Unwired Business: Cases in Mobile Business

Stuart J. Barnes
University of East Anglia, UK

Eusebio Scornavacca
Victoria University of Wellington, New Zealand

IRM Press
Publisher of innovative scholarly and professional information technology titles in the cyberage

Hershey • London • Melbourne • Singapore

Acquisitions Editor:	Michelle Potter
Development Editor:	Kristin Roth
Senior Managing Editor:	Amanda Appicello
Managing Editor:	Jennifer Neidig
Copy Editor:	Joyce Li
Typesetter:	Cindy Consonery
Cover Design:	Lisa Tosheff
Printed at:	Yurchak Printing Inc.

Published in the United States of America by
IRM Press (an imprint of Idea Group Inc.)
701 E. Chocolate Avenue, Suite 200
Hershey PA 17033-1240
Tel: 717-533-8845
Fax: 717-533-8661
E-mail: cust@idea-group.com
Web site: http://www.irm-press.com

and in the United Kingdom by
IRM Press (an imprint of Idea Group Inc.)
3 Henrietta Street
Covent Garden
London WC2E 8LU
Tel: 44 20 7240 0856
Fax: 44 20 7379 0609
Web site: http://www.eurospanonline.com

Library of Congress Cataloging-in-Publication Data

Unwired business : cases in mobile business / Stuart Barnes and Eusebio Scornavacca, editors.
p. cm.
Summary: "This book provides practical case studies of the planning, implementation and use of mobile and wireless data solutions in modern business"--Provided by publisher.
Includes bibliographical references and index.
ISBN 1-59140-664-1 (hardcover) -- ISBN 1-59140-665-X (softcover) -- ISBN 1-59140-666-8 (ebook)
1. Wireless communication systems--Economic aspects. 2. Mobile communication systems--Economic aspects. 3. Electronic commerce. I. Barnes, Stuart, 1971- II. Scornavacca, Eusebio, 1975-
HE9713.B37 2005
658.8'72--dc22

2005020634

British Cataloguing in Publication Data
A Cataloguing in Publication record for this book is available from the British Library.

All work contributed to this book is new, previously-unpublished material. The views expressed in this book are those of the authors, but not necessarily of the publisher.

Unwired Business:
Cases in Mobile Business

Table of Contents

Section IV. Mobile Applications in Healthcare

Section V. Mobile Technologies in International Markets

Preface

Homo sapiens is by nature a very mobile animal, striving to cover new ground and push existing boundaries. However, we have certain patterns of habitual geographical movement in our lives—between home, work, study, entertainment, shopping, family, friends, and so on. This equilibrium is punctuated only by extraordinary movements, such as visiting a client overseas or going on holiday (Pica & Sørensen, 2004).

Many activities that humans perform are dependent on communication and information. Information is key to decision making, whether for a customer to buy a certain product or for a manufacturer to procure a specific quantity of raw materials, or any other activity where information can determine outcomes. Communication is an important channel for conveying information as well as fulfilling roles of social interaction, purposeful group decision-making, and many other functions.

Until very recently, the combination of mobility, information, and communication was rather staccato; not only would an individual need to move from A to B to do something, but he/she would also need access to C, a point at which he/she could obtain information or communicate in a meaningful way to complete tasks at B. For example, C could be a fixed-line telephone, a networked personal computer, or simply a person to talk to face-to-face. In this situation significant value could be added by information and communication at the point of need—what if B and C were at the same location? This would require technology for mobile communication.

With well over a billion handsets worldwide, mobile phones have been one of the fastest adopted consumer products of all time (Chen, 2000; de Haan, 2000; Emarketer, 2002; Kalakota & Robinson, 2002). According to a study by Telecom Trends International (2003), global revenues from m-commerce—that is, transactions over mobile networks—could grow from $6.8 billion in 2003 to over $554 billion in 2008.

Alongside mobile phones, distributed network computing has been a significant technology trend. This has put more computing power directly in the hands of networked individuals. Beyond organisational, academic, and military networks, the trend spread rapidly to the general globalisation of distributed networking in the 1990s, spearheaded by the Internet. In 2005, it is estimated that there are more than a billion users of the Internet.

Although developing along separate paths, mobile communications and the Internet have started to converge. The products of the partnership between mobile devices and the Internet are sophisticated wireless data services, centering on mobile data access and electronic messaging on mobile devices (Yoo, Lyytinen, & Yang, 2004). The market for these services is diverse, and the most commonly cited applications are in the business-to-consumer (B2C) and business-to-employee (B2E) segments. Such applications are built on some fundamental value propositions, such as ubiquitous access to information, the personal nature of devices and customization, and contextual properties of the device and user, such as time, location, personal preferences, and the task at hand. In the consumer space, the wireless applications have included person-to-person messaging, e-mail, banking, news, games, music, shopping, ticketing, and information feeds. In the business space, applications include sales force automation, navigation, tracking, field force automation, wireless telemetry, and the mobile office (McIntosh & Baron, 2005; Scornavacca, Barnes, & Huff, 2005).

More broadly speaking, mobile (m-) business is likely to have a tremendous impact on organisations, as wireless technologies and applications begin to challenge the existing processes, strategies, structures, roles of individuals, and even cultures of organisations. Here, m-business is defined as the use of the wireless Internet and other mobile information technologies for organisational communication and coordination, and the management of the firm. Indeed, by 2004, cost savings could permit wireless business services around the world to generate annual value of up to $80 billion, and at least as much value could be created if corporations used wireless services to improve their current offerings or to deliver new ones (Alanen & Autio, 2003).

Features of the Book

This book aims to provide a source of high-quality, practical case studies of the planning, implementation, and use of mobile and wireless data solutions in modern business. The case studies are selected both as exemplars of wireless and mobile solutions and as typical cases in a variety of areas of common development. The book provides a number of insightful analyses of business applications of mobile technologies that help the practitioner understand the nature of the technology and how its value can be best harnessed in a wide variety of organizational settings. The focus is to present how these emerging technologies can help business to create a strategic advantage in the market, typically by becoming more efficient, effective, and profitable. The examples should provide ideas and points of reference for managers as they seek to devise and implement mobile applications for business advantage.

Since the book is proposed to be an imprint of Cyber Tech Publishing, naturally it is aimed at practicing managers. In particular, it is aimed at managers who would like to better understand the implications of wireless and mobile technologies for today's organisations. By providing examples and analyses, it provides a source of ideas for managers to take with them to their workplace. In addition, the book also has value as a source of cases for academics and students. Thus, the cases could possibly be used as the basis of classroom discussion.

This text has arisen from extensive investigation into the impacts of wireless technologies in a variety of areas of business and organisation, each highly dependent upon recent technological developments. It has also arisen from a review of the available professional and academic literature on this and related topics, based on experience, and in the context of recent developments in the field. The chapters of this book illustrate the wide array of business opportunities afforded by mobile business. They describe and discuss the important strategic, managerial, and technological issues that follow in the wake of an organisation deciding to embrace wireless technologies. Chapters have been created to bring a balance of conceptual and practical issues, focusing on recent and emerging trends. Where possible, the book examines wireless issues from an international perspective, pointing to specific examples from around the globe.

It is, of course, impossible to cover all aspects of this emerging topic. The focus of this book has been on attempting to cover a selection of the core, recent, or possibly more important areas of m-business, with reference to different markets, technology foundations, applications, services, and impacts for organisations. The implications are that whilst technological aspects are covered in some detail, this is always in a mode accessible to the manager.

Structure of the Book

This book's 18 chapters are structured into five sections, each emphasising different but interrelated aspects of the m-business landscape.

Section I. Consumer Applications of M-Business

The first section examines the impact of mobile communications on relationships in the consumer marketspace. The mobile medium provides significant potential for businesses to augment existing consumer products or services or even provide new ones tailored to the mobile context. This section examines case studies in some of the most popular or promising areas of consumer application development. In particular, it includes applications such as mobile ringing tones, banking, gaming, alerts, and the use of barcodes for information transfer.

Section II. Mobile Marketing

Following on from the last section, one region within the consumer space that provides significant potential is mobile marketing. The individual nature of mobile devices, along with the recognition of time, space, and personal characteristics, provides an unprecedented platform for one-to-one marketing. This section examines the nature and potential of mobile advertising, including successful instances of application, as well as issues of permission and acceptance among consumers.

Section III. Organizational Applications of M-Business

The mobile applications that are currently reaping the biggest rewards are those operating within organisations. Many organisations have gained significant return on investment from their B2E mobile solutions. This section picks up on this important topic by examining specific organisational cases in a variety of industrial contexts. These include examples of wireless sales force automation (wSFA) in the food industry, a mobile knowledge management system (mKMS) in a university environment, as well as other mobile applications in the paper industry and in the supply of heavy machinery.

Section IV. Mobile Applications in Healthcare

One sector of organisational application that is worthy of specific attention is healthcare. This is an area where numerous solutions have been created and deployed for the improvement of patient care. In this section, we examine the current use and future potential of mobile health applications by utilising a number of case examples. This includes working applications in a variety of contexts, such as a handheld solution in emergency services, a clinical messaging facility, and mobile information systems for residential care.

Section V. Mobile Technologies in International Markets

The final section examines the advance of mobile technologies and markets in an international context. The examples provided in this section focus specifically on developments in Finland, Japan, and South Korea. These cases will be of interest to other countries contemplating their own trajectories for future development in mobile communications.

As you will now be aware, m-business is a complex and diverse subject. It is not simply concerned with technological issues, but it also incorporates aspects of strategic management, marketing, operations management, and behavioural science, among others. Such an interdisciplinary perspective is critical if the subject domains are to be fully understood. Recent examples of m-business offerings that overestimate technology and underestimate consumers exemplify this point. For this reason, we advocate a broader management viewpoint. The issues debated here are far too important to be left to the technologists; although technology is an important enabler, the vision, strategy, and management of the evolution of m-business belongs to managers. To reap the real rewards of m-business, management competence is paramount.

We hope you find this book of interest and that it raises some important issues relevant to consideration in your organisation, study, or research. By harnessing the power of m-business, your organisation could become the next to step into the wireless world.

References

Alanen, J., & Autio, E. (2003). Mobile business services: A strategic perspective. In B.E. Mennecke & T.J. Strader (Eds.), *Mobile commerce: Technology, theory and applications* (pp. 162–184). Hershey, PA: Idea Group Inc.

Chen, P. (2000). Broadvision delivers new frontier for e-commerce. *M-Commerce*, October, 25.

de Haan, A. (2000). The Internet goes wireless. *EAI Journal, April*, 62–63.

Emarketer. (2002). One billion mobile users by end of Q2. Retrieved May 23, 2003, from *www.nua.ie/surveys/index.cgi?f=VS&art_id=905357779&rel=true*

Kalakota, R., & Robinson, M. (2002). *M-business: The race to mobility*. New York: McGraw-Hill.

McIntosh, J.C., & Baron, J.P. (2005). Mobile commerce's impact on today's workforce: Issues, impacts and implication. *International Journal of Mobile Communications, 3*(2), 99–113.

Pica, D., & Sørensen, C. (2004). *On mobility and context of work: Exploring mobile police work*. Paper presented at the 37th Hawaii International Conference on System Sciences, Big Island, HI.

Scornavacca, E., Barnes, S., & Huff, S. (2005). *The emergence of mobile business as a research discipline: Past present and future*. Paper presented at the European Conference on Information Systems, Regensburg, Germany.

Telecom Trends International. (2003). M-commerce poised for rapid growth. Retrieved October 27, 2003, from *www.telecomtrends.net/pages/932188/index.htm*

Yoo, Y., Lyytinen, K., & Yang, H. (2004). *The role of standards and its impact on the diffusion of broadband mobile services: A Korean case*. Paper presented at the Austin Mobility Roundtable, Austin, Texas.

Acknowledgments

We would like to thank all the authors who contributed to this edited book.

In addition, we would like to acknowledge the support we received from our families—without them this publication would not be possible.

Stuart J. Barnes
Eusebio Scornavacca

Section I

Consumer Applications of M-Business

Chapter I

Mobile Innovation and the Music Business in Japan: The Case of Ringing Tone Melody ("Chaku-Mero")

Akira Takeishi, Hitotsubashi University, Japan

Kyoung-Joo Lee, Hitotsubashi University, Japan

Abstract

This paper examines the development process and successful factors of the ringing tone melody downloading service, or "Chaku-Mero," in Japan. Chaku-Mero, arguably the most successful mobile (m)-commerce business in the world, is a mobile Internet service in which a subscriber could download from a wide selection of music melodies his/her favorite for a fee to get it ring when the mobile phone receives a call or message. This chapter describes the process of how this business has evolved from pre-mobile-Internet phase of related business; examines the structure of the business; and analyzes why some content providers have been more successful than others. Some implications for the prospects of mobile Internet businesses for music and other cultural contents will be provided.

Objectives

The primary purpose of this chapter is to examine how the Chaku-Mero business in Japan has been launched and developed successfully to create about $1 billion (currently US$1=110 yen) market within 4 years. "Chaku-Mero" is an abbreviation of "chakusin merodi" in Japanese, which stands for ringing tone melody. A Chaku-Mero service user downloads a digitalized music file through the Internet for a fee and has it played as a ringing tone for the mobile phone handset. This download service accounts for the largest market among Japan's paid mobile Internet services, thus making the world's largest mobile Internet market and the most successful mobile (m)-commerce business. Furthermore, the Chaku-Mero business is arguably one of the most successful cases in the world of paid cultural content business on the Internet, be it fixed or mobile.

This chapter attempts to give brief answers to the following questions: What evolutionary paths has the Chaku-Mero business taken to form the current business model? What has made it possible to create the largest mobile Internet market within a short period of time? How do firms compete in this business?[1] Although this chapter still remains preliminary and descriptive, the research results provide some valuable implications for future directions of mobile Internet services not only in Japan but also in other countries. In a broader context, this chapter constitutes a part of a larger research plan that will examine interactions between the music business and technological innovations, including the impact of Internet technologies.

Overview of Chaku-Mero Service and its Brief History

What is Chaku-Mero Service?

The current Chaku-Mero service is an Internet-mediated content downloading service. A subscriber accesses a content provider's Web site by his/her mobile phone with an Internet browser, and downloads a selected music melody file into the handset. The downloaded file is saved and used as the handset's receiving tone melody.

The music data of Chaku-Mero is coded in a simplified Musical Instrument Digital Interface (MIDI) format, which is a digital format designed for music replaying. MIDI file is basically signal data that orchestrate musical tones and control their volume and length. MIDI file is smaller in memory capacity than other recorded music files on the MP3 or WAVE digital format. Chaku-Mero's file format is a simplified version of the MIDI format specially tailored for mobile handsets' small memory capacity. The two dominant mobile formats currently used are Compact MIDI (adopted by NTT DoCoMo, KDDI) developed by Faith, and SMAF (Synthetic Music Application Format, adopted by J-Phone, KDDI) developed by Yamaha. Content providers prepare a Chaku-Mero file by

(1) creating a music melody file in MIDI format, (2) transforming it into a mobile MIDI file by an authoring tool, and (3) uploading it to Internet servers. Authoring tools are available for each format and provided by the file format developers such as Yamaha and Faith.

Users download a melody file through Internet-browsing mobile phones, and the music synthesizing processor chip in the handset reads the file and replays the melody when a phone call or e-mail message arrives. Major manufacturers of mobile music synthesizing chips include Yamaha, Rohm, and Qualcomm (installed in its CPUs). They supply chips to handset manufacturers.

Downloading service is charged in various ways. Some providers charge monthly subscription fee for limited or unlimited melody downloads, while others collect transaction-based fee for each download, depending on marketing and competitive strategies of service providers. For instances, users can download 11 melodies at $3.00 a month, or they can download without limit at $1.00 a month but only from the most popular 100 melodies. The discrete payment for every download purchase is also available (except for providers for NTT DoCoMo), paying, for example, 10 cents for each melody. Moreover, free download services are also widely diffused as effective marketing means for advertisement and promotion.

A Brief History of the Chaku-Mero Service

It was in December 1999 that the current model of Internet-mediated ringing tone melody download service was first commercialized on NTT DoCoMo's i-mode platform. However, the creation of Chaku-Mero was not an overnight incident. Previous experiences and resource accumulations both in related markets and technologies had existed as precursors of the Chaku-Mero service. Particularly critical are (1) the fact that many users had already enjoyed Chaku-Mero in different ways before the mobile Internet became available, and (2) the fact that related technologies, data, and know-how had already been accumulated in the karaoke business.

The earliest version of Chaku-Mero was "handmade" by individual users as they directly set music tone by pushing their handset's dial buttons. A variety of Chaku-Mero guidebooks were published and a growing number of mobile phone users came to enjoy Chaku-Mero by themselves. The first guidebook of this kind was published in July 1998 and a million copies were reportedly sold. In those days, a substantial number of users manually input their favorite melodies by pushing handset buttons following the instructions of Chaku-Mero guidebooks and magazines. The instructions were tailored for each handset's different technological specifications.

To give customers melody files, rather than let them push buttons tediously by themselves, J-phone introduced a first version of melody download service named "Sky Melody" in November 1998, 1 year before today's Internet-mediated download service started (a service similar to the Sky Melody was also provided at that time by Astel, a PHS [Personal Handy Phone System] mobile service carrier). In the Sky Melody service, which is still in operation, a subscriber calls up the carrier's automatic responding system and selects a favorite melody following the system's operation guidance. The selected melody in binary code is sent to the handset as a form of short message after the user hangs up the phone. In this model, J-phone provides the automatic responding system as well as music data, and could make revenue by increasing subscriber's communication time to call and select the melody. This system was realized by modifying the short message service system to handle the melody download service. Sharing the basic infrastructure with the existing short message system, the business was very successful and brought profit to J-phone.

The first melody download service on the mobile Internet was introduced on NTT DoCoMo's i-mode service platform. However, it took some time for the service to start after DoCoMo launched i-mode in February 1999. A related service that first appeared on i-mode was a Web site where Chaku-Mero instructions were shown to subscribers, an Internet version of Chaku-Mero guidebooks. The site, named "Club GIGA," was provided by GIGA Networks for free. Initially, NTT DoCoMo's staff members were skeptical about the demand for such service. However, consumer reaction betrayed the initial skepticism when Club GIGA attracted more than 3.5 million page views a month, and such popularity never ceased even after the site started charging for the service. In September 1999, the number of service subscribers reached 100,000 and 9.1 million page views were recorded.

Users' strong interests in Chaku-Mero had thus been well indicated by the success of Sky Melody and Club GIGA, when the current model of Chaku-Mero download service through the Internet was launched in December 1999. At this point, another precursor joined to contribute to the development of Chaku-Mero business—the karaoke business.[2]

At the beginning, there were three i-mode Chaku-Mero service sites, and all of them were provided by karaoke service companies: GIGA, which was mentioned above, XING, and SEGA. These providers and many other early entrants in the Chaku-Mero business were leading firms in the wired karaoke service industry. The wired karaoke service is a service to provide karaoke music online to karaoke places where users enjoy karaoke. Musical data files are provided through fixed lines and played in the terminal (sing-along machine) installed in karaoke places. Through the direct channels, these companies distribute newly released popular songs very efficiently to a large number of karaoke places around Japan. After the first wired karaoke technology was introduced in 1992, this online system gradually became dominant and diffused widely in the karaoke business. In the growth process, wired karaoke service companies accumulated a huge stock of digitalized music files on the MIDI format. Indeed, NTT DoCoMo adopted a simplified MIDI format for the i-mode Chaku-Mero service because accumulated MIDI music data in karaoke could be readily used for the Chaku-Mero service. It should also be noted that Japan has led the world in MIDI technologies, driven by long-term development efforts by Yamaha and other companies.

To summarize, the Internet-mediated Chaku-Mero download business was founded on the two critical bases, one on ample digital music data resources and related technologies, which had been accumulated in the karaoke industry, and the other on the established needs of mobile phone subscribers to enjoy Chaku-Mero in the pre-mobile-Internet era.

Initially Chaku-Mero was composed in just two or three musical chords because of technological limitations. Now, due to technological advances in handset music processing chips, driven by the so-called "Wa-on Sensou (chord war)," 40-chords music replays have come to be realized for most advanced handsets. Stimulated by such improvement in sound quality (more chords), a growing number and variety of service providers and melodies available, and fierce market competition among providers, the market has grown rapidly.

As of summer 2004, the total number of official sites for the three carriers (NTT DoCoMo, KDDI, and J-Phone) reached more than 500, and Chaku-Mero commanded the largest sales in every carrier's mobile Internet services. Chaku-Mero claimed about 40–60% of the whole mobile Internet sales for each carrier. It was estimated that in FY 2003 (from April 2003 to March 2004) the three carriers' total revenue from (i.e., subscribers' total payment for) paid mobile Internet services was approximately $2 billion, and the revenue from Chaku-Mero sites was $1 billion, a very rapid growth from zero within 4 years.

Profile of Chaku-Mero Users

Chaku-Mero is the most popular and widely used among various mobile Internet services, regardless of age and sex.

Table 1 shows the survey results (about 1,000 respondents) on mobile Internet usage patterns in 2002. Similar to other mobile Internet services, young consumers were more active in using Chaku-Mero downloading. Yet, in all ages and both sexes, Chaku-Mero was the most popular mobile Internet service, with the exception of the group aged between 40 and 69, who preferred weather forecast and news to Chaku-Mero. When users were grouped by frequencies of mobile Internet usage, there was no difference; for both heavy and light users, Chaku-Mero was the most popular.

Downloading Chaku-Mero into the handset has been established as the first action of a typical consumer when he or she gets a new mobile phone handset with browser. A growing and strong demand for the Chaku-Mero service seems to reflect subscribers' needs to customize their own handset and to satisfy the desire of appealing to friends and surrounding people with their own characteristic ringing melody. Also, many users download a number of their favorite melodies every month, saving them and building up their own collections. The Chaku-Mero service not only satisfies the functional needs to have customized ringing tone, but it also fulfills the desire to enjoy music and to boast about new melodies to friends.

Many new usages of Chaku-Mero have also been explored. For example, you can send a melody with an e-mail message to your friend to celebrate his/her birthday. Or, as it has recently become possible to download human voices, you can use the voice of your favorite celebrity as a morning call.

Table 1. Mobile Internet site genre usually used by cutomers

		Chaku-Mero DL	Weather Forecast	Screen Savers DL	Traffic Information	General News	Sports Information	Game DL	Restaurant Information
Total		54.1	32.5	31.7	27.2	20.8	17.0	12.7	11.7
Man	Total	48.2	33.7	27.6	25.1	22.6	25.6	12.6	9.5
	age 12-19	84.4	15.6	56.3	3.1	3.1	9.4	34.4	0.0
	20-29	58.6	25.7	34.3	27.1	27.1	30.0	12.9	7.1
	30-39	37.0	37.0	19.6	21.7	23.9	30.4	8.7	17.4
	40-69	21.6	52.9	7.8	39.2	27.5	25.5	2.0	11.8
Woman	Total	60.0	31.3	35.9	29.2	19.0	8.2	12.8	13.8
	age 12-19	88.9	14.8	59.3	7.4	11.1	7.4	25.9	3.7
	20-29	63.4	37.8	41.5	42.7	15.9	9.8	13.4	15.9
	30-39	47.3	29.1	25.5	25.5	23.6	3.6	7.3	16.4
	40-69	48.4	32.3	19.4	19.4	25.8	12.9	9.7	12.9
Heavy User		51.7	38.1	30.5	32.2	30.5	21.2	12.7	11.0
Intermediate User		57.8	32.5	35.5	25.9	19.3	14.5	15.7	10.2
Light User		50.9	26.4	27.3	23.5	12.7	16.4	8.2	14.5

Note: Survey conducted in July 2002 (with 1039 respondents) % of the respondents who usually use the site genre.

Heavy, Intermediate, and Light indicate the frequency level of mobile internet service usage. Source: Video Research, Mobile Phone Usage Situation.

Structure and Competition of the Chaku-Mero Business

Structure of the Chaku-Mero Business

The basic structure of the Chaku-Mero business is depicted in Figure 1. The official content provider sites, which are approved by each carrier and listed on the carrier's official menu pages, provide melody download service. Users subscribe to their favorite sites, and download music melodies on a monthly fee basis or by discrete payment on each download (NTT DoCoMo allows only monthly paid content subscription). Each month, on the monthly bill, carriers charge and collect the service fees from subscribers on behalf of the service providers. Carriers gain their own business revenue by charging (1) data communication time (the packet communication fee, for instance, costs around 6–7 cents for a Chaku-Mero download) to subscribers for their site access, and (2) a fixed

Figure 1. Basic structure of Chaku-Mero

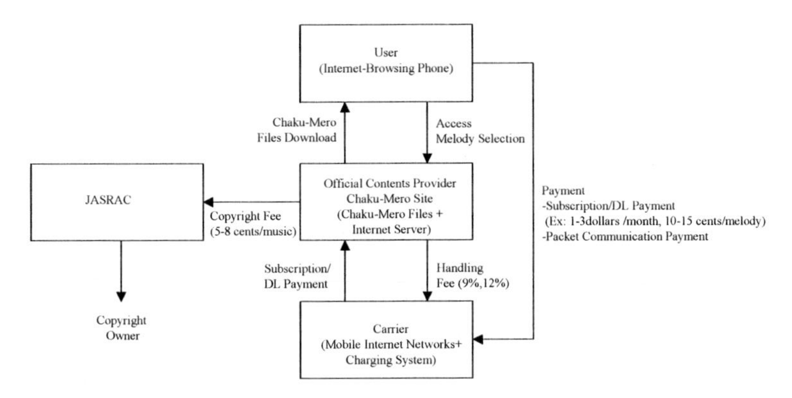

rate of handling fee to content providers (DoCoMo and KDDI charge 9% and J-phone 12% as the handling fee of the collected revenue for the provider). And the service providers pay the copyright fees to the Japanese Society for Rights of Authors, Composers, and Publishers (JASRAC). Five cents per melody is charged for melodies shorter than 45 seconds, and 8 cents for melodies longer than 45 seconds.

It is important to point out that the music copyrights for Chaku-Mero are collectively and solely managed by JASRAC in Japan. As JASRAC has taken charge of copyright procedures for most of the ringing tone melodies, Chaku-Mero service providers could use most of popular tunes rather freely as long as they pay the fees to JASRAC. A historical background of such smooth copyright coordination for Chaku-Mero could be traced back to institutional arrangements made for the karaoke business and the Sky Melody service. For the Chaku-Mero service, the copyright negotiations with JASRAC were conducted by the Association of Musical Electronics Industry (AMEI) with the leadership of Yamaha, Faith, and major service providers. They agreed on the current copyright rule in late 2000, about 1 year after the first Chaku-Mero download service was introduced.

In the beginning, as Chaku-Mero became popular, a large number of general sites (or sites not approved by the carrier) for Chaku-Mero sprung into existence, offering a wide variety of free melody download services. To counter this undesirable trend, JASRAC vigorously monitored and charged copyright violations, and eventually cooled down the movement. JASRAC thus made an important contribution to the Chaku-Mero business by preventing the spread of improper free services, in addition to the provision of smooth copyright coordination. This evolution pattern contrasts with the U.S. situation in which an earlier widespread of free music download services such as NAPSTER on the fixed Internet has disturbed the creation of paid online music download services, having established the first image of "free music on the Web."

When we look at other countries, we find that their situations are more problematic. As shown in Table 2, for example, in some countries copyrights are managed in a more complicated manner with different rules and/or there is no institution that could manage necessary copyrights all together for one-stop shopping, thus causing high costs of copyright coordination or allowing the spread of illegal sites. Japanese service providers

Table 2. Institutions for Chaku-Mero-related copyrights in some countries

Country	Institutions for Copyrights Coordination	Characteristics, Problems, etc.
Japan	JASRAC	As JASRAC collectively controls almost all the copyrights, it could efficiently coordinate copyrights for Chaku-Mero providers.
U.S.A	ASCAP,BMI, etc.	Because of multiple numbers of copy right institutions, copy right coordination process is complex.
Germany	GEMA, etc.	The copyright rule requires Chaku-Mero service providers to pay 0.2 Euro for the composer as they transform the original music into Chaku-Mero, the distribution cost is high.
Singapore	COMPASS	Because there is no effective collective copyright institutions, coordination procedurals are difficult.
Korea	KOMCA	As there is almost no collective copyright institutions, inappropriate Chaku-Mero business is wide spread.

Source: Nihon Keizi Shinbun, 2003/02/08

often complain about the absence of efficient copyright procedures as a major obstacle for their business overseas. This fact, in effect, indicates that smooth copyright coordination has substantially contributed to the success of Japan's Chaku-Mero business.[3]

Competition Surrounding Chaku-Mero Business

Table 3 lists major Chaku-Mero sites and their sponsoring companies. Largest content providers in the market include Xing, GIGA Networks, Dai-ichikosho, SEGA, Yamaha, and Dwango. The largest service provider is Xing, which has about 7 million subscribers and had revenue of $150 million in FY 2003. The common characteristic of leading companies like Xing, GIGA, Dai-ichikosho, and SEGA, as mentioned before, is that they all have diversified into the Chaku-Mero business from the wired karaoke industry. In the late 1990s, these companies were eagerly searching for new businesses to compensate for their saturated karaoke market. Their abundant stock of MIDI files prepared for the karaoke business leveraged their venture into the Chaku-Mero business. These former karaoke service providers had already built up capability to swiftly arrange and make MIDI files for the newly released songs and lost no time in gripping the new chances with briskly prepared melody menus in the early rapid-growth period of Chaku-Mero business. They succeeded, as a first mover, in capturing a large number of users. A larger amount of subscribers would give the provider some competitive advantages. First, there are some switching costs for customers; once accustomed to a particular site they do not frequently change to another. Second, the carrier's official menu lists the service sites in the order of the number of accesses/subscription. Being at the top on the first page in the list would bring the provider more accesses than being at the bottom of the subsequent pages.

Another feature of two top-ranked Chaku-Mero providers, Xing and GIGA, is that they have focused their business on the services with NTT DoCoMo. By launching very early Chaku-Mero sites for NTT DoCoMo, which first created the mobile Internet market, they have built up large subscriber base and gained competitive advantages. On the other

Table 3. Major Chaku-Mero sites

Carrier	Major Chaku-Mero Site	Content Provider	Parent/Related Company
NTT DoCoMo (87)	PokeMeroJOYSOUND	Xing (Wired Karaoke)	Brother (Machinery)
	Chaku-Mero GIGA	GIGA Networks (Wired Karaoke)	Richo (OA Equipment)
	Mero DAM (Karaoke DAM)	Daiichikosho (Wired Karaoke)	Dai-ichikosho
	SEGAKara	SEGA Music Networks	SEGA (Game Software)
	Yamaha MerocCha!	Yamaha (Music Equipment)	Yamaha
	40MeroMix	Dwango (GameSoftware)	Dwango
KDDI (108)	Meropa	MTI (Mobile Contents)	Music.Co.Jp (Digital Music Distribution)
	Mero-Chaku Club	Infocom (IT Solution)	Infocom
	N-Melody Town	NEC (Electronics)	NEC
	Mero DAM (Karaoke DA)	Daiichikosho (Wired Karaoke)	Dai-ichikosho
	Oricon Super Sound	Oricon Entertainment	Oricon (Music Information)
	Ymaha MerocCha!	Yamaha (Music Equipment)	Yamaha
J-Phone (49)	40MeroMix	Composit (IT Solution)	Dwango (Game Software)
	N-Melody Town	NEC (Electricity Equipment)	NEC
	MecchaMero	Infocom (IT Solution)	Infocom
	Oricon Super Sound	Oricon Entertainment	Oricon (Music Information)
	SEGA Kara (Melody)	SEGA Music Networks	SEGA (Game Software)
	Taito Sky-On Rakuen	Taito (Game Software)	Taito

Note: Number in () with the carrier shows the number of official Chaku-Mero sites as of December 2002

Source: Carriers' Web sites and an interview with Yamaha.

hand, some of leading Chaku-Mero service companies for KDDI and J-Phone provide their services to multiple carriers' platforms. In other words, the relationships between service providers and carriers are mixed, including both open and closed ones. However, the general trend moves toward the more open structure, and recently the DoCoMo-only service providers have begun to establish their services with other carriers.

One key factor in the Chaku-Mero competition is the scope and quality of the service menu. All the leading service companies offer a wide variety of melodies in their menu. For instance, Xing, the market leader, provided 10,200 melodies as of March 2003, while adding 200 new melodies every month. Differentiated from these "full-line" service providers, some providers specialize in unique music genres (for instance, theme songs of popular TV programs and animations, guitar sounds, and jazz arrangements, to name a few).

Another critical competitive dimension in the Chaku-Mero business is the ability to arrange melodies fine-tuned for different carriers and different mobile phone handset models. First of all, service providers need to arrange the melody files to cope with each carrier's different format of Chaku-Mero files such as Compact MIDI and SMAF, each of which has a couple of versions. What makes the situation even more complicated is that each handset model has a different synthesizer chip and speaker. Older handsets could play only three chords, while the latest ones can play 40 chords, for instance. To maintain the sound quality of Chaku-Mero and satisfy users, each melody should be rearranged and tuned according to the file format and handset specifications, a condition that significantly increases the variations of music files needed for optimal sound.

Therefore, the same melody often needs to be encoded on 20–30 different file formats. The capability to efficiently and quickly provide melody files fine-tuned for individual handset models to play better musical sound would significantly affect competitive positions in the Chaku-Mero business.

Other critical competitive factors include the number of chords available, cheap pricing, and a better Web site design to allow subscribers to find favorite melodies quickly and thus economize on time and communication fees.

Future of the Chaku-Mero Business: The Music and Mobile Innovation

Today and Tomorrow of the Chaku-Mero Business

The Chaku-Mero business took a head start and has been growing very rapidly until now. About 4 years have passed since NTT DoCoMo launched the first mobile Internet service. During this period the Chaku-Mero service, as the largest business, has driven the development and growth of the mobile Internet. Although its share in the mobile Internet market tends to gradually decrease because new services such as games have been introduced, the Chaku-Mero market has so far kept expanding and still maintains the top position.

However, it is also true that the pace of market growth has slowed and the business does not appeal as much as before. An important revenue source for Chaku-Mero is those subscribers who have just purchased new handsets for the first time or for replacement. The market for mobile communication subscribers, however, has entered the maturation phase and keeping a high pace of growth as in the past will not be feasible. The estimated annual growth rate of the Chaku-Mero service recorded 94% in FY 2001, 40% in FY 2002, and 20% in FY 2003, indicating market saturation. The number of new entrants in the business and intensified competition lead to price cuts, making the business less profitable. The "chord war," which has stimulated the demand for better quality melodies, now likely comes to an end, now that 40-chord processors have become available on the market. The shared industrial recognition is that the subsequent advances in this dimension would not significantly raise the current consumer satisfaction level.

As mobile technologies advance into the Third Generation (3G), new services related to Chaku-Mero have become available, such as "Chaku-Koe" service (providing voices of people, including celebrities, as a ringing tone), "karaoke" service (providing melody, words, and graphical image of a song to enjoy karaoke with mobile phones), and "Chaku-Uta" service (providing a piece of song as a ringing tone). The future of the Chaku-Mero business would depend on how attractive these new services are and how much they create new market demand.

Among these new services, we are particularly interested in the Chaku-Uta service, which was introduced in December 2002 for KDDI's 3G mobile Internet platform (au). Whereas in the Chaku-Mero service people download melodies, in the Chaku-Uta service people can download a part of original songs with vocals and full instrumentation. This new service seems to have made a good start. Now more than 30 sites have been opened and the revenue from the Chaku-Uta service reached $45 million in FY 2003. The service has become a killer application for KDDI's 3G service and helped KDDI gain market share against NTT DoCoMo.

Until now, no service has been realized for downloading music, rather than melody, by way of the mobile Internet channel (although DoCoMo provides a music distribution service on its PHS platform [M-stage], it does not operate by way of Internet and the demand has not been strong). The music download service on the mobile Internet has not been realized because voluminous file size necessary for full music demands larger download and playing capacity, longer time, and high costs. Furthermore, music download services need to handle complicated copyright concerns of various parties.

However, since Chaku-Uta uses just a small part of music, it could overcome the problems with capacity, download speed, and consumer cost (nevertheless downloading a Chaku-Uta still costs 10 times as much as Chaku-Mero, including communication fees. Typically, to download a Chaku-Mero costs 15–20 cents, while Chaku-Uta costs $1.50–$2.00). In addition, the Chaku-Uta service simplified copyright problems by limiting the scope of the music usage. Only a short part of the whole song is used, and the downloaded file, which is made in the format called "ez-movie," is protected for subsequent copies and not removable from the mobile handset.

Another characteristic of Chaku-Uta is that the service has been provided predominantly by recording companies, which have not been successful in the Chaku-Mero business. A download service of original music involves matters and concerns that go beyond what JASRAC could cover and manage. Thus, while leading Chaku-Mero service providers (mostly karaoke companies) have not entered the market, Labelmobile, a company that has been jointly established by Japan's major recording companies and maintains necessary copyrights of many popular songs, became the first Chaku-Uta service provider. This is different from what happened in the Chaku-Mero business.

Ironically, the recording companies have not been able to receive benefits from the lucrative Chaku-Mero business. They have watched a rapid growth of the market and business with envy. In the Chaku-Mero business, the copyrights and attached fee revenue go only to music composers as the business uses just a short melody of music. This business structure has frustrated recording companies, which have risked huge financial commitments to make hit-songs to be used for Chaku-Mero. To them, Chaku-Mero providers are free-riders, who just enjoy the delicious fruits yielded by recording companies without committing any investment and risk to create popular songs. To make their own presence in the Chaku-Mero business, recording companies jointly launched a Chaku-Mero site, "Reco-Choku," operated by Labelmobile, in addition to other sites provided by individual record companies. However, their business performance has not been notable. Chaku-Mero is a melody just arranged for ringing tone and quite different from the original songs. Recording companies do not have distinctive capability to prepare simplified MIDI files fine-tuned for individual handset models and compete against leading service providers. Actually, recording companies have experienced the same frustration as when the karaoke businesses were created and developed.

Facing decreased CD sales, recording companies have recently been in a slump. Japanese CD production peaked at $5.4 billion in 1998 (the second largest in the world next to the

United States), and then fell to $3.6 billion in 2003. In the meantime, the karaoke business has a $8.1 billion market, though matured, and the Chaku-Mero business has a $1 billion market, still growing. It is a critical issue for leading recording companies as to how to establish their presence in wired and wireless online Internet music businesses. In this context, the Chaku-Uta service is regarded as an important business opportunity, where they could have some competitive advantage. The new service is also expected to serve as a new promotion means for newly released songs.

How much does the Chaku-Uta service to download music, instead of melody, create market demand? What business could recording companies command in this field? Answers to these questions surrounding Chaku-Uta would cast critical implications on the future of the music business on the Internet.

Music Business and Mobile Innovation

Technological innovations in music creation, production, recording, playing, distribution, promotion, and advertising change the structure of the music business and industry, and give birth to new popular music genres and styles of performance. The history of the music business has repeatedly witnessed such patterns of interactions between technological changes, business structure, and music itself.[4] What would be the impact of the mobile Internet innovations on the music business and music itself? To think about this question, the case of Chaku-Mero in Japan would be of great interest as one of few examples of successful music content download business on Internet.

This chapter has briefly examined how the Chaku-Mero business has developed, and provides some findings and implications. First, the Japanese Chaku-Mero business was made possible by some favorable conditions:

1. Needs for Chaku-Mero were already established among mobile phone users.
2. Technological know-how and digitalized music resources were accumulated in the wired karaoke business.
3. Institutional foundation was established for smooth and efficient coordination of copyright issues.

It is often argued that the success of the Japanese mobile Internet business could be explained by young people's unique market needs peculiar to Japan, or by delayed diffusion of fixed Internet networks. However, our observation in this chapter shows that such argument would lose explanatory bases. Chaku-Mero enjoys popularity across generations as shown before. Also, the Chaku-Mero service occupies the top rank in popularity among mobile Internet services in foreign countries where mobile Internet services have been launched recently. Chaku-Mero seems to have a firm market potential in many countries. The Chaku-Mero service is a mobile-only business to use melodies as ringing tones, and should be distinguished, though related, from music download on the fixed Internet. Thus whether customers have experiences in fixed Internet or not has little to do with the success of the Chaku-Mero business in Japan. Rather, the major industrial success factors of Japanese Chaku-Mero seem to be the long accumulated music data resources and technological competence (in the karaoke business and MIDI technologies) and institutional arrangements (JASRAC).

The question still defies a quick answer if the success of Chaku-Mero would lead to another success story in Internet music download services. Copyright management and control for original music pieces are much more complicated than for Chaku-Mero. The whole music file is still too heavy to be downloaded even in the 3G mobile networks. Positioned in the "middle" of simple melody and full-track song, the outcome of the Chaku-Uta service seems interesting and suggestive for the future. Yet, it is still questionable if the experiences in Chaku-Uta could be applied to full music download services on the Internet, both fixed and mobile.[5] Expectations for music and graphical content services grow as 3G and 4G mobile technologies come to be realized, but we still must jump over many hurdles to achieve success.

References

Takeishi, A. (2004). Digitalization and the Evolution of Music Business: Interactions of Technology, Business and Music. *Hitotsubashi Business Review, 52*(1), 78–94.

Takeishi, A., & Lee, K.J. (2004). Mobile Music Business in Japan and Korea: Copyright Management Institutions as a Reverse Salient. Working Paper WP#04-02, Institute of Innovation Research, Hitotsubashi University.

Endnotes

1. Data sources for this chapter include published information, industry data on Internet Web sites, and interviews with mobile communication carriers (NTT DoCoMo, KDDI, and J-Phone) and content providers (Xing and Yamaha). We would like to thank these companies and interviewees for their kind cooperation. Note that this chapter was originally written in 2003. Although we have updated wherever possible, some descriptions may be outdated, as is always the case in rapidly changing industries. This research has been financially supported by the Mobile Innovation Research Program at the Institute of Innovation Research, Hitotsubashi University.
2. Karaoke is the activity of singing to specially recorded music for fun. Originated in Japan, karaoke has increased in popularity since the 1980s among Japanese people.
3. Takeishi and Lee (2004) compared mobile music businesses in Japan and Korea.
4. See Takeishi (2004).
5. See Takeishi (2004). Takeishi and Lee (2004) examined changes in mobile music businesses and institutional factors behind them in Japan and Korea, each of which has taken a different evolutionary path.

Chapter II

Strategic Implications of M-Banking Services in Japan

Eusebio Scornavacca, Victoria University of Wellington, New Zealand

Stuart J. Barnes, University of East Anglia, UK

Abstract

This chapter explores the state of the art of mobile (m-) banking in Japan. In Japan, in early 2005, there were more than nearly 100 million users of mobile Internet services. A brief discussion about the main characteristics of Japanese banking practices is accompanied by an overview of this country's mobile market. This is followed by a detailed analysis of the mobile Internet services of three major Japanese banks—Mizuho, Sumitomo Mitsui, and UFJ—and the development of a strategic framework for m-banking. The chapter concludes with a discussion about the future of mobile banking.

Introduction

In 2003, the number of mobile phone users around the world passed the 1 billion mark (Emarketer, 2002). The proliferation of mobile Internet devices is creating an extraordinary opportunity for e-commerce to leverage the benefits of mobility (Barnes, 2004; Chen, 2000; Clarke, 2001; Dholakia & Rask, 2000; Durlacher Research, 2002).

Mobile e-commerce, commonly known as m-commerce, is increasing the overall market for e-commerce by expanding beyond the traditional limitations of the fixed-line personal computers (Anwar, 2002; Barnes, 2002a, 2002b, 2004; Bayne, 2002; Durlacher Research, 2002; Lau, 2003; Newell & Lemon, 2001; Sadeh, 2002; Siau and Shen, 2003).

One of the first commercial applications of the mobile Internet was mobile (m-) banking. M-banking can be defined as a channel whereby the customer interacts with a bank via a mobile device, such as a mobile phone or personal digital assistant (PDA). The emphasis is on data communication, and in its strictest form, m-banking does not include telephone banking, either in its traditional form of voice dial-up, or through the form of dial-up to a service based on touch-tone phones (Barnes & Corbitt, 2003).

One interesting aspect of m-commerce is the differing pattern of development and use compared to "traditional" e-commerce. While the United States leads the world in almost every aspect of traditional e-commerce, it has struggled to make significant inroads into m-commerce. In this market, the leader is Japan (Barnes & Huff, 2003). The Japanese market is at least 2 years ahead in comparison with the West. By the end of February 2005, there were almost 100 million users of mobile Internet services in Japan, 25.5% based on third- generation (3G) networks. The main providers of mobile Internet services in Japan are NTT DoCoMo (i-mode) with 43 million subscribers, KDDI (EZweb) with 17.5 million, and J-phone/Vodapone (J-Sky) with 13 million subscribers. Each of these players also offer 3G network services for their costumers, and as of February 2005, KDDI's Au 3G has 16.8 million users, J-Phone 3G has 0.36 million, and NTT DoCoMo's Freedom of Mobile Access (FOMA) has 8.9 million users (Mobile Media Japan, 2005).

This chapter aims to explore the state of the art of mobile (m-) banking in Japan. The following section provides an overview of the country's mobile market. This is followed by a brief description of the main characteristics of Japanese banking practices. Subsequently, a detailed analysis of the mobile Internet services of three major Japanese banks is presented, and analysed using a strategic m-banking framework. The chapter concludes with a discussion about the future of mobile banking.

Mobile Technology in Japan

This section provides some background on the diffusion of wireless Internet services in Japan. In particular, it explores the key determinants for the high level of penetration, some recent examples of devices, and the general pattern of usage for wireless services.

The Remarkable Diffusion of the Mobile Internet in Japan

In terms of technology adoption and diffusion, Japan provides an interesting case study. In this market, a number of complex forces have combined to create the strong platform for wireless Internet services that we see today. Such forces do not necessarily exist elsewhere—as evidenced by a number of wireless industry analysts—leading to considerable debate surrounding the possible transfer of similar service offerings to overseas markets.

Determinants of wireless Internet diffusion in Japan have included the following (Barnes & Huff, 2003):

- The poor penetration of wireline Internet services, stifled by the high cost of access. Before the introduction of wireless Internet services, only 3% of the population had Internet access.
- Market dominance created by centralised market structures and vertical integration. This has been used by DoCoMo to leverage the supply chain and provide superior i-mode devices, particularly in 1999 and 2000. NTT and NTT DoCoMo have also established a high brand trust.
- Japanese culture. Adoption and diffusion has been enhanced by the Japanese cultural tendency towards group conformity, fascination with small devices, and youth and young adult appeal. Seeing devices used in public has accelerated this process.
- Innovative choices of technology platform. The early selection of packet-switched networks and compact hypertext markup language (cHTML) provided a compelling foundation for service development. Instant, affordable data communications and an easy, familiar development environment created early momentum. The simple and intuitive interface of devices reduced complexity, as did DoCoMo's management of content providers.

Recent Example of Japanese Mobile Device

As a result of its innovativeness, Japan has the most advanced consumer devices available for the wireless Internet in the world. Figure 1 shows one of the many devices available on the Japanese market in January 2005 (NTT DoCoMo, 2005). The F900i takes SXGA (960 × 1,280 pixels) photos using a built-in CCD camera with 1.28 mega-pixel effective resolution and 1.23 mega-pixel recorded resolution. Also it has an external memory slot card—miniSD compatible. The F900i can connect to a PC via USB cable to synchronize schedules and phone books. The F900i is equipped with a Macromedia Flash browser handling Flash applications of up to 100K. Java(R)-based i-appli applications offer a 400K scratch pad plus up to 100K for content/archiving. It also has videophone capability of up to approximately 100 minutes.

Usage Patterns for Wireless Internet Services in Japan

Recent surveys of wireless Internet usage have underlined the peculiar nature of youth and young adult market. Data from the Telecommunications Carriers Association of Japan (Telecommunications Carrier Association of Japan, 2002), suggests that downloading ringing tones is the most popular area of usage, with around one-third of all traffic, as users download tones to personalize their individual phones. This is closely followed by entertainment, with a quarter of all traffic, and a closely related segment,

Figure 1. Japanese mobile phone

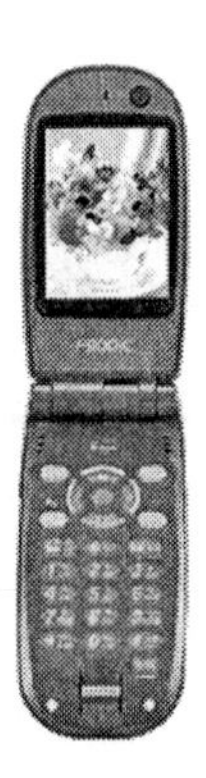

games and horoscopes, with a fifth. Clearly, the overarching emphasis of services is on entertainment, fun, and personalization. Transactions accounted for only 5% of all traffic.

This finding is also supported by another survey conducted by Infoplant at the beginning of 2003 (NE Asia, 2003). This study examined Internet sites registered under "i-menu" (i.e., DoCoMo-authorized sites) and found that 95.5% of the 36,666 respondents bookmark their favourite sites in the personalised "My Menu." Among the bookmarked sites, favourite destinations were sites providing ringing melodies and wallpaper images, followed by news and weather updates. Transaction-oriented sites, such as mobile banking and trading, were not a popular destination of users.

Banking in Japan

Commercial banks in Japan tend to have a strong national base, and are allowed to create nationwide branch networks. The activities they may conduct are diverse; for example, they engage in other financial services such as underwriting of securities and insurance.

Notwithstanding, Japan possesses very peculiar banking practices compared to most Western countries. There is no doubt that in Japan cash is still the most widely used form of payment among individuals. It is not uncommon to see people buying high-value products or services—of $3,000 or more—using cash. The use of debit cards is still in an embryonic stage and personal cheques are rarely issued and seldom used. Therefore, direct debit is widely accepted as the preferred method of payment for utility and telephone bills, mortgages, and many other payments. In addition, the recent growth of personal banking software has encouraged transactions, such as account transfers, via the Web. While credit cards are gaining popularity as a method of payment for individuals, most Japanese are very cautious about using credit cards for Internet shopping (Sumitomo Mitsui Banking Co., 2003).

Automated teller machines (ATMs) in Japan do much more than dispense cash and take deposits. In addition to the services normally offered in Western countries, most ATMs in Japan are capable of domestic wire transfers and accepting time deposits. However, an interesting aspect of the Japanese ATM network is the traditional "curfew." Typically, the customer is only able to withdraw money from the bank's ATM from Monday to Friday 7:00 to 23:00 hours, and 9:00 to 21:00 hours on Saturday, Sundays, and holidays. On weekdays from 8:45 to 18:00 no fees apply for ATM use, but outside of these times a ¥105 fee is charged. Bank customers are also charged by the nature of the transaction (intrabranch, interbranch, or interbank), and the bigger the size of the local bank's ATM network, the lower the charges associated with interbank transactions. Another interesting feature is the localised nature of services; most of the ATMs are linked to one specific bank, and usually do not accept transactions from a different network.

Internet banking operations by Japanese banks have largely been limited to their domestic markets (Pyun, Scruggs, & Nam, 2002). They have also been relatively late to appear compared to other developed markets, for some of the reasons mentioned above. The first cyberbank of Japan—JNB—was established in October 2000 by a consortium consisting of Sakura Bank, Fujitsu Limited, Nippon Life Insurance Company, and Sumitomo Bank. JNB operates over the Internet without a single physical branch. JNB offers savings accounts, time deposit, money transfer and consumer loans, mutual fund accounts, mortgages, fund transfer, and insurance (Fujitsu, 2001). Physical withdrawal can be made at over 100,000 ATMs across Japan.

Given the peculiar banking system in Japan, and the late entry of Internet banks into the market, it is interesting to note the phenomenal expansion of mobile banking services. In a relatively short time, mobile banking has exploded onto the Japanese mobile market. The next section provides more details on m-banking services offered in Japan, examining the value added by the mobile banking services available.

Mobile Banking Services in Japan

M-banking services in Japan have a strong emphasis on mobile phones. Only recently have a few Japanese banks made m-banking services available to clients through PDAs. Typically, the services available via PDAs are identical to those available through mobile phones. Since PDAs are not used by a significant percentage of m-banking customers, and the services available usually do not differ from one mobile device to another, this chapter will focus on the services provided through mobile phones.

According to NTT DoCoMo (2003) there are over 1,370 i-mode–compatible sites under the mobile banking category. For most Japanese financial institutions, mobile banking is a "must have" channel extension. This section examines the mobile banking channel in detail. It is divided into three parts. First, we explain how customers access banking services through a mobile phone in Japan. Second, we provide case studies of three of Japan's largest banks: Mizuho, Sumitomo Mitsui, and UFJ. Finally, we summarise the available services. It is worth noting that one of the authors worked in Japan as a researcher from April 2001 to March 2003 and has considerable experience in using these services.

Figure 2. Access to mobile banking in Japan

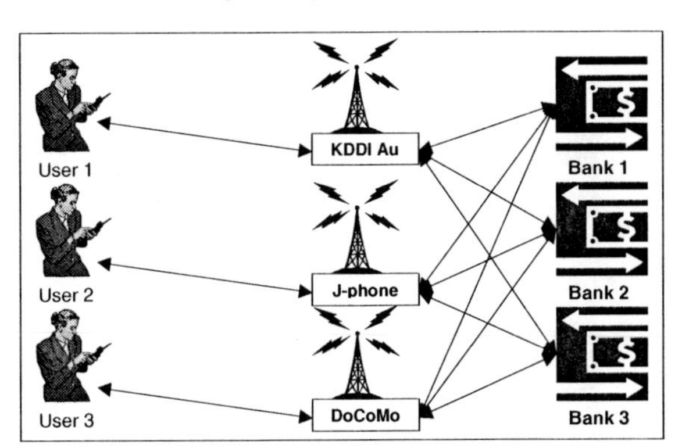

Accessing M-Banking Services

The wireless service providers in Japan—the three main telecommunications operators—do not charge extra fees for using mobile banking services. The user pays the standard fee for sending and receiving information through the handset. Notwithstanding, the banks may charge a fee for some specific services. Most of the banks require the customers to register in order to gain access to m-banking services, although some banks only require registration for advanced features. Figure 2 shows how a mobile phone user can gain access to m-banking services. Most of the banks have Web sites that can be viewed from any device, independently of the wireless service provider. A few applications—such as foreign exchange simulations that run on DoCoMo's Java-based "i-appli" platform—can only be accessed from suitably equipped i-mode devices. For example, suppose that the wireless service provider of User 2 is J-phone. User 2 would be able to access any of the mobile banking services of banks where she holds an account. However, if the bank offers some advanced services that run on "i-appli," she would not be able to access those from the J-Sky platform (J-Phone, 2003).

Figure 3 provides an example of the user of mobile device interacting with an m-banking service. In this scenario, the user of a DoCoMo device is checking her account balance, which takes seven clicks. In step 1, the user selects the "i-menu" from the main i-mode menu. There, in step 2, the content menu list is selected. A wide selection of content is presented—such as weather, news, and so on—and a mobile banking service—"DoCoMo bank"—is selected (step 3). In step 4, "DoCoMo bank" is displayed and "DoCoMo direct" is selected. Consequently, a login page is displayed, where the user must enter his/he ID number and password (step 5). Step 6 presents the user with three options: account balance (selected), money transfer (*furikomi*), and direct debt (*furikae*). In step 7, it shows that the user holds accounts at two branches. After selecting a branch, the balance is displayed in step 8 with an option to return to the menu.

The navigation process from the start-up menu to the bank's homepage varies according to the wireless service provider. For example, if a person wishes to access Mizuho bank mobile services via a particular service provider:

Figure 3. Example — m-banking on i-mode (NTT DoCoMo, 2003

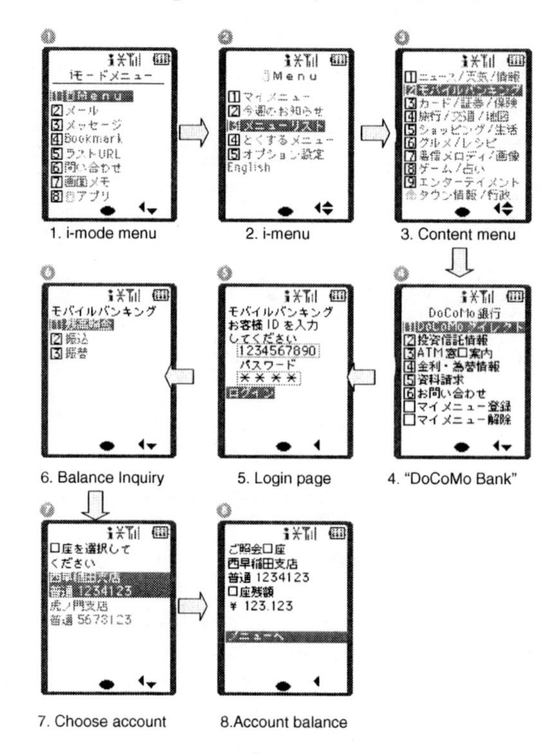

- On a KDDI/Au device the navigation process will be:

  Top menu → Daily life information → Banking, card information → Mizuho bank

- On a J-phone device it will be:

  J-Sky menu → Banking → City bank → Mizuho bank

- And via DoCoMo it will be:

  i-menu → Menu list → Mobile banking → City bank → Mizuho bank

The m-banking services available will vary according to each bank.

Examples of M-Banking Services at Three Banks

In this section, we examine the specific services offered by three Japanese banks. Banks were selected according to the importance of the banks in the Japanese market and data availability on m-banking. According to the *Financial Times* (Financial Times, 2003), Mizuho, Sumitomo Mitsui, and UFJ are among Japan's largest banks. Let us consider each of these in turn.

Mizuho

In Japanese, "Mizuho" means "a fresh harvest of rice," and the phrase "mizuho country"—meaning "fruitful country"—is used to refer poetically to Japan. In April 2002, the Mizuho Financial Group, through a corporate demerger and merger process, reorganized the operations of Dai-Ichi Kangyo Bank, Fuji Bank, and Industrial Bank of Japan into Mizuho Corporate Bank and Mizuho Bank. Mizuho Corporate Bank primarily serves large corporations, financial institutions and their group companies, public sector entities, and overseas corporations including subsidiaries of Japanese companies. Mizuho Bank's main customer base consists of individuals, domestic small and medium-sized or middle-market enterprises, and local governments. Mizuho Securities and Mizuho Trust & Banking also became direct subsidiaries of Mizuho Holdings through corporate demerger, creating a new management structure with four core legally separate subsidiaries (Mizuho Bank, 2003).

The m-banking services provided by Mizuho can be accessed from any Internet enabled device, independent of the wireless service provider. Figure 4 illustrates the login page and the "Mizuho Direct Card" with user ID number, expiration date, and general support information. The bank provides a variety of services via the mobile Web site:

- Check account balance
- Transaction inquiry
- Check previous transactions
- Funds transfer
- Create a transaction statement
- Request cancellation of service
- Check status of service requests
- Payment of bills
- Time deposits
- Foreign currency information
- Change password
- Account information
- Product information
- Branch locations
- Loan information

The bank has combined phone, Internet, and mobile banking services into a single strategy called "Mizuho direct."

Figure 4. Mizuho Bank's login page and user card (Mizuho Bank, 2003)

Sumitomo Mitsui

Sumitomo Mitsui Banking Corporation (SMBC) is a wholly owned subsidiary of Sumitomo Mitsui Financial Group, Inc. (SMFG), a holding company established in December 2002 through a share transfer. SMBC is relatively large; the capital stock of SMBC is worth more than ¥560 billion. The SMBC Group offers a broad range of financial services centred on banking. The Group is also engaged in leasing, securities, credit cards, investment, mortgage securitisation, venture capital, and other credit-related businesses. SMBC employs over 24,000 staff and has a network of 437 domestic branches (Sumitomo Mitsui Banking Co., 2003).

In a similar fashion to Mizuho, SMBC has combined phone, Internet, and mobile banking services into a single strategy called "One's direct." Again the m-banking services provided by SMBC can be accessed from any Internet-enabled device, independent of the wireless service provider. The m-banking Web site offers similar services to Mizuho, including account balance, transaction inquiry, previous transactions, funds transfer, transaction statement, service cancellation requests, status of service requests, bill payment, time deposits, foreign currency information, password change, account information, product information, and branch locations.

UFJ

The UFJ Group was formed in April 2001 as a result of the establishment of UFJ Holdings, integrating Sanwa Bank, Tokai Bank, and Toyo Trust. UFJ Group has been aggressively building a framework capable of supporting a broad range of financial services. In January 2002, UFJ Bank was established through the merger of Sanwa Bank and Tokai Bank, and Toyo Trust was renamed UFJ Trust. Other companies in the group are also in the process of merging, and the UFJ brand is starting to become known in all sectors of the financial marketplace. The UFJ Group focuses on two strategic market segments: retail, especially housing loans and consumer loans, and medium-sized companies. The UFG Group has capital worth over ¥1 trillion (UFJ, 2003).

Like the other banks mentioned above, UFJ bank provides a standard bundle of services via its m-banking Web site: account balance, transaction inquiry, previous transactions, funds transfer, transaction statement, service request cancellation, service request status, bill payment, time deposits, foreign currency information, password change, account information, product information, branch locations, and foreign exchange investment simulation with graphics (which is restricted to i-mode handsets with i-appli). Like the other banks, UFJ has an explicit multichannel strategy, combining phone, Internet, and mobile banking services into a single strategy called "UFJ direct." Most of the m-banking services provided by UFJ can be accessed from any Internet-enabled device, independent from the wireless service provider.

Summary of Available Services

The case studies of the three major Japanese banks are remarkably similar. Each of three banks approached in this study have a very similar portfolio of m-banking services. In addition, they have all recently embarked upon a similar, multichannel strategy that combines telephone banking, Internet banking, and mobile banking services.

To demonstrate the similarity in service provision, Table 1 provides a list of the services offered by each bank. Only two services—loan information and foreign exchange simulation—appear to be unique to a particular bank, albeit different ones. Interestingly, there is no evidence of a special partnership between wireless service providers and the banks; the Web sites of each of the banks could be accessed from any Internet-enabled device, independent of the wireless service provider.

Table 1. Mobile banking services provided by the banks

Service List	Mizuho	Sumitomo	UFJ
Check an account balance	*Yes*	*Yes*	*Yes*
Transaction inquiry	*Yes*	*Yes*	*Yes*
View the last transactions made	*Yes*	*Yes*	*Yes*
Transfer funds	*Yes*	*Yes*	*Yes*
Request a transaction statement	*Yes*	*Yes*	*Yes*
Cancel a service request	*Yes*	*Yes*	*Yes*
Check the status of service requests	*Yes*	*Yes*	*Yes*
Pay a utility bill or credit card	*Yes*	*Yes*	*Yes*
Time deposit	*Yes*	*Yes*	*Yes*
Foreign currency information	*Yes*	*Yes*	*Yes*
Change password	*Yes*	*Yes*	*Yes*
Find account information	*Yes*	*Yes*	*Yes*
Find product information	*Yes*	*Yes*	*Yes*
Find branch location	*Yes*	*Yes*	*Yes*
Loan information	*Yes*	No	No
Information about telephone banking	No	No	No
Simulation of foreign exchange (i-mode only)	No	No	*Yes*

A Strategic Perspective on Japanese M-Banking Services

Now that we have a better understanding of the nature of m-banking services in Japan, let us consider the strategic implications of these services for both banks and customers, particularly in relation to other markets.

Barnes (2003c) examined the strategic implications of m-banking services in a number of technology-related scenarios (see Figure 5). Within Barnes' strategic model, he categorises the potential development of m-banking services in different markets according to the penetration of mobile telecommunications and personal computer technology:

a. Reliance on traditional banking in less developed markets;
b. Extension to multiple banking channels in developed markets;
c. Online banking as a complementary banking channel; and
d. Adoption of m-banking services in emerging and underdeveloped markets.

Within this framework, Japan was given as an example of a market where the potential of m-banking was high, due to the strong penetration of wireless data services in advance of PC access.

While m-banking does present considerable potential, it is now clear that the situation is more complex. In particular, the adoption path for technology-based banking services is quite different from Europe or the United States. Traditional banking services (step 1 in Figure 5) in Japan have been quite localised, low-tech, and not oriented to the consumer (especially in the early to mid-1990s). As explained in section 3, this is evidenced by the persistence of cash for high-value transactions, the quite separate service offerings of

Figure 5. Strategic model for m-banking (Barnes, 2003c)

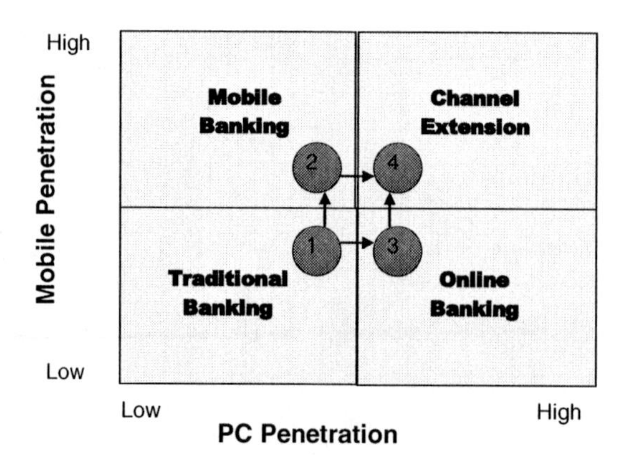

banks, typically without collaboration, and the rigid nature of services, for example, ATM curfews.

The development of remote banking services was initially spearheaded by m-banking (step 2), which emerged before online banking. The development of these services provided a very different set of strategies. Services became customer oriented, and banks clambered to provide similar services on the wireless Internet, particularly out of fear of losing customers through poor service provision. This has been despite the slim usage patterns of services. The services tended to be stand-alone and independent of other banks or operators.

More recently in Japan, the wireline Internet has begun to make inroads (step 3). The rollout of broadband is likely to provide a strong impetus to further adoption. In this climate, online banking has started to develop. In terms of the technical platform, this provides a superior service offering (Barnes, 2003c), and a complementary channel to other methods of access. In the future, we are likely to see a more developed multichannel strategy (step 4) pursued throughout all banks, and there is evidence to suggest that this is already beginning to take hold in the strategies of the banks examined in this study.

The strategic benefits of Japanese m-banking are another aspect of service development worthy of further investigation. To this end, Barnes (2003c) provides a simple framework for evaluating the direct and indirect strategic benefits from m-banking. In all, there are five main types of benefits. Table 6 applies the strategic benefit framework to the

Table 2. The strategic benefits of Japanese m-banking services

	Mizuho		Sumitomo Mitsui		UFJ	
Indirect benefits						
Cross-selling	✓✓	Mizuho has been more proactive than the other banks, for example, in cross-selling loan products.	✓	There is evidence of cross-selling for products.	✓	There is evidence of cross-selling for products.
Cheaper customer acquisition	✖	There is no evidence to suggest that customers have joined the bank via the m-banking service.	✖	There is no evidence to suggest that customers have joined the bank via the m-banking service.	✖	There is no evidence to suggest that customers have joined the bank via the m-banking service.
Customer loyalty	✓	Mizuho developed the service to deter switching of customers and extend loyalty.	✓	Sumitomo Mitsui developed the service to deter switching of customers.	✓	UFJ developed the service to deter switching of customers.
Direct benefits						
Customer benefits	✓✓	The compelling value proposition presented to customers from a personal, mobile channel has created benefits. The cost of services is a drawback. Usage by customers is low.	✓✓	The compelling value proposition presented to customers from a personal, mobile channel has created benefits. The cost of services is a drawback. Usage by customers is low.	✓✓	The compelling value proposition presented to customers from a personal, mobile channel has created benefits. The cost of services is a drawback. Usage by customers is low.
Cost reduction	✖	M-banking services are seen as an expensive necessity for Mizuho. Although the potential for cost reduction is there, the critical mass of users is not.	✖	There is no evidence of cost reduction. Developing m-banking services has been an expensive necessity for Sumitomo Mitsui.	✖	There is no evidence of cost reduction. Developing m-banking services has been an expensive necessity for UFJ, especially as it has tried to provide innovative services on the i-appli platform.

Japanese example, examining whether such benefits are apparent in each of the case studies.

As Table 2 demonstrates, the main beneficiaries of m-banking services in Japan have been the users or consumers. The swift replication of mobile service offerings among banks has created a relatively level playing field, while at the same time adding the cost burden of banks, which have typically not gained the critical mass of usage to return benefits. In addition, there is evidence to suggest that technologies such as online and mobile banking can encourage greater transaction volumes for a typical user, which, although reducing cost per transaction, can increase the overall cost of services (Olazabal, 2002).

Summary and Conclusions

Mobile banking was one of the first commercial applications of the mobile Internet. Currently, in Japan, most mobile data devices have access to a multitude of m-banking services. Indeed, there are now more than 1,300 mobile Web sites for Japanese financial institutions. Mobile banking has proven to be a viable channel for consumer banking with exciting possibilities of growth.

This chapter has examined mobile banking in a Japanese context. The three major Japanese banks presented in this study have a very similar strategy and offer a similar range of services. The banks provide services to customers independently of the wireless service providers. However, as the above analysis has shown, mobile banking is fast becoming a channel of extension of electronic banking—part of a multichannel strategy. These results can provide some useful insights and implications for other markets where mobile Internet-based services are following or could follow similar trajectories.

Among the wireless service providers in Japan, DoCoMo has taken the most aggressive strategy for mobile banking and mobile payments. In 2001, DoCoMo launched the c-mode service in cooperation with Coca-Cola; c-mode allows i-mode users to buy drinks and other items (e.g., ringtones, graphics, or tickets to various types of entertainment) from vending machines. DoCoMo also offers—in a partnership with IYBank—a service that lets IYBank customers withdraw and deposit money at ATMs in convenience stores and supermarkets using their handsets instead of a cash card. In May 2003, DoCoMo launched the DoCommerce payments service, enabling both 2G and 3G users to shop using their mobile phone with just a single user ID and password. Initially, 10 virtual shops were available on the DoCommerce site, selling items such as clothing accessories, perfume, and healthcare. An account aggregation feature enables DoCommerce users to check—simultaneously and on one single screen—the balances of their various financial accounts held with banks and credit card companies. More than 15 financial institutions, including VISA and MasterCard, are now part of the service (NTT DoCoMo, 2005).

The next 2 years will provide important evidence regarding the future of m-banking in Japan. The general trend towards acquisition and merger could well move over into wireless financial services. In particular, there is considerable speculation regarding future integration between wireless service providers and financial institutions. Another development, the advanced use of subscriber identity module (SIM) cards for personal and financial data, will further enable data integration and provide an important building block towards the establishment of a true "mobile wallet."

References

Anwar, S.T. (2002). NTT DoCoMo and m-commerce: A case study in market expansion and global strategy. *Thunderbird International Business Review, 144*(1), 139–164.

Barnes, S.J. (2002a). Under the skin: Short-range embedded wireless technology. *International Journal of Information Management, 22*(3), 165–179.

Barnes, S.J. (2002b). The mobile commerce value chain: Analysis and future developments. *International Journal of Information Management, 22*(2), 91–108.

Barnes, S.J. (2003c). Pocket money: Banking on mobile devices. *e-Business Strategy Management, 4*(4), 263–271.

Barnes, S.J. (2004). *m-Business: The strategic implications of wireless communications.* Oxford: Butterworth-Heinemann.

Barnes, S.J., & Corbitt, B. (2003). Mobile banking: Concept and potential. *International Journal of Mobile Communications, 1*(3), 273-288.

Barnes, S.J., & Huff, S. (2003). Rising sun: i-mode and the wireless Internet. *Communications of the ACM,* in press.

Bayne, K.M. (2002). *Marketing without wires: Targeting promotions and advertising to mobile device users.* London: John Wiley & Sons.

Chen, P. (2000). Broadvision delivers new frontier for e-commerce. *M-Commerce, October*, 25.

Clarke, I. (2001). Emerging value propositions for m-commerce. *Journal of Business Strategies, 18*(2), 133–148.

Dholakia, N., & Rask, M. (2000). *Dynamic elements of emerging mobile portal strategies: M-commerce is all about personalization, permission and specification.* RITIM Working Paper, RITIM.

Durlacher Research. (2002). Mobile commerce report. Retrieved July 10, 2002, from *www.durlacher.com*

Emarketer. (2002). One billion mobile users by end of Q2. Retrieved May 27, 2003, from *www.nua.ie/surveys/index.cgi?f=VS&art_id=905357779&rel=true*

Financial Times. (2003). SMFG shores up finances with issue plan; Japan's number-two takes measures to deal with potential problems on capital adequacy ratio. *FT2*, 17 February, p. 20.

Fujitsu. (2001). Japan Net Bank—case study. Retrieved May 27, 2003, from *http://crm.fujitsu.com/en/case-study/japanljapan-net*

J-Phone. (2003). J-Phone. Retrieved May 31, 2003, from *www.j-phone.com/scripts/japanese/top.jsp* (in Japanese)

Kalakota, R., & Robinson, M. (2002). *M-business: The race to mobility*. New York: McGraw-Hill.

KDDI/Au. (2003). Sehin rain apu. Retrieved June 1, 2003, from *www.au.kddi.com/seihin/index.html* (in Japanese)

Lau, A.S.M. (2003). A study on direction of development of business to customer m-commerce. *International Journal of Mobile Communications, 1*(1/2), 167–179.

Mizuho Bank. (2003). Mizuho ginko. Retrieved June 1, 2003, *www.mizuho-fg.co.jp* (in Japanese)

Mobile Media Japan. (2005). Japanese mobile Internet users. Retrieved February 2, 2005, from *www.mobilemediajapan.com*

NE Asia. (2003). I-mode users like sites for ring melody, wallpaper data most: survey. Accessed June 1, 2003, from *http://neasia.nikkeibp.com/wcs/leaf?CID=onair/asabt/news/236966*

Newell, F., & Lemon, K.N. (2001). *Wireless rules: New marketing strategies for customer relationship management anytime, anywhere*. New York: McGraw-Hill.

NTT DoCoMo. (2003). DoCoMo Net. Retrieved June 1, 2003, from *www.nttdocomo.co.jp* (in Japanese)

NTT DoCoMo. (2005). Sehin rain upu. Retrieved January 20, 2005, from *www.nttdocomo.com* (in Japanese)

Olazabal, N.G. (2002). Banking: The IT paradox. *McKinsey Quarterly, 1*, 47–51.

Pyun, C.S., Scruggs, L., & Nam, K. (2002). Internet banking in the U.S., Japan and Europe. *Multinational Business Review, Fall*, 73–81.

Sadeh, M.N. (2002). *M-commerce: Technologies, services, and business models*. London: John Wiley & Sons.

Siau, K., & Shen, Z. (2003). Mobile communications and mobile services. *International Journal of Mobile Communications, 1*(1/2), 3–14.

Sumitomo Mitsui Banking Co. (2003). Sumitomo Mitsui Banking Corporation. Retrieved May 30, 2003, from *www.smbc.co.jp* (in Japanese)

Telecommunications Carrier Association of Japan. (2002). *Market report*. Tokyo: TCAJ.

UFJ. (2003). UFJ. Retrieved May 30, 2003, from *www.ufj.co.jp* (in Japanese)

Endnotes

- An earlier and shorter version of this paper appeared as Scornavacca, E., & Barnes, S.J. (2004). M-banking services in Japan: A strategic perspective. *International Journal of Mobile Communications, 2*(1), 51–66.

Chapter III

Buongiorno! MyAlert: Creating a Market to Develop a Mobile Business

Guillermo de Haro, Instituto De Empresa, Spain

José María García, Instituto De Empresa, Spain

Abstract

In 1999 Jorge Mata, vice president of Broadvision and former expert in interactive solutions for Banco Santander and McKinsey, decided to leave everything to create MyAlert. The company was born on the basis of offering the same Internet services on the new and growing mobile devices. With a strong financial capitalization after raising more than 50 million euros during the bubble burst, in 4 years the company figures were in the black, and the journey had led to the creation of the European sector of mobile data services market and the European leader in that sector. As Charles Darwin emphasized, if a being wants to survive in a shifting environment, it must evolve at least as fast as the medium itself: Buorngiorno! MyAlert ruled the change.

Introduction

This case helps to understand, through the history of Buongiorno! MyAlert (García & de Haro, 2003), how the mobile data services market works, who are the main players, and what are their business models, explaining clearly the *initial development of the mobile data services market* in Europe.

MyAlert was a pioneer in the sector. It was the first company to have the idea and to define a business model. It was the first to appear in the mobile services market, leveraging on its own technology platform. Also, it was the first in raising enough money to make this market grow on the basis of products and services development. Once the market was launched, the firm consolidated the business model by merging with Buongiorno.

Continuous design and development of products, applications, and business models was a key. As Nomura pointed out, "MyAlert is positioned to move into several areas of the value chain at low cost." The flexible strategy of the company made it possible to lead a changing market, influenced by many major players. Value chain analysis and strategic diagnosis on the environment and its evolution is provided. Also product development and business models are explained.

Another key was the market-driven evolvement of the offering, maintaining a close look at customers, competitors, and environment. Firm changed on a market basis, not on a technological basis, for example, focusing on SMS instead of WAP when it was "flavour of the month."

Beginning from a "sweet spot" of the value chain with technology as their competitive advantage, by 2004 the company had evolved looking for new business, consolidating and looking for new markets.

The Company Evolution

MyAlert

The idea was simple: to take advantage of the "alerts"—data messages sent through the mobile phone network—to create services like those being launched in the Internet. MyAlert started as a "portal of alerts": users defined via the Web which information alerts they were interested in receiving via a text message delivered to their mobile phone. In the midst of the Internet boom in Spain and Europe, Jorge pioneered the extension into the mobile phone's mailbox.

Jorge Mata, telecommunications engineer, MBA from New York University, 4 years at McKinsey & Co, VP at Banco Santander developing Internet and GSM banking solutions, and VP at Broadvision, had the idea. Time to market was critical: he gave up his job and also an important stock-option plan and explained his vision to Pehong Chen, Broadvision's founder. Chen decided to invest in the project. MyAlert was created in March 1999 with 500,000 euros capital.

Jorge recalled later, "That could seem a lot of money, but when you make strong expenditures in technology and staff, it runs away quickly. I got top engineers, we got an office and we tried to make the money last as much as we could . . . but in September 1999, we had nothing." In October 1999 a new capital increase of 4 million euros was subscribed mainly by Banco Santander and BBVA. Jorge attracted qualified top talent, but the bulk of the recruits happened in the technology area, reaching 35 employees in 1999, most of them engineers devoted to developing the platform.

Market Launch

The first versions of MyAlert's Internet portal were operational in July 1999 with services such as top headlines of the day or soccer match scores, based on data feeds provided by Europa Press, a leading Spanish news agency which also took a stake in the company. Due to agreements with other content and service providers the number of services increased: information about travel and services as innovative as an alert to remind you to quit smoking.

The company became the leader within that market segment and began to identify potential business services for the companies it was partnering with, starting to develop customized services for corporate customers. For instance, the recruitment consultant Adecco could contact candidates in record time by delivering them employment offer alerts in real time. Traffic in the portal increased hand in hand with product development, surpassing in less than a year 200,000 registered users without a substantial advertising expense.

MyAlert went for all European major markets replicating what was already working in Spain: launching advertising supported free services in a way that allowed growing and developing a local presence and then starting offering business services based on its proprietary technological platform. France was first in October 1999, and by July 2000 Italy, Germany, and the United Kingdom were covered through organic growth. For R&D a development center was based in Madrid, and investments were made in companies in Bulgaria (low costs and high productivity) and Finland (control stake in Future121, specialized in WAP and 3G developments).

The team reached 80 members and consolidated internationalization but the market demanded further growth. In May 2000 MyAlert successfully closed a new financing round, bringing in an additional 48 million euros from top-tier investors such as Nomura, Brokat, 3i, or the original shareholders. Investors found highly valuable the proprietary technology platform. MyAlert's market value went from 14.2 million euros by December 1999 to 163.8 million euros a year later.

But the NASDAQ Index fell dramatically in 2000. Markets went deeply pessimistic, the economy fell into recession, and new technologies took the worst part of the crisis. Suddenly MyAlert was forced to confront its business model with the new scenario.

Buongiorno! MyAlert

After the financing round two challenges were yet to be solved. First, how to sustain the speed required for growth, and second, how to move from now unacceptable "new economy" standards to a positive P&L as required by the "old economy"?

Some subjects became important. First, competition escalated and a substantial chunk of the company's expected revenues came from advertising and mobile commerce, both sectors seriously hurt by the crisis. The company had reached a 5-million-euro turnover, one of the highest within this sector, but it was presenting sizable losses. Financials were acceptable due to the investment community supporting the strategic need for short-term strong losses to buy market share, but the financial markets were now requiring the revenue base to cover the cost base in every venture.

Second, the targets of growth and leadership required reaching in all major markets a leading position, something impossible to achieve through organic growth: it was the hardest option from a practical standpoint but could also seriously put into risk the company's bottom line. Another option was a sell out to a strategic partner, following the Finnish IOBox example, bought out by Telefónica for 230 million euros—even when its yearly turnover was 60,000 euros.

MyAlert decided to seek for a "twin soul," a company sharing the same objectives and ambition. By September 2001 the company agreed to merge with BuonGiorno!, Italian leader in personalized e-mail alerts and newsletters, creating a 260-employee company, with nearly 30-million-euro yearly revenues and 20 million subscribers (a yearly total of more than 3,000 million messages via Internet or mobile phone). Curiously both companies had exhibited an impressive story in raising funds: together more than 85 million euros.

Complementarity in their core skills was remarkable: MyAlert's strengths were in the mobile data services market; Buongiorno! was the leader in e-mail marketing services, having reached an 80% penetration over the total PC base in Italy.

Restructuring to reap synergies, decrease costs, and speed up the path to profitability was simplified by the complementarities in the strengths: Buongiorno! has advertising, marketing know-how, and Italian leadership, while MyAlert has ASP services technological excellence and reference in Spain.

Mauro del Río, founder of Buongiorno!, was president and Jorge Mata was vice president. Andrea Casalini, former CEO of EDS for Italy, was the new CEO. One year later 34 million subscribers were reached and the first positive EBITDA achieved.

The Mobile Data Services Market in Europe

MyAlert and Buongiorno! grew in the midst of the two main business driving forces at the turn of the 20th Century: the development of mobile telephony as a new mass communication channel and the Internet as a universal information network.

The mobile telephony market developed in an accelerated way during the 1990s. Initial marketing was targeted to corporate users, soon becoming a mass market. In 2004 the total number of mobile telephone handsets worldwide was estimated at no less than 1,000 million, with more than 450 million users in Europe (EITO, 2004), surpassing the number of fixed telephony handsets in several countries.

A key factor for this development in Europe was the adoption of a shared digital standard, GSM, since 1992 (Huidobro, 1996). This standard helped a faster market penetration and a substantial acceleration in its innovation curve creating a pan-European market. Development of new functionalities or handset designs, and aggressive marketing offers from the telecom operators, such as subsidizing the cost of the handsets speeded up the market growth, with rates exceeding a yearly 60%. In Spain figures changed from less than a million users in 1995 to 7 million in 1998 or 15 million in 1999 after the entry of a third telecom operator (Amena). By 2004 more than 36 million users had a mobile phone.

The Internet was another force allowing the launching of new businesses. Its interactivity and universal access provoked an accelerated development of the Internet user base. By August 2004, worldwide total users reached almost 800 million (according to the Internet World Stats).

The development of both technologies revolutionized the economy and changed global society, altering social usages or creating new sectors and corporations.

Mobile Data and Value-Added Services Markets

"Mobile data services" refers to the delivery of information messages—as opposed to voice messages—through the mobile telecom networks.

Two distinctive businesses may be included: the delivery of messages from one user to another ("peer-to-peer" or "P2P") and the value-added services provided by a third party ("value-added services" or "VAS"). Both use the same technology platforms and communication networks, but they are different businesses in terms of industry players and value split. P2P services are mainly a simple communication service provided by the mobile telecom operator; VAS allow the users access to every sort of digital alerts supplied by content and service providers, similar to Internet browsing.

The mobile data services business started taking advantage of the Short Messaging Service (SMS) standard technology. Its enthusiastic usage by the youngest mobile customers—which stand for 45% of the European mobile users—made them a new revenue model. According to analysis, the European market in 2002 implied more than 10,000 million monthly text messages, and an average of 35 text messages a month per user.

Most of the revenue was coming from P2P services, although VAS were increasing. New significant submarkets appeared, such as mobile gaming (around 500 million euros in Europe; Frost & Sullivan, or ring tones' downloading (1,500 million euros in Europe; Strand Consult. Expected growth trend favors a much higher relative increase of VAS, which will exceed the P2P messages share in 2005 according to Strand. In that year, data mobile services will account for 33% of total ARPU, and 17% will come from VAS.

Another aspect of this market is its growth rate. Telefónica Móviles, already exceeding the 15% share in 2002, expected text messages to become 30% of its total turnover by 2005. These are startling figures considering that by 2000 the mobile data revenues were only 4% of the total mobile telephony market in Spain.

Evolution of the Mobile Data Services Market

How to enable such a spectacular increase in sales? The main agents focused on data services after the SMS boom. Nevertheless and because of the SMS limitations, new technology platforms with a broader range of features were developed to guarantee future growth (DBK, 2003).

By 2000 the mobile telecom industry started to create the so-called "Mobile Internet" (Lautenschläger & Schmidtke, 2000). The new model would be similar to the Web navigation: mobile portals translating browsing experience to the mobile phone screen interface. They failed as WAP protocol was unable to provide an appealing enough user experience (slow, no attractive contents, etc.).

Growth was fuelled by new services such as the adoption of Japanese i-Mode, an alternative technology allowing basic services (weather forecast and horoscopes) and advanced messaging features (e-mail and image delivery via color screens). So it seemed necessary to develop new technologies with faster and higher data transmission capabilities to enable a substantial increase in the number of mobile services provided (Lamont, 2001).

In Europe the industry decided to launch a new standard: UMTS (Universal Mobile Telecommunications System). The development process became much slower than expected (Gómez & González Martínez, 2001). In the meantime the industry kept on milking GSM or using other transitional technologies (2.5G), such as HSCSD (High Speed Circuit Switched Data), which works at 56Kbps and only requires software updates, GPRS (General Packet Radio Service), which works up to 115Kbps but requires new hardware routers, or EDGE (Enhanced Data rates for GSM Evolution), which is closer to 3G and able to reach 384Kbps. Furthermore, network equipment and handset manufacturers faced difficulties in developing in time the new hardware to put in place the UMTS mobile networks infrastructure.

At this very moment in 2002 MMS (Multimedia Messaging Service) messages were launched, which improved P2P applications (enhanced e-mail, image delivery, audio and video delivery, etc.), browsing and downloading applications (news, horoscopes, adult entertainment, games, etc.), and massive participation applications (voting, quizzes, etc.).

Industry expected a new increase in the usage level of its customers but this new message format required new handsets. Camera phones, phones that allow music downloading, personal organizer phones, or handsets geared towards gaming were also launched (Funk, 2004).

This phenomenon reflects the absolute need for the industry to keep its high growth rate on the basis of handset renewal and additional features. NTT DoCoMo managed to bring to the Japanese market 3 million camera phones in 6 months.

Main Players

To make a data mobile service work, participation of different players is necessary. Buongiorno! MyAlert is a specialist or "pure player" in the production of VAS. The other agents may be grouped in four areas: mobile telephony operators, technological infrastructure providers (hardware, software, and services), media, and interactive media.

VAS providers were the smaller in size. They identified a growing market niche and launched their services, directly to end users (downloading of logos), or developing tailored solutions for the other type of players involved such as telecom operators and infrastructure providers (developing information services customized to their customers), mass media (voting services to interact in a TV program), or interactive media (mobile versions of their offering).

Pure players are heterogeneous in terms of business lines (some specialize in ring tones, logos, and TV voting; others are mobile marketing specialists; and some others cover the whole spectrum of services) and in their background and business approach (some have strong technological foundations, other come from the advertising field).

The number of companies in this scene was high. Easy access to the necessary technology was possible once the market exploded, no longer being a barrier to compete. However, top players were concentrated, and in 2002 80% of the total SMS traffic in Spain was split among five or six companies, when there were more than 40 companies competing. Despite its pan-European development, market remains highly local. Top positions are usually taken by companies tied to the top local telecom operators and media or by "early movers" within that country.

In Europe relevant companies may be divided into four groups:

1. International companies, relevant position in several countries, covering a broad range of services, sound technological foundations: Buongiorno! MyAlert or iTouch.
2. International companies with presence in several countries, single product line: KiWee for consumer services or 12Snap for business services.
3. Relevant companies for a single market: MoviListo in Spain.
4. Companies tied to a main player: Vizzavi (Vodafone) or Terra Mobile (Telefónica).

Their business depended on share captured from mobile operators' total revenue and marketing budget that their customers channeled through them. Telcos kept a share of

Table 1. Players in the mobile data services market

PLAYER	CHARACTERISTICS	SAMPLE PLAYERS
Mobile telephony operators	Own the network, provide voice and data telecommunication services	Vodafone, Telefónica Móviles, TIM
Infrastructure providers	Produce and provide the necessary technological infrastructure and services (hardware, software, services) to enable telecom services and network intelligence leading the way to the provision of VAS	Hardware: Nokia, Ericsson, HP Software: Nokia, Microsoft, Oracle, Symbian Services: IBM, Accenture
Media groups	Produce content and entertainment which may be enhanced by mobile telephony- enabled services	Vivendi, Bertelsmann, Endemol, PRISA
Interactive media	Produce interactive content and entertainment which may be enhanced by mobile telephony-enabled services—mostly Internet portals	Terra, Wanadoo, Yahoo!, eBay, Amazon

revenues while the rest was split among the agents involved in the production of the service (content providers, associated TV programs, etc.). In Spain and Italy it was usually 50%, in Japan DoCoMo was only charging 9%.

VAS providers seemed to have placed themselves in a high-growth hot spot. The Economist Intelligence Unit emphasized, "Will survive those who exploit today's technologies and successfully drive the transition to deliver other applications that consumers will pay for." And the report added that Buongiorno! MyAlert was at the forefront of the international wireless sector.

Buongiorno! MyAlert Business Model

After the merger product portfolio was conformed by technology and mobile marketing services provided by MyAlert and the interactive marketing services provided by Buongiorno. The business model was based on three revenue lines: advertising, business services, and consumer services. But when revenues were generated by a corporate customer it was called a "business service," and "consumer services" if they came from an end user.

Business Services

Services sold to corporate customers like advertising or interactive marketing. Also custom made applications to allow customer companies perform these services.

In the beginning mainly e-mail newsletters and mobile alerts' sponsorship, on the basis of its database with more than 34 million subscribers and its CRM tools which allowed segmentation and campaign hit rates higher to traditional media. A business based on volume: marketing impacts through new channels.

Lately appeared marketing services geared towards brand building or loyalty: interactive games, direct promotions and even interactive market surveys. The company developed a new business model: "Digital Marketing Project," consulting-like projects with higher margins and project size.

Head & Shoulders® launched a campaign in association with the movie *Men in Black*, giving the chance to win a Smart car prize to those sending messages to the 5556.

Another business services was the provision of some infrastructure and technology services to corporate customers. Thus CRM, e-mail marketing and SMS delivery tools were available through ASP (Application Service Provider) agreements, with relevant revenues.

Consumer Services

They are paid by the end user, who is billed by the mobile telco. Interactive games, voting, downloading of ring tones and logos, and P2P communication services such as SMS-based chats. The company was continuously investing in the development of new applications such as MMS-supported games or group messages.

Model based on the revenue split of the price paid by the end user. Telcos charged each user a previously set price (in Spain usually in the 0.3–0.9 euros range). This revenue is broken down between the telco, the content provider, and the service aggregator or VAS provider. Telcos retain 50% of revenues. If product was fully developed and managed by Buongiorno! MyAlert, it could keep the other 50%, otherwise it depended on the agreement among partners involved in the service development. KPMG estimated an average of 12% for the content provider and 38% for the VAS provider.

Organization Structure and Operations

After the merger a new organizational structure was defined and conformed by geographical markets rather than product units. International structure was defined along the already-existing country units: Italy, Spain, France, United Kingdom, and Germany/Austria. Each branch was run by a Country Director leading a sales managers team. Only Italy was structured per product line.

Country business units were supported by centralized staff units at the new Italian headquarters. There were two types of staff: purely corporate departments (Administration or Finance) and operational departments (IT, Content Development, and Customer Relationship Management). If a new game was needed the project manager would make a request to IT and this unit would make it work within the technological platform. This area was also in charge of the management and administration of B!3A. Content Development was accountable for negotiating and managing third-party content licensing terms, and the development of Buongiorno! MyAlert's own proprietary content. The CRM unit helped in the personalization of advertising and marketing campaigns.

The launching of a new service was centralized in the sales managers, whose role evolved into that of "product managers." In consumer services, the launching decision was made by each sales manager. With the new service defined the IT team started developing or customizing the necessary support systems. Meanwhile, the Content Development team acquired or licensed the appropriate contents required. In a parallel process, the service is added to the mobile operators' network infrastructure. The sales manager also decides on the appropriate marketing and is accountable for the profitability and final results of the service. In business services a similar process is followed with a consulting-like business approach.

This process differs from MyAlert's pre-merger product development methodology. Formerly the decisions about new products were taken by top management, based on the ideas and products developed by the more than 100 engineers in charge of MAGO (ideas such as m-auction services). In the next step these technological solutions were enhanced by features and content ideas developed by the Marketing department (for instance, defining specific alerts, such as the nonsmoking example). Finally only those new products successfully marketed by the sales managers were incorporated to the technological platform.

Buongiorno! MyAlert's Technology

Andrea Casalini stated that "Buongiorno! MyAlert has two distinguished technological platforms. MAGO is like a jumbo jet: highly advanced technology, but very costly maintenance. B!3A resembles a fighter aircraft: fast and cheaper to maintain, but with more limited features." At merger time, these platforms were able to process more than 250 million monthly e-mails and 2,000 SMS per second, respectively. B!3A was created to manage very large Internet end-user communities, and MAGO was geared toward the management of wireless users communities.

MAGO

MAGO had allowed creating and sending every type of highly personalized alerts to end users, through different channels (SMS, e-mail). A system supported a personalization engine, several system management tools, client and end user interfaces, and connectivity devices with the different networks. From the start it was conceived with a prospective vision to serve as the basis for future development. A highly scalable architecture, it is robust and flexible, based on the compliance of industry standards which would allow future growth and multiple-user acceptance. Its architecture was structured around CORBA, an open object communications standard. Its software was coded in object-oriented languages (J2EE, C++), JavaScript, and XML/XSL. The operating system was HP/UNIX and the primary database is Oracle Parallel Server RDBMS.

Other additional services were security (PKIs), commerce gateways, or event managers (www.buongiorno.com). Part of these applications and services were developed internally within MyAlert's development team, while others were based on adapting third-party developments.

The system included management tools and interfaces developed to allow customer companies access to the system: the foundations of the ASP business model. Easy to use and configure applications supported in a Web interface and application performance metrics are incorporated to increase the value for customers. The end-user interface was also developed in Web format easily customizable to customers.

The delivery engine integrated communication channels with end users through own-developed gateways, managing mobile networks (SMS, WAP, etc.), Internet, even a UMS (Unified Messaging System) integrating fax, voice and e-mail, among others. Because of its international ambition it was designed to support GSM, GPRS, TDMA, and

CDMA standards. The company developed its own virtual network to link mobile operators' message centers. This network was a competitive advantage and barrier to entry for any competitor.

By 2002 due to the fine welcoming of its technology the company started "packaging" its technological platform into a software license agreement for its usage by third parties.

Buongiorno! and the Merger

B!3A was created only to enable the massive delivery of e-mail messages: the Buongiorno! branded daily bulletins. Afterwards delivery through platform was offered to corporate customers via ASP services. It was claimed that wherever there was a PC in Italy, there was a BuonGiorno! newsletter.

B!3A was a simpler platform, based on the Linux open-source operating system and developed on Java, built with the purpose of achieving the required performance at minimal cost. For this reason B!3A had its own application server, and could not easily work with other standards, which made impossible its license to third parties.

Performance was the main driver, as it should manage a high number of subscribers and e-mail messages to be delivered (in late 2002 the company servers delivered up to 400 million messages monthly by sending the company's different newsletters to its 34 million subscribers). This implied requirements in terms of scalability, to cope with continuous increases in traffic volumes, and flexibility, because of the seasonality and peak operation highs. The must was having a technology that could easily operate in an extremely short time, rather than a safer and sounder solution implying costly development and maintenance.

Also MAGO user interfaces were based on a Web format, accessible with an identifier and a password, without requiring software installation. The management tool allowed multichannel campaign management (wireless and e-mail) based on a single database, with CRM capabilities and campaign reporting generation. More than 500 corporate customers used this tool.

After the merger, both platforms continued delivering their services and carrying on "independent lives." Even for the same service, the technology platform used could be different depending on the country.

Buongiorno! MyAlert simultaneously marketed both platforms to its corporate customers with license agreement, ASP business model, or even customized developments to generate their own corporate platforms. This turned into a new separate business unit which focused on software licensing and technology consulting.

B!Digital Technologies

By the end of 2002 B!Digital Technologies was a separate P&L-focused business unit in break even. It brought together some of the Group's business services, such as software licensing and technology consulting, and R&D efforts.

Born with a technical team of 40 engineers (half based in Bulgaria), sales headquarters in London, and operating center in Madrid, and with an interesting customer portfolio (Hutchison 3G and the Chinese Government), was based in the software license of the MobileCast Messaging technological platform, the new commercial name for the latest evolution of MAGO. For licensing purposes an API was added allowing its modification by their clients for new developments and interfaces with their own back-end systems (billing, CRM, provisioning, etc.).

Core customer was Hutchison 3G, which selected the platform to launch its 3G alert and messaging services in seven countries. Another customer was Hellas Online which licensed also Infotainment applications, enabling it to offer in its home market services comparable to those currently offered in Spain and Italy by Buongiorno! MyAlert.

A second line of business was technology consulting professional services, for example, the creation of an intelligent traffic system for the Chinese Government, enabling alerts to the authorities about traffic signs and about congestion and routes.

B!Digital Technologies has continued development efforts towards MobileCast.

Conclusions

The Business Model

SMS alerts potential to create information and interaction services was huge. It was feasible the adoption by end users (Montero, 2000). The mobile market was under an explosive growth so the product was an open door to new services. As a pioneer the leadership was easy to achieve due to lack of competence. The matching of the entrepreneur and the project was perfect: Jorge had the experience (technology, business development), knowledge (in the new interactive channels), and ambition (refused his stock options). But there was a black spot: the lack of a clear and profitable business model from the beginning.

One of the main factors for the success and leadership of the company has been flexibility, the ability to change with the environment. To analyze this we will focus in the value chain and its evolution. The value chain of any industry is a conceptual tool that defines the activities needed for the creation and deliverance of a product or service. The methodology is to "chain" graphically the different functions that get to the final product, so it is easy to understand all the steps needed to make that product or service available.

The original value chain was as follows:

- Content: Development of the content of the alerts (the soccer info with the results of a match in real time)
- Packaging (or VAS): Integration of the content with the technical and operational requirements to send it (the soccer result in the text of an SMS)

- Marketing and Sales: To achieve new users (TV campaigns for the soccer alert)
- Transfer: Net providers who send the alert through the mobile networks to the end user (Vodafone sending the soccer message to a Vodafone end user)
- Infrastructure Providers of VAS: Development of the technology (software, hardware, services, etc.) to design and package the alerts (a tool to customize the soccer clubs' alerts to each end user)
- Infrastructure Providers of Transfer: Development of the technology to deliver the alerts (the telecommunication equipment created by Ericsson and sold to Vodafone)

The value chain helps us to understand the business of the company (links where it is operating) and the opportunities available. The attractiveness of one part can differ dramatically from others. When the company was created in 1999 the Transfer link was an opportunity for telcos like Vodafone: launching new services with a marginal cost they could amortize the investment made in the network. New entrants had no network to amortize, so they focused in other links. MyAlert developed two business lines that covered different links.

One was a Portal of Alerts to Mobiles. A user registered to an alert, MyAlert developed it with its technology (proprietary) and the content needed (provided by a third party or developed by MyAlert), and sent it via the telco that operates in the user's mobile phone (negotiating with all telco providers). This business implied the VAS, and Marketing and Sales links, and sometimes the Content: also the Infrastructure Provision, but for internal operation of MyAlert only.

The other was as a service provider, developing customized solutions for companies that wanted to have their own alert services. MyAlert designed and packaged the alert and was responsible for sending it, leaving to the client company the Content, and Marketing and Sales, and acting only in the VAS link. Here the Infrastructure Provision is commercialized occasionally as a service, not as a product.

An important fact is that MyAlert acted as an Infrastructure Provider not being its business aim, but because no one in the market had developed the infrastructure needed to send alerts. Thus the company put a stake on the technology as a competitive advantage to confront competitors that want to get into the market.

Figure 1. Value chain

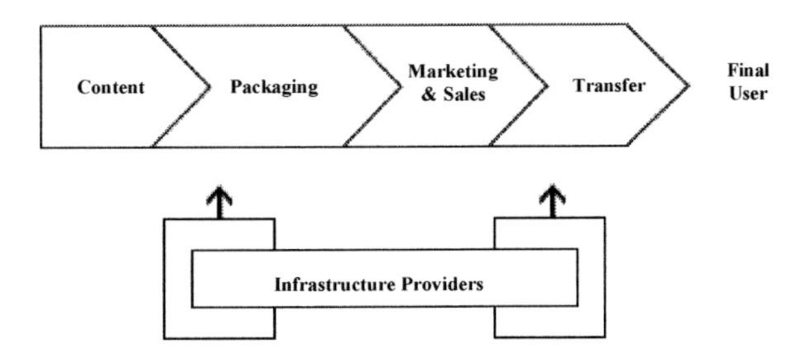

As we pointed out before, the profitability of the business model depended on the announcers who paid MyAlert. The costs of sending an alert are the same, regardless of whether the ad was sold or not, and the telco has to be paid for each SMS (alert). If the number of users increases the problem grows because advertisement income cannot cover costs.

MyAlert survived because of flexibility, correcting the business model to increase income (Gual & Ricart, 2001). Thus it became a provider of technology consultancy services and a development services provider to others in an ASP model.

The Merger

This was another good example of the strategic flexibility as a key success factor. By 2000, after the crash, MyAlert noticed that its position was strong but vulnerable, so consolidation was a good way to confront the future.

From the options valuated a merger among equals was the better option if the companies where complementary. Organic growth could not keep up with the growing rates needed for this environment, less having problems to generate cash flow. Selling the company was another possibility, but the only main had been to "make cash" and leave the project.

Buongiorno! was a good mate because of its high complementarities:

- Both were strong, MyAlert technologically, Buongiorno! in marketing services
- MyAlert income came from business services (mainly technological), Buongiorno! from advertisement
- Both were pioneers in developing direct marketing services, MyAlert for mobile, Buongiorno! for e-mail
- MyAlert was a Spanish leader, Buongiorno! was an Italian leader, and both had international presence
- MyAlert had the technology to access the mobile market, and B! the marketing and sales capabilities. The main indicator of the success: black numbers by 2002.

New Business Model

The new commercial focus forced continuous development of new products and services, possible due to the new structure based upon Product Managers. Formerly product development was developed in the R&D department with less flexibility.

In the first business plan m-commerce was expected to be 30% of the income by 2002, technology 15%, and advertising 55%. The real figures in 2002 where totally unexpected: 50% of advertising, 25% of technology, and 25% of consumer services, without any income from m-commerce.

Those deviations are normal. If prediction of new products' success is difficult, it is much more difficult if it is related to the adoption of new trends by end users. But the company was capable of reformulating its original plans to refocus towards more profitable areas. Most of the 2002 income was not expected in the original plan. Nomura Equity Research pointed out in 2000 that "MyAlert is positioned to move into several areas of the value chain at low cost."

Competitive Advantage Evolution

A competitive advantage helps a company be successful in a market on the basis of its superior conditions to those of its competitors. Those conditions come from the following:

1. Offer Advantages, due to processes or systems of the company in the design, development, and operation of products and services (lower costs due to scale economies).
2. Demand Advantages, due to the particular characteristics of the demand of products or services (relationships with clients or net economies that makes it difficult for them to change to another provider in the future).
3. Advantages related to the control of assets, so the competitive advantage in one business is related to the leadership in another business or in the influence power over the providers of other business (Microsoft leadership in operating systems market helped success in the text processing market).
4. Advantages related to innovation, when a real innovation in business term is developed and launched successfully.

At the beginning the competitive advantage was based upon the innovative product and the leadership, but it was vulnerable from the other perspectives because existing advantages were replicable by competitors. Thus to maintain the first-mover advantage acquired by being innovative in a new niche market, it had to keep on innovating quicker and better than its competitors. In the mean time, it needs to also develop offer and demand advantages to reinforce the original advantage and make it sustainable.

The merger favored this, first consolidating the leadership position. The merger implied lower costs (scale economies), developing offer advantages, and increased the number of clients: demand advantages. Not so clear and influential seem the associated assets (the extension of products from MyAlert to Buongiorno!) or the innovation (the product portfolio was not more innovative when combined).

When Buongiorno! MyAlert reached profitability the competitive original advantage had been consolidated and its position reinforced due to growth and consolidation. Its size was important in comparison with other pure players, but mean in comparison with other players like Vodafone.

When markets evolve strategic assets should change: the firm created new competitive advantages. When the market was born, business key factors were flexibility in products, and consequently good access to technology. While market is expanding keys turn into low costs and scale.

Technology

Is the property of technology important? The company played in the technology business and also used technology for the business. It changed from a technology-based company to a consolidated business model, but by 2003 it had presence in a totally technological business (B!Digital Technologies), and another two (Business Services and Consumer Services) requiring technology but not implying its development.

In terms of strategy and resources assignment, should the technological business line be prior? What is the importance of the technology development business for the rest of the business of the company? We will analyze the business lines attractiveness to define the strategic priorities.

Business lines are defined answering three questions: Who is the client (who pays, B2C or B2B?); What is the nature of the services sold (marketing/advertising or technological); and What is the business model (standard services—paid per unit; standard products—selling license; services ad hoc).

ASP and Software Licenses provided a quite similar service, but differences were in the way they were provided. A license implies a software package "as is," according to the industry standards. ASP implies a periodical payment for the operation of a technological platform.

To evaluate the attractiveness business lines can be reduce to three: business services (advertising, digital marketing projects, and ASP), consumer services, and software and technology consultancy. We decided to include new possibilities that the mobile market

Table 2. Business lines

Nature of Service	Business Model	B2B	B2C
Advertising/ Contents	Standard Services	Advertising Impacts Selling	Consumer Services
	Ad Hoc Services	Digital Marketing Projects	
Technology	Standard Services (paid per unit)	ASP	
	Standard Products (selling license)	Software Licenses	
	Ad Hoc Services	Technological Consultancy	

evolution could provide (e.g., 3G). Evaluation was made aligning capabilities of the company with the following criteria: market growth, competitors, and profitability.

First consideration is profitability. Consumer services is a margin business based upon the infrastructure already available with profitability was under 10%. Business services is a line with bigger margins but also bigger fixed cost, similar to consultancy. Software and technology consultancy is about 10% of income. Some aspects were difficult to valuate due to data availability, but according to available data, business services had fewer competitors and high margins in some of the products (e.g., Digital Marketing Projects), and also more volume in the future with consumer services. To that extent, the software and technology consultancy business will be in a second priority level.

Second consideration is the influence of technology development in the business development of the other lines. By 2003 the market had changed and was looking for profitability, once demonstrated it was possible, and there was more technological availability. Basic technologies offerings increase and the development rate slows down. To add up the company did not integrate technologies after the merger, evidencing that the less sophisticated B!3A can compete with the advanced MAGO once some of the functionalities are available (mainly gateways to mobile networks). Need for technology was a must at the beginning due to a lack of market products necessary to develop the services. After the merger R&D efforts in new platforms were abandoned, focusing in developing new products and services as a response to market changes.

We consider that technological development was not a priority for the company, not even conditioned by the needs of the other business lines. Nevertheless further analysis could be interesting because of cases like Amazon, the world leader in electronic commerce also considered as a leader in developing software for creation and operation of Internet shops.

Table 3. Business priorities for the future

Business Line	Market Attractiveness	Capabilities
Business Services	High Profitability Attractive Growth Moderated Competence	• Marketing • Projects
Consumer Services	Medium Profitability High Growth High Competence	• Marketing • Projects
Software and Technological Consultancy	Required to Invest in R&D Attractive Growth Moderated Competence	• Proprietary Technology • Product
Future Products (3G)	Uncertain and High Expectations Risk of Getting Out of Business	• Needs?

Final Conclusions

The company was able to become a leader in a newly created market thanks to technological development and strategic flexibility. Now less focused on technology, changing with the environment as the market consolidated. Facing new challenges like development of new products and services, to increase size of the business, and looking for new markets all over the world to complete Jorge Mata's vision: *"MyAlert aims to become the world leader in mobile commerce."*

2002–2004

Industry presented major changes. In Spain the five main companies—Movilisto, Netsize, Terra Mobile, Buongiorno! MyAlert, and Gsmbox—represented a market share of 57.6% (DBK, 2003) by 2002. By 2004 Buongiorno! MyAlert acquired Gsmbox and I-Touch acquired Movilisto.

Telcos began developing content to position their business in another link of the value chain. Telefonica launched the E-mocion service, via WAP and i-mode, and Vodafone Life was launched all across Europe.

UMTS has come with the first services in Europe, video download and videoconferences. Market was growing at a 30% rate, with a progressive evolution towards multimedia terminals, the basis for new services, like Java Games or video chatting. Market size in Europe was almost 1,300 million euros, and expected to double in 3 years.

Nowadays Buongiorno! (B!) is the leading provider with nearly 400 employees, 56 million euros in revenue in 2003, and traded on the Milan stock exchange since July 2003, and is the only European company with presence in every European country.

References

DBK, Informes Especiales. (2003, October). *Contenidos y Servicios para Telefonía Móvil.* Madrid.

European Information Technologies Observatory (EITO). (2004). European Union.

Funk, J.L. (2004). *Mobile disruption.* Hoboken, NJ: John Wiley & Sons.

García, J.M., & de Haro, G. (2003). *Buongiorno! MyAlert Case.* Madrid: Instituto de Empresa.

Gómez, F., & González Martínez, A. (2001). *Telefonía Móvil Digital.* Madrid: Anaya Multimedia.

Gual, J., & Ricart, J.E. (2001). *Estrategias empresariales en Telecomunicaciones e Internet.* Madrid: Fundación Retevisión-Auna.

Huidobro, J.M. (1996). *Telefonía Fija y Móvil*. Madrid: Thompson Paraninfo.

Lamont, D. (2001). *Conquering the wireless world: The age of m-commerce*. UK. Capstone Publishing.

Lautenschläger, G., & Schmidtke, B. (2000). *Móviles: SMS, WAP y compañía (guía rápida)*. Madrid: Data Ibérica de Software S.L.

Montero Pascual, J.J. (2000). *Competencia de las comunicaciones móviles: de la telefonía a Internet*. Valencia, Spain: Tirant lo Blanc.

www.buongiorno.com

Chapter IV

Customer Perceptions Towards Mobile Games Delivered via the Wireless Application Protocol

Clarry Shchiglik, Victoria University of Wellington, New Zealand

Stuart J. Barnes, University of East Anglia, UK

Eusebio Scornavacca, Victoria University of Wellington, New Zealand

Abstract

The rapid uptake and increased complexity of mobile phones has provided an unprecedented platform for the penetration of mobile services. Among these, mobile entertainment is composed of a variety of services such as ringing tones, games, gambling, and so on. Games are predicted to replace ringing tones as the main driver of mobile entertainment. This chapter contributes to the development of the mobile game industry by understanding corresponding consumer perceptions towards wireless application protocol (WAP) games. A series of focus groups were conducted to gain in-depth qualitative insight of perceptions towards WAP game services. The results indicate a number of clear areas for the delivery of successful WAP game services. WAP games were perceived as lacking complexity, but at the same time, were also seen as possessing several beneficial qualities. The chapter concludes with some recommendations and predictions regarding the future of WAP games.

Introduction

The past decade has seen mobile phones emerge as one of the fastest adopted technologies of all time (Chen, 2000; de Haan, 2000). Mobile phones present traditional e-commerce with an abundance of possibilities for new services and paradigms (Barnes & Huff, 2003; Siau & Shen, 2003). Commonly referred to as mobile (m-) commerce, the conduct of e-commerce through Internet-enabled mobile phones allows the delivery of services beyond a traditional fixed-line connection (Barnes, 2003; Bergeron, 2001; Sadeh, 2002). As a result, m-commerce possesses greater ubiquity and thus market size than previous notions of e-commerce. According to a study by Telecom Trends International (2003), global revenues from m-commerce are projected to grow from US$6.8 billion in 2003 to over US$554 billion in 2008.

In addition to their rapid adoption and proliferation, mobile phones have progressively improved in terms of their capabilities and network connectivity. These advances have created a strong impetus for mobile entertainment, a service that has gained considerable attention for its potential to rapidly drive the adoption of m-commerce. According to the ARC Group (2001), the market for mobile entertainment will reach 1.6 billion global users by 2006, creating extraordinary opportunities to leverage m-commerce revenues.

Mobile entertainment represents a variety of services, including ringing tones, games, gambling, and many others (Baldi & Thaung, 2002). Currently, ringing tones comprise the greatest market share of mobile entertainment. However, games are predicted to overtake ringing tones within the next few years (Strategy Analytics, 2003). The significance of mobile games is already evident in some markets. In markets such as Japan and South Korea mobile games have become a killer application for m-commerce (Datacomm Research, 2002; Datamonitor, 2002). Screen Digest (2004) forecast the global market to grow more than six times to $6.4 billion by the end of the decade.

Even though mobile games are a relatively recent phenomenon, there are now a variety of these services available in most developed markets (Vodafone, 2005). The nature of these games is heavily dependent on the boundaries created by device, network, and application. Most of the games currently available can be categorised within three mainstreams: messaging, downloadable, and online (Nokia, 2003a). In the future, the vision is for mobile games to be colour interfaced, real time, multiplayer, and location sensitive (Choong, 2003). These are qualities that present opportunities for all three game types, but particularly for online games. The most common type of online games available in developed markets are currently based on the wireless application protocol (WAP). WAP games have the ability to provide synchronous multiplayer gaming to a global audience, to be played using location-based services, and to be easily customised to user preferences and profile. However, because WAP games are reliant upon online connectivity, they are susceptible to the limitations of current mobile networks and are as a result typically of a start-stop nature, not too dissimilar to turn-based games.

Experience in the gaming market has shown that while a game's brand may initially be able to attract consumers, it will not guarantee the success of a game. In the long-run people embrace games that deliver them value (mGain, 2003). Furthermore, games delivered over the mobile network operate in a different paradigm to those of the traditional wired Internet, as dictated by differences in infrastructure and user behaviour. Therefore, it is interesting to observe that a good part of the problem with the initial wave of unsuccessful WAP games was due to a lack of understanding of consumer needs and expectations and how these can be met over the mobile medium. Consequently, just as with any other mobile service, it is fundamental to have appreciation of corresponding consumer perceptions in order to achieve successful deployment of these types of games (Barnes, 2002; mGain, 2003).

Accordingly, there exists a stream of research dedicated to understanding consumer behaviour and perceptions towards m-commerce (Chae & Kim, 2001; Landor, 2003; Lau, 2003; Samtani, Leow, Goh, & Lim, 2003; Vrechopoulos, Constantiou, Sideris, Doukidis, & Mylonopoulos, 2003). However, none of these studies have specifically focused on the field of mobile games. By focussing on mobile games, distinctive features that would otherwise be unobserved through broad m-commerce research can be exposed.

This paper aims to contribute to the development of a better understanding of consumer perceptions towards WAP games. In order to achieve this goal, four focus groups, each consisting of six participants, were conducted. The next section provides a background to mobile games. This is then followed by discussion regarding the research methodology. Subsequently, an examination and analysis of data gathered is provided. Finally, we conclude with a discussion of limitations, future developments, and further research.

Background on Mobile Games

The potential market for mobile gaming is huge; worldwide, there are already over 1 billion mobile phone users, a large proportion of whom maintain mobile phones capable of gaming, and this figure is set to grow (eMarketer, 2003). These devices also provide a challenge for service provision since they restrict mobile games to small displays and limited controls. Nevertheless, they also possess advantages over other digital gaming media, most notably, mobile phones are multifunctional as opposed to specialist devices such as a radio, thus, consumers habitually possess them wherever they go—they are both ubiquitous and networked. By working around limitations and utilising advantages, mobile games are able to deliver a very different gaming experience. For the purposes of this paper, mobile games are defined as games played on mobile phones that are either embedded or at some stage require the use of wireless connectivity, excluding any games that are reliant upon cartridges.

Beyond simply allowing gaming at anytime and anywhere, mobile games can be massively multiplayer and can exploit information gathered such as players' location and proximity to one another to create a new concept in mobile entertainment (Datacomm Research, 2002). The use of location-based services has created an ability to play virtual games in a "real-world" context. Two pioneering games that illustrate the current diversity of location-based mobile gaming are BotFighters (It's Alive, 2005) and TreasureMachine (Unwiredfactory, 2005).

BotFighters allows players to create a robot that is housed in their mobile phone, by choosing the robot's armour, shield, and eyes, which they then set upon other robots by sending text "attack messages" to the central game server. Those messages are then relayed to their local game opponents in the form of beeps. The game has become so addictive that players have been known to play for many hours in order to defeat opponents (Kharif, 2001). An involved player, who plays on average 30 minutes a day, will pay somewhere between US$5 to US$10 per month in addition to regular mobile phone charges.

TreasureMachine releases clues to guide players to a predefined location. Whenever a player believes they physically stand on the right spot, they "dig" for the treasure using their mobile phone. The first player to "dig" for the treasure at the predefined location wins. Players are charged a small fee for each clue they receive and digging attempt (Unwiredfactory, 2001).

BotFighters and TreasureMachine are the world's earliest location-based games (It's Alive, 2000; Unwiredfactory, 2001). Partly due to their release being at a time when interpreted language (e.g., Java)-enabled mobile phones were not widespread these games were made available over both short message service (SMS) and WAP platforms to increase circulation. SMS and WAP both offer a distinctive array of capabilities for mobile games.

SMS together with multimedia message service (MMS) applications form a gaming category that can be classified as messaging mobile games. The means of interaction amongst these games is analogous to other data communications. To initiate game-play an SMS or MMS message is sent to a game server. The player then receives a reply message consisting of instructions. From this point onwards, messages are sent back and forth consisting of commands from the player as well as status and directions from the game server until the game is concluded. Games that are particularly well suited for this medium include trivia, combat, and strategy. Messaging games can be played either as single or multiplayer, and are able to feature location sensitive game play.

WAP games are always played on a game server through the use of a microbrowser and the mobile Internet. Thus, along with constraints created by device, the limitations in mobile networks restrict the dynamics and interactivity of WAP games. However, WAP is designed to accommodate these limitations by bridging the gap between wired and wireless environments. Originally, termed WAP 1.0 and written in the WML programming language, its latest version, WAP 2.0, has now progressed to employ the more advanced xHTML and to adopt more recent Internet standards. WAP 2.0 attempts to optimise the usage of higher bandwidths, packet-based connections, and improved device capability, while at the same time providing backward compatibility to preexisting WAP content (WAP Forum, 2002). The most significant advancement for games based on this platform is that WAP 2.0 recognises the capabilities of users' devices, such as screen size and colour in order to maximise performance potential and bring increased consumer satisfaction. Furthermore, WAP games are easily customisable to user preferences and profile. Genres suited for this medium include role player, casino, and trivia games. These games are typically of a start-stop nature.

In addition to WAP, the Japanese mobile service i-mode also provides a form of online games. I-mode originally differentiated itself from WAP 1.0 by being based on cHTML, a subset of HTML. However, the release of WAP 2.0 has signalled the unification of cHTML and xHTML. The unique feature of i-mode is that it has been able to provide a form of online gaming not previously seen with older versions of WAP. For example, one i-mode service links mobile phones to video arcade games. This i-mode service comple-

Figure 1. Connection between i-mode and video games (NTT DoCoMo, 2003)

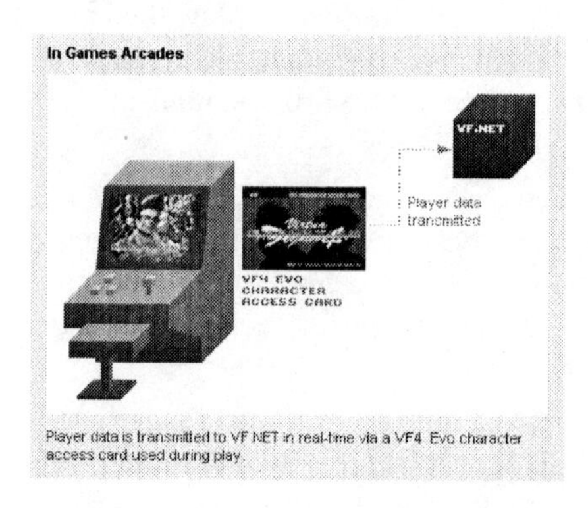

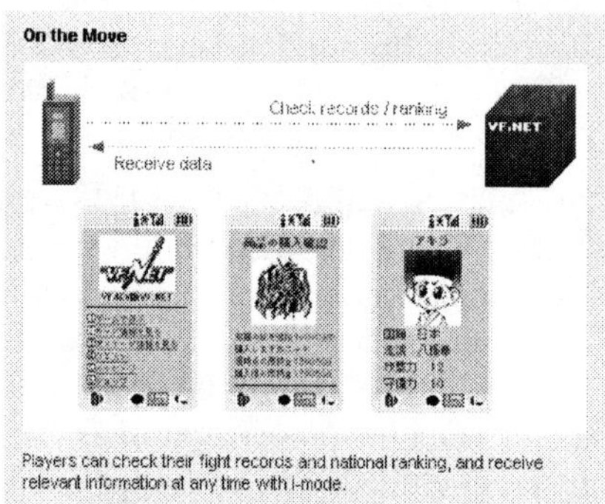

ments the video games by allowing a number of functions to be played over mobile phones. For instance, as shown in Figure 1, the arcade game Virtual Fighter 4 allows players to check their fighting match history, national rankings, customise their characters, search for arcades with Virtual Fighter 4, and communicate with other players nationwide (NTT DoCoMo, 2001).

The forecasted growth among each of the various mobile game formats displays some disparity. As depicted in Figure 2, messaging and WAP games are heavily out of favour, perhaps even fading away towards the year 2008. At the same time, downloadable games are destined to offer the greatest potential for growth.

Downloadable games are made possible by way of technologies such as J2ME (Java 2 Micro Edition) and BREW (Binary Runtime Environment for Wireless). As their name suggests, they are downloaded into devices and can be played repeatedly without the need for any further network activity. Embedded mobile phone games are essentially also included in this category. At present, due to mobile network limitations and a lack of standardisation, synchronous multiplayer capability is largely restricted to short-ranged embedded technologies such as Bluetooth, while asynchronous multiplayer functions such as the uploading of high scores is facilitated by mobile networks. Already there exists a comprehensive range of branded downloadable games, including The Lord of the Rings, Tiger Woods PGA Tour, and Pacman. Downloadable games are arcade-style and sufficiently advanced to contribute to the ubiquitous network of gaming. In Japan, consumers are able to play portions of Sony Playstation console games over Java-

Figure 2. Global revenues by mobile game format (2003–2008) (Strategy Analytics, 2003)

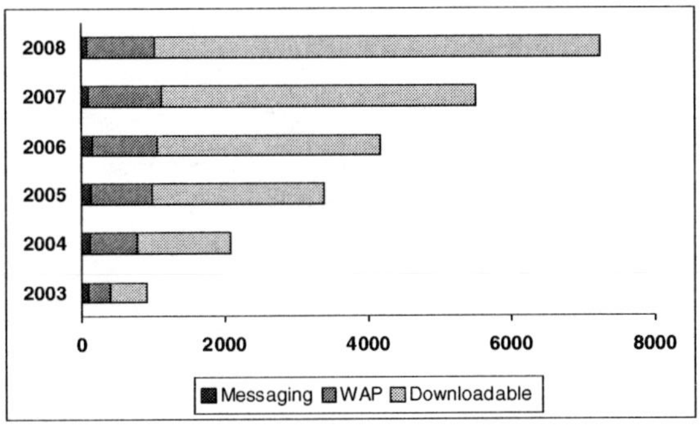

enabled i-mode mobile phones (NTT DoCoMo, 2001); i-mode phones plug into a Playstation console permitting games to be later played while on the move.

The following sections present the methodology used in this research and the respective consumer perceptions towards WAP games.

Methodology

The use of qualitative methods of research can be found in many disciplines. Within the domain of Information Systems, qualitative research has become increasingly used in the past decade (Myers, 1997). Focus groups are especially valuable when there is a need to obtain qualitative data filled with vivid and rich descriptions. Focus groups generate data through bilateral communication between the moderator and participants. From listening to people share and compare their different perspectives a wealth of in-depth insight can be uncovered regarding opinions and attitudes (Morgan & Stinson, 1997). Figure 3 illustrates the research design undertaken.

Participant Selection

A total of four focus groups, each consisting of six participants, were conducted during June and July 2003 in Wellington, New Zealand. In line with recommended procedure, participants were selected and grouped according to their previous experience with the Internet, mobile phones, computer games, and WAP in order to gain insight into consumer perceptions of WAP games from a variety of perspectives (Morgan & Stinson, 1997). Although participants in all four groups were high Internet users, groups were unique with respect to participants' level of usage of the other three technologies. With reference to Table 1, the focus groups were assembled as follows: WAP users (group A), computer game and mobile phone users (group B), mobile phone users (group C), and

Figure 3. Research design

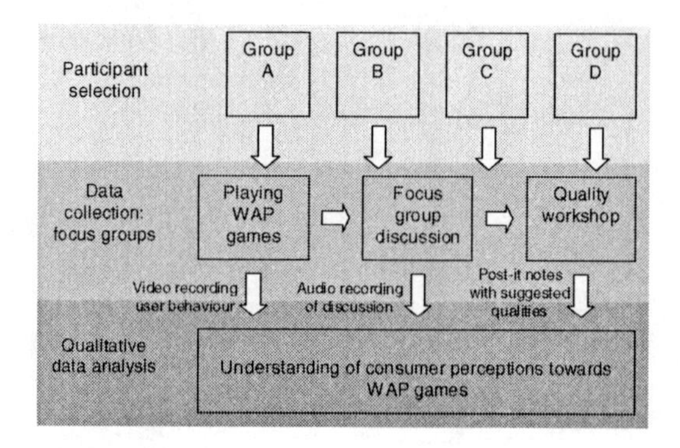

Table 1. Focus group experience levels

Group	Internet	Mobile phone	Computer Games	WAP Experience
A	High	High	High	Yes
B	High	High	High	No
C	High	High	Low	No
D	High	Low	Low	No

Figure 4. Key mobile phone attributes

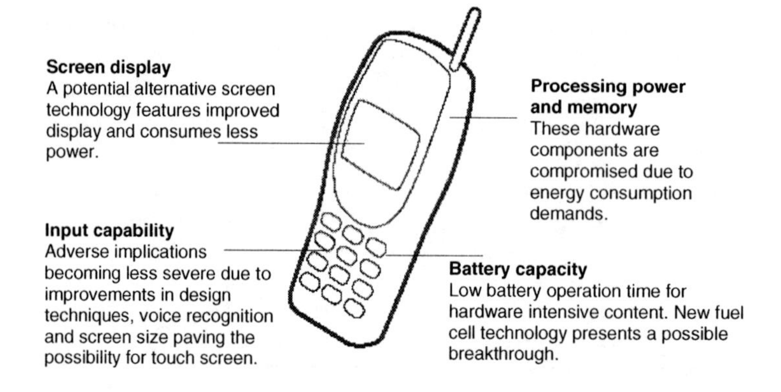

Internet-only users (group D). Based on demographic statistics all participants were aged between 21 and 35 (Chang, 2003; Datamonitor, 2001; Lipp, 2002).

Focus Groups

At the start of each focus group, participants were presented with a mobile phone and worksheet. The worksheet included the purpose of the focus group, instructions on how

Table 2. Games played by focus groups

Group	Multiplayer	Role player	Classic
A	Boy Racer	Ecowarrior	Rock, paper, scissors
B	Shark Hunter	Swingers	Tic Tac Toe
C	Trivia: Trivia Racer	Wipe Out	Hangman
D	Trivia: Woodbine	Gone Fish'n	Slots

to operate the participant's respective mobile phone and the proceedings of the focus group. Each participant used a unique model of mobile phone associated with the service of Telecom New Zealand. To increase the level of consumer perceptions uncovered, a variety of activities were conducted in each focus group. Focus groups ran for between 90 and 120 minutes and consisted of three stages:

- Playing WAP games
- Focus group discussion
- A quality workshop

Playing WAP Games

Participants were provided with approximately 45 minutes to play and build familiarity with three designated games. To gain understanding of consumer perceptions with respect to a range of WAP games, each focus group was designated a unique set of three games consisting of one game from each of the following genres: multiplayer, role player, and classic. This interaction was recorded onto digital videocassette. A total of 12 games were played, represented by a unique set of three games for each of the four focus groups. Table 2 details the specific games played by each focus group. All games were played using WAP 1.2.1 over a CDMA 2000 1x mobile network. At the time of focus group exploration, these were the most advanced mobile technologies available in New Zealand—a more recent version of WAP has since been made available to the market.

Focus Group Discussion

Once participants had played the three designated games they were asked open-ended questions requiring them to draw upon their experience. A conscious effort was made to ensure that questions were clearly formulated, neutral, appropriately sequenced, and easily understood. Focus groups were recorded on cassette tape. Indicative questions used at each focus group included:

- What problems did you have?
- What did you find easy?

- What did you like?
- What did you not like?
- Would you play the games again? If yes, why? If not, why?

Furthermore, follow-on questions evolved from several sources:

- Drilldown questions on specific issues expressed by participants to clarify thought and gain a deeper understanding
- Comments provided by participants, which fuelled questions on related issues to allow a broader comprehension
- Observations from participants playing WAP games in stage one of the focus group

Questions were not preconceived for specific focus groups. However, a number of distinct follow-on questions arose amongst focus groups. Most noticeably, groups consisting of experienced game players (groups A and B) raised concerns regarding a lack of game functionality and media richness, while groups consisting of inexperienced game players (groups C and D) tended to raise issues concerning perceptions of control and challenge in game play. This trend is consistent with existing knowledge on the behaviour of mobile game players (Andersen, 2002).

Quality Workshop

For the final stage of the focus group, a quality workshop was run to uncover perceived qualities of an excellent WAP game. Bossert (1991) recommends a three-stage process for such workshops. Participants were first provided with post-it notes and asked to answer the following question: "What are the qualities of an excellent WAP game?" Participants were instructed to work in silence, write one quality per post-it note and encouraged to also write a brief explanation and rationale for the proposed quality.

Once all participants felt they had exhausted their ideas, participants were then sorted into two groups of three to combine their post-it notes into affinity groups. Participants initially performed this task in silence, moving post-it notes around and creating headings as felt appropriate. Finally, participants worked as one group to develop one combined affinity group of demanded qualities.

Data Analysis

Results presented in this section represent derivatives of the data collected during all three stages of the four focus groups. Initially, data gathered from stage one, user behaviour, and stage two, focus group discussion, were organised. First, the observations concerning the qualities of an excellent WAP game were coded openly and qualitatively (Strauss & Corbin, 1990). Following this, these observations were recoded axially (Strauss & Corbin, 1990), converging into categories: user friendliness, media richness, interactivity, price, rewards, responsiveness, functionality, multiplayer, personalisation, and enjoyable. Within these results, areas were identified that could improve current WAP game offerings.

User Friendliness

The user-friendliness quality consisted of the highest number of identified observations. First, participants felt it was difficult to use input keys. This may have been due to participants' lack of experience with the mobile phone they were using. Although presented with instructions, with the vast number of models of mobile phones available, it is likely participants were using phones with which they lacked familiarity. Second, the need to scroll down through text was seen as annoying. To remove the need to repetitively scroll, particularly in storytelling games, one participant suggested that audio could be used as a substitute. Third, an important need was recognised for WAP games to have high-quality instructions as participants were often confused and needed a longer-than-anticipated length of time to understand game play. One possible explanation for this could be due to participants' misguided preconceptions of the nature of WAP games. These preconceptions can be attributed to a number of sources: previous experiences of mobile gaming derived from playing embedded games that have a much higher level of interactivity; an association of Internet-based technologies consisting of colour graphics and moving pictures; and connotations derived from the titles of WAP games. In addition to misguided preconceptions, another need for game instructions could be that the design of a game meant it was inherently difficult to learn to play. For example, one participant identified the display in a WAP game as not conducive to explaining the need to scroll down; the lack of a scroll bar in WAP meant the user was left perplexed. Also, it was interesting to observe comments that revealed how in a real-world scenario participants would have given up playing a WAP game after experiencing initial difficulty.

Media Richness

Lack of colour, sound, and animation induced one participant to comment how WAP games were reminiscent of early handheld games played 20 years ago. This was a feeling that typified the expressions voiced relating to media richness. Some participants indicated that until the media richness in terms of colour, graphics, screen size, and sound develops to a sufficient level, they will avoid WAP games. Additionally, the ability of a mobile phone to vibrate was identified as a possible characteristic to exploit to enhance WAP gaming.

Price

Relative to the entertainment value obtained, participants considered WAP games overpriced. WAP games are charged at US$0.04 per screen, which aggregates to approximately US$6 for every 30 minutes of game play. The high price was suggested as a factor that would make participants reluctant to play WAP games, especially since embedded games are a "free" alternative.

Rewards

A very strong consensus across all focus groups emerged regarding rewards. Participants believed there was a lack of some form of reward structure. They saw rewards as a way of acknowledging skill and enthusiasm. Participants also believed that by providing rewards an incentive would be created to play WAP by increasing entertainment value in the form of indirect gambling and recompensing for the steep price paid to play games. Furthermore, it was expressed that rewards do not necessarily have to be financially based; for example, a reward may simply comprise the acquisition of entertainment value from playing an enjoyable game.

Responsiveness

Participants invariably provided adverse comments concerning the mobile network. The network was found to be slow and unreliable. "Network not responding" anecdotally appeared to be the most common WAP page cited. Participants felt that under alternative circumstances, they would have been inclined to give up playing due to the delays. A fast connection speed was desired to facilitate less frustrating and more interesting game play.

Interactivity

User involvement was identified as a desirable quality evidenced by participants' disappointment at the lack of control and range of tasks they could perform in determining the outcome of some WAP games. After experiencing games that were found to be determined entirely by luck as opposed to skill, participants felt they would tend to stay away from games that failed to provide a challenge. In contrast, games that captivated the user by providing an interesting storyline or requiring an appropriate level of thought or skill when taking action were perceived favourably. Games that lacked any degree of complexity became uninteresting very quickly. The length of time participants were willing to play a WAP game was dependent on the game's level of complexity.

Functionality

Games were seen as lacking a range of functions. Participants had anticipated the ability to record high scores, develop a personal gaming history, and to progress on to higher levels once a game had been mastered. The absence of functions of this nature restricted the level of value participants could derive from playing WAP games.

Multiplayer

Games that required participants to compete against others were found to have the potential to make the focus group atmosphere exciting and lively. The ability to play other people was seen as a quality that favourably differentiated WAP games to previous mobile gaming experiences. Games that were multiplayer and at the same time immersed the user provided the most enjoyment. Additionally, it was envisaged that value would ensue from being able to talk to other players, an observation indicating perceived benefits of an online community.

Personalisation

Participants appreciated how certain games were customised by featuring the names of nearby localities. Conversely, adverse reactions were created by games that were considered gender biased or incompatible with personal morals.

Enjoyable

Participants expressed how they desired WAP games to be entertaining, exciting, and stimulating. When these variables were not met, games were described as boring, simplistic, too long to complete, too hard, and slow to respond.

While the results above discuss findings of WAP games holistically, Table 3 presents a summary of positive and negative perceptions across the different focus groups in relation to each specific game genre examined. The title of each game is provided in Table 2. Although some perceptions displayed in Table 3 overlap with those discussed earlier in the data analysis section, observations discussed below concern perceptions that are peculiar to specific game genres. Compared to the other two game genres, multiplayer games were enjoyed for their ability to provide competition against other people. However, at the same time, multiplayer games caused some confusion. Adverse comments from participants related to a lack of awareness that other people were playing and an inability to identify which player they were. Role player games gained positive perceptions from participants with respect to narrative qualities of a captivating storyline, customisation to player details and humour. Although, narrative qualities of role player games were enjoyed, the method in which they were delivered through scrolling and reading was perceived unfavourably. Classic games were only able to attract negative perceptions. Their simplicity failed to attract participant's interest.

Table 3. Positive and negative perceptions in relation to game genres

Genre	Focus group	Positive perceptions	Negative perceptions
Multiplayer	A	Enjoyed for the competitive nature created by being able to play against other people. Game exercised memory skills.	Lack of ability to store personal gaming history as well as other personal gaming details. Network speed affected ability to compete. Overinflated expectations due to connotations of the game's name.
	B	Ability to play against other people.	Although there are a variety of actions a player may choose, game is entirely up to chance. Participants developed an incorrect understanding of the game even when they felt they had gained a reasonable level of familiarity.
	C	Ability to play against other people. Requirement to think when playing.	Participants were not always aware they were competing against another person. Difficult to decipher which player is which. Too simplistic.
	D	Ability to play against other people. Requirement to think when playing.	The game disallowed answers if too much time has passed, but despite a quick response by a player time would pass as a result of slow network speed. Difficult to decipher which player is which.
Role player	A	Engaging storyline. Storyline indirectly provided instructions by describing possible player actions.	Need to continually scroll down through text was irritated participants. So many instructions to remember note taking was at times required.
	B	–	Only one action can be taken when playing.
	C	Game customised by naming nearby localities. Players rewarded by seeing a picture.	Requires only one button to be pressed intermittently to successfully complete. Confuses participants by featuring animation, yet requiring one button to be pressed. Pictures appeared that offended participants.
	D	Provided humour.	Does not require the player to produce any thought.
Classic	A	–	Gets boring quickly. Lack of reward even when lucky enough to win.
	B	–	Network failure prevented testing.
	C	–	Complex words used. Poorly adapted for mobile phone screen size.
	D	–	Failed to attract interest.

Further insight into understanding mobile games can be gained by relating the results discussed above to mobile game usability guidelines established by Nokia (2003b, 2003c). By providing a series of recommendations on game design based on usability principles, Nokia (2003b, 2003c) offered a number of guidelines that engender games with greater user satisfaction. Since these guidelines are strictly focussed on usability and are dedicated for downloadable games, there are a number of disparities to the findings

Table 4. Relevant guidelines specified by Nokia for WAP games

Variables identified from focus groups	Game development guidelines specified by Nokia (2003b, 2003c)
User friendliness	Use natural controls. Provide a clear menu structure. Offer help when it is needed – if it is relevant to the situation it is more likely it will be remembered. The user must always understand his/her current status.
Media richness	Go easy on the sounds (e.g., use sound for feedback, but the game must also be playable without the sounds). The same applies for phone vibration. Use a consistent colouring scheme within the game and the navigational structure. Things that look similar should behave similarly.
Price	—
Rewards	Provide a high score system as a reward.
Responsiveness	Make the game launch in reasonable time, the faster the better. To reduce perceived launch time, make something happen on the screen or give the user something to do.
Interactivity	The user needs to feel he/she is in control of the situation. Must be challenging enough for advanced users, but easy enough for beginners to stay motivated.
Functionality	Provide save and pause. Special levels are wonderful features that bring more variety to play. However, if the game rules for these levels are different, players must understand this.
Multiplayer	Given the difficulty of chat via mobile devices, players may benefit from a system where they can automatically offer, accept, and break alliances or similarly signal their intentions by in-game actions rather than with words. Utilise systems to match players of equivalent skill against one another, that is, an inexperienced player will be put off a multiplayer game if they are convincingly defeated at their first attempt by a skilled player. Alternatively, balance a game by employing a system that provides players of different skill an equal chance of victory, for example, lower-skilled player begins with a head start.
Personalisation	It is important to understand the user's preferences. The game must preserve all the data the user has entered. Including their name, options selected before playing and options selected during the game.
Enjoyable	-

of this chapter. Nevertheless, there are also areas in which they are aligned. Moreover, in some instances the findings presented are extended. A selection of Nokia's (2003b, 2003c) usability guidelines are provided in Table 4.

Findings from the focus groups can also be interpreted to identify key mobile phone attributes that affect the quality of a player's gaming experience. As shown in Table 4, these key attributes are classified according to screen, input capability, processing power and memory, and battery. Screen size and display hamper user friendliness and media richness. Improvements in the mobile phone screen will bring benefits such as reduction in the scrolling of text, increased sensory gratification, and easier-to-read content. OLED (Organic Light-Emitting Diode) technologies show strong promise for producing improved screen display at reduced power consumption (mGain, 2004). From one perspective, as WAP games are reliant on the mobile Internet, improvements in network connectivity will enhance their interactivity and dynamics. To improve the current situation, the goal of such games should be to minimise network traffic by performing as much processing as possible in the game server and mobile phone individually. However, restrictions of the processing power and memory of mobile phones may cause undue reliance upon mobile networks and confine the range of game designs that may be delivered (Nokia, 2004). WAP games were found to drain battery power significantly within a short period of time. Although mobile phones may operate for hundreds of hours in standby mode, their operation time for gaming appears to be severely reduced due to power requirements for screen display and processing. The severity of this limitation not only concerns game-playing time, but also forces a substantial sacrifice in talk time and the usage of other mobile phone capabilities. With the promise of OLED operating at reduced power consumption there is the possibility that screen size could increase with a lengthened operation time. Nevertheless, processing power requirements remain a concern for battery consumption. At present, input capability lacks intuitiveness for the WAP gaming context. Also, the small size of the keypad would likely become a more noticeable concern with games of greater interactivity. Improvements in mobile phone design techniques, voice technology, and touch-screen capability are likely sources for the betterment of the mobile phone input interface.

Overall, participants responded to WAP games with mixed reactions. Although WAP games were perceived as lacking complexity, they were also seen as possessing several beneficial qualities. When combining these qualities with destined enhancements to design and the introduction of increasingly sophisticated technologies, WAP games were generally regarded as a promising medium of gaming. In addition to lacking complexity, price was also seen as a significant deterrent.

Conclusion

The advancement of mobile devices, networks, and applications and the ability to provide an innovative form of gaming previously unattainable is suggestive of the future direction of mobile gaming. These advances will undoubtedly enhance integral game elements such as media richness and responsiveness. However, even before this takes place, by uncovering a number of consumer perceptions towards WAP games by way of focus groups, we have been able to identify areas of design that could improve current offerings. These areas include the introduction of functionality to allow storage of personal gaming details, such as statistics or preferences, improvements in user friendliness, offering rewards to compensate for high costs, recognise skill and add excitement, as well as the important role of a game title to aid the instruction of a game and guide expectations. Another key finding of this chapter was the significant entertainment value gained by consumers when playing multiplayer WAP games that engaged players by providing an appropriate level of challenge. Value was also gained when games were personalised and offered a captivating storyline. Furthermore, the results indicate the importance of user friendliness, responsiveness, and appropriate pricing. Without sufficient user friendliness and responsiveness, consumers will become frustrated and inclined to abandon a game at a very early experimental stage of playing, while appropriate pricing is critical to entice demand.

There were a number of limitations identified in the focus groups. The mobile phones used in this research all featured monochrome screens. In addition to this, participants played WAP games within a restricted length of time and area of space. As a result participants gained an unrepresentative exposure of the speed and reliability of the mobile network and perceptions that would arise from a more longitudinal-based study, such as the importance of game longevity, were obstructed from observation. A further consequence due to area constraints was an inability to examine location-based games. Furthermore, participants were asked to play games that may have belonged to a gaming genre they find adverse, creating artificial, experiment-type conditions. Accordingly, in order to develop a better understanding of consumer perceptions towards WAP games and other mobile games platforms, further research must be undertaken in areas such as human-computer interaction (HCI), mobile networks, and user contexts. Undoubtedly, this is an area with great potential in which a deep understanding of its domain will be a key factor for successful consumer adoption.

References

Andersen (2002). *Mobile multimedia study*. Brussels: European Commission Directorate.

ARC Group. (2001). *Mobile entertainment*. Boston: ARC.

Baldi, S., & Thaung, H.P. (2002). The entertaining way to m-commerce: Japan's approach to the mobile Internet – a model for Europe? *Electronic Markets*, *12*(1), 6–13.

Barnes, S. (2002). The mobile commerce value chain: Analysis and future developments. *International Journal of Information Management*, *22*(2), 91–108.

Barnes, S. (2003). Location-based services: The state-of-the-art. *e-Service Journal*, *2*(3), 57–70.

Barnes, S., & Huff, S. (2003). Rising sun: i-Mode and the wireless Internet. *Communications of the ACM*, *46*(11), 78–84.

Bergeron, B.P. (2001). *The wireless web: How to develop and execute a winning wireless strategy*. New York: McGraw-Hill.

Bossert, J.L. (1991). *Quality function deployment, a practitioner's approach*. Milwaukee, WI: ASQC Quality Press.

Chae, M., & Kim, J. (2001). Information quality for mobile internet services: A theoretical model with empirical validation. *Proceedings of the Twenty-Second International Conference on Information Systems*, Louisiana (pp. 43–53).

Chang, N. (2003, March 4–5). Enorbus opening a new world of wireless entertainment. *Proceedings of the 2003 Game Developers Conference*, San Jose, CA.

Chen, P. (2000). Broadvision delivers new frontier for e-commerce. *M-Commerce, October*, 25.

Choong, A. (2003). Nokia aims to N-Gage gamers. Retrieved January 7, 2004, from *http://news.zdnet.co.uk/business/0,39020645,2129974,00.htm*

Datacomm Research. (2002). *Winning business strategies for mobile games*. Chesterfield, MO: Datacomm Research.

Datamonitor. (2001). *Asia Pacific wireless gaming*. Tokyo: Datamonitor.

Datamonitor. (2002). *Best practice in Asia-Pacific mobile gaming*. Tokyo: Datamonitor.

de Haan, A. (2000). The internet goes wireless. *EAI Journal, April*, 62–63.

eMarketer. (2003). Mobiles outnumber fixed-line phones. Retrieved January 28, 2004, from *www.emarketer.com/news/article.php?1002522&trackref=edaily*

It's Alive. (2000). It's Alive launches world's first location-based mobile game. Retrieved January 26, 2005, from *www.itsalive.com/page.asp?t=press&id=61*

It's Alive. (2005). BotFighters. Retrieved January 26, 2005, from *www.botfighters.com*

Kharif, O. (2001). Excuse me I've got to take this game. *BusinessWeek Online*. Retrieved January 26, 2005, from *www.businessweek.com/bwdaily/dnflash/jul2001/nf2001072_760.htm*

Landor, P. (2003). Understanding the foundation of mobile content quality a presentation of a new research field. *Proceedings of the 36th Hawaii International Conference on Systems Sciences*, Hawaii (pp. 1–10).

Lau, A.S.M. (2003). A study on direction of development of business to customer m-commerce. *International Journal of Mobile Communications*, *1*(1/2),167–179.

Lipp, D. (2002). *Marketing to win mobile gaming*. Presentation, Telecom New Zealand, Wellington.

mGain. (2003). Benchmarking literature review, European Commission, Brussels, Deliverable D6.2.1. Retrieved January 26, 2005, from *www.mgain.org/mgain-wp6-d6214.pdf*

mGain. (2004). Emerging and future mobile entertainment technologies, European Commission, Brussels, Deliverable D4.2.1. Retrieved January 26, 2005, from *www.mgain.org/publications.html*

Morgan, D., & Stinson, L. (1997). *What are focus groups?* Alexandria, VA: American Statistical Association.

Myers, M. (1997). Qualitative research in information systems. Retrieved January 24, 2004, from *www.qual.auckland.ac.nz*

Nokia. (2003a). Introduction to the mobile games business, Version 1.0. Retrieved March 9, 2004, from *www.forum.nokia.com/main/0,6566,050,00.html*

Nokia. (2003b). Nokia series 40 J2ME™ game usability guidelines and implementation model, Version 1.0. Retrieved March 9, 2004, from *www.forum.nokia.com/main/0,6566,050,00.html*

Nokia. (2003c). Overview of multiplayer mobile game design, Version 1.1. Retrieved March 9, 2004, from *www.forum.nokia.com/main/0,6566,050,00.html*

Nokia. (2004). Multiplayer game performance over cellular networks, Version 1.0. Retrieved March 9, 2004, from *www.forum.nokia.com/ndsCookieBuilder?fileParamID=4195*

NTT DoCoMo. (2001). *DoCoMo iMode service information.* Tokyo: NTT DoCoMo.

NTT DoCoMo. (2003). VF net corporate alliance. Retrieved January 26, 2005, from *www.nttdocomo.com/corebiz/imode/alliances/vfnet.html*

Sadeh, M.N. (2002). *M commerce: Technologies, services, and business models.* London: John Wiley & Sons.

Samtani, A., Leow, T.T., Goh, P.G.J., & Lim., H.M. (2003). Overcoming barriers to the successful adoption of mobile commerce in Singapore. *International Journal of Mobile Communications, 1*(1/2), 194–231.

Screen Digest. (2004). *Wireless gaming: Operator strategies, global market outlook and opportunities for the games industry* (Report). London.

Siau, K., & Shen, Z. (2003). Mobile communications and mobile services. *International Journal of Mobile Communications, 1*(1/2), 3–14.

Strategy Analytics. (2003). Download platforms to drive $7 billion wireless games opportunity. Retrieved January 28, 2004, from *www.strategyanalytics.com/press/PR00058.htm*

Strauss, A., & Corbin, J. (1990). *Basics of qualitative research.* Newbury Park, CA: Sage.

Telecom Trends International. (2003). M-commerce poised for rapid growth, says Telecom Trends International. Retrieved October 23, 2003, from *www.telecomtrends.net/pr_MIIS-1.htm*

Unwiredfactory. (2001). The launch of the world's first location-based competition. Retrieved January 26, 2005, from *www.unwiredfactory.com*

Unwiredfactory. (2005). TreasureMachine. Retrieved January 26, 2005, from

www.unwiredfactory.com/pdf_documents/TreasureMachine.pdf

Vodafone. (2005). Vodafone.com. Retrieved January 26, 2005, from *www.vodafone.com*

Vrechopoulos, A., Constantiou, I., Sideris, I., Doukidis, G., & Mylonopoulos, N. (2003). The critical role of consumer behaviour research in mobile commerce. *International Journal of Mobile Communications, 1*(3), 329–340.

WAP Forum. (2002). Wireless application protocol 2.0 – technical white paper. Retrieved December 23, 2002, from *www.wapforum.org*

Endnotes

- An earlier and shorter version of this paper appeared as Shchiglik, C., Barnes, S.J., & Scornavacca, E. (2005). Mobile entertainment: A study of consumer perceptions of games delivered via the wireless application protocol. *International Journal of Services and Standards, 1*(2), 155–171.

Chapter V

Barcode Applications for M-Business

Eusebio Scornavacca, Victoria University of Wellington, New Zealand

Stuart J. Barnes, University of East Anglia, UK

Abstract

One pertinent area of recent m-commerce development is in methods for personal transaction and information transfer. Several companies around the world have begun to use barcodes for the provision of m-commerce services. This chapter provides background on the enabling technological platform for providing such services. It then continues with three cases where mobile barcodes have been used—in Japan, New Zealand, and the UK. Subsequently, these are used as the basis for a discussion and analysis of the key business models, and strategic implications for particular markets. The chapter concludes with predictions for the market and directions for future research.

Introduction

With well over a billion mobile handsets worldwide, wireless technologies are enabling e-businesses to expand beyond the traditional limitations of the fixed-line personal computer (Bai, Chou, Yen, & Lin, 2005; Barnes, 2003; Barnes & Huff, 2003; Barnes & Vidgen, 2001; Bergeron, 2001; Chen, 2000; Clarke, 2001; de Haan, 2000; Emarketer, 2002;

Figure 1. Technology convergence between OCR and mobile phones

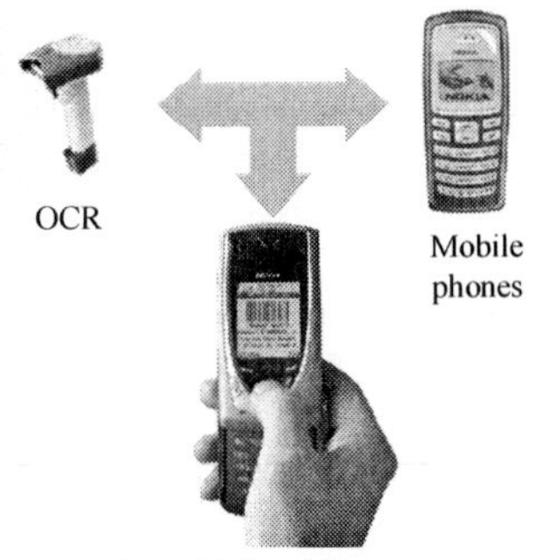

Kalakota & Robinson, 2002; Sadeh, 2002; Yuan & Zhang, 2003). According to a study by Telecom Trends International (2003), global revenues from m-commerce could grow from $6.8 billion in 2003 to over $554 billion in 2008.

As each mobile device is typically used by a sole individual, it provides a suitable platform for delivering individual-based target information, purchasing goods or services and making payments (Barnes, 2003; Barnes & Scornavacca, 2004; Bayne, 2002; Kannan, Chang, & Whinston, 2001; Newell & Lemon, 2001; Scornavacca & Barnes, 2003). One recent development in m-commerce is the application of barcodes solutions (NTT DoCoMo, 2003). This revolutionary development benefits from the convergence of two widespread technologies: optical character recognition (OCR) and mobile telephony (see Figure 1). Through barcodes, users of mobile phones can, for example, purchase tickets, receive coupons, access information, make payments, and interact with point-of-sale information systems (Airclic, 2003; AsiaTech, 2003; Bango, 2003; Barcode 1, 2003; Ecrio, 2003; NTT DoCoMo, 2003; Wireless Newsfactor, 2002).

This chapter aims to explore the applications of barcodes in mobile commerce. The following section provides a brief explication of barcodes and OCR technologies. This is followed by three case studies of the application of barcodes in m-commerce: in Japan, New Zealand, and the United Kingdom. Subsequently, we explore the business models being employed, and analyze the strategic implications of these models for different m-commerce markets. The chapter concludes with a discussion about the future of mobile commerce and directions for further research.

Technological Foundations: Optical Character Recognition and Barcodes

OCR technology has been used in business for nearly half a century. OCR devices transform specially designed marks, characters, and codes into a digital format. The codes can contain any kind of information, such as alphanumeric codes, date, time, names, and so on. The most widely used application of optical code is the barcode, and this has become used in a vast range of organizations, including supermarkets and retail stores (point-of-sale [POS] systems), libraries, hospitals, schools, and factories (production and supply chain management) (Uniform Code Council, 2003).

There are several varieties and standards for barcodes. The symbol found on most retail products around the world is based on UPC/EAN standards (Figure 2a). The Universal Product Code (UPC), developed by the Uniform Code Council (UCC), was the first barcode symbol widely adopted in the world. In 1973, the grocery industry formally established UPC as the standard bar code symbology for product marking in the United States. European interest in UPC led to the adoption of the European Article Numbering (EAN) code format in 1976 (EAN, 2003). Today, EAN International is a global not-for-profit organization that creates, develops, and manages jointly with the UCC open, global, multisectoral information standards, and the EAN/UCC standards. All businesses must apply for membership in order to be assigned a unique company identification number for use on all its products (Uniform Code Council, 2003). There are now five versions of UPC and two versions of EAN. The Japanese Article Numbering (JAN) code has a single version identical to one of the EAN versions. UPC and EAN symbols are fixed in length, can only encode numbers, and are continuous symbologies using four element widths (Barcode 1, 2003). The barcode used on books, for example, is generated based on the International Standard Book Number (ISBN).

In terms of business use, barcodes have been promoted largely as a machine-readable "license plate," where each label provides a unique serial number coded in black and white bars linking to a database entry containing detailed information. More recently, end users have sought to code more information; making the barcode a portable database rather than just a database key (Barcode 1, 2003). One good example is the Data Matrix barcode, a two-dimensional (2-D) matrix code (Figure 2b). A Data Matrix symbol typically stores between one and 500 characters. The symbol is also scalable, from a 1-millimeter square to a 14-inch square. Theoretically, a maximum of 500 million characters to the inch is possible, but the practical density will be limited by the resolution of the printing and reading technology used (Barcode 1, 2003).

Figure 2. Examples of barcodes

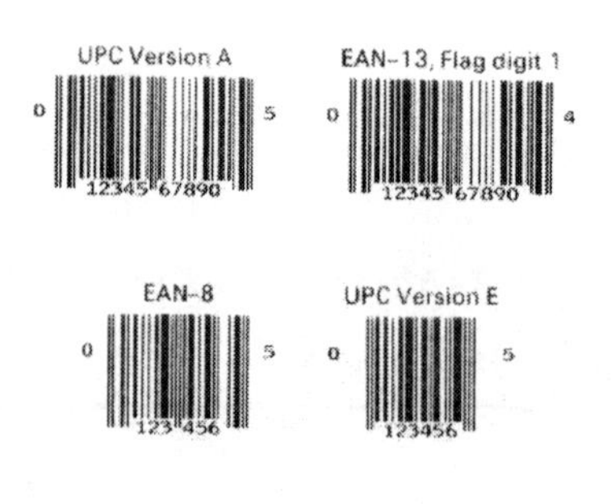

(a) EAN/UPC barcodes

(b) Data Matrix 2-D barcode

Barcodes have been widely used and accepted in the business world for decades. Given the existence of a solid infrastructure, it is interesting to note the development of barcode applications for mobile e-commerce (Airclic, 2003; AsiaTech, 2003; Bango 2003; Barcode 1, 2003; Ecrio, 2003; NTT DoCoMo, 2003; Vodafone, 2003; Wireless Newsfactor, 2002). Given a solid platform, there is potential for these applications to become commonplace in the mobile world. The next section provides more details about specific barcode applications in m-commerce, examining the value added by this technology.

Case Studies of Barcode Use in M-Commerce

In this section, we examine specific mobile barcode applications offered by companies in Japan, New Zealand, and the United Kingdom. We have focused on exemplars of existing applications that are being offered to and used by consumers in each of the respective markets. Let us consider each of these in turn.

NTT DoCoMo (Japan)

NTT DoCoMo is Japan's leading mobile communications company with more than 50 million customers. The company provides a wide variety of leading-edge mobile multimedia services. These include i-mode, the world's most popular mobile Internet service, providing e-mail and Internet access to over 43 million subscribers; and FOMA (Freedom of Mobile Access), launched in 2001 as the world's first third-generation (3G) mobile network service based on W-CDMA (Wideband-Code Division Multiple Access) (NTT DoCoMo, 2005).

Combien?

Since May 2002, NTT DoCoMo customers have been able to pay monthly mobile phone bills at convenience stores using a 2-D Data Matrix barcode on the screens of their mobile phones. The service is available nationwide in over 2,000 Am/Pm, Lawson, and Mini-stop convenience stores. Since the service eliminates the need for a hard-copy invoice, customers can pay their bills whenever they visit a participating convenience store. There is no fee for the service, although users must pay a transmission charge to download the barcode (via the i-mode portal). To prevent double payment, the barcode is valid only on the day when it is downloaded. A dedicated scanner is used by the merchant to read the 2-D barcode, after which the customer pays the bill. The procedure for downloading the barcode is presented in Figure 3.

This application allows customers to benefit from the large network of convenience stores in Japan. Alongside, they are able to pay their mobile phone bills 24 hours a day,

Figure 3. NTT DoCoMo's Combien? (NTT DoCoMo, 2003)

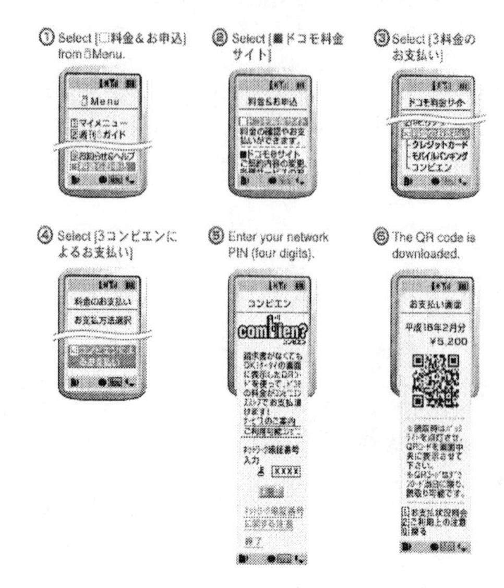

7 days a week. For the operator, NTT DoCoMo, the main benefit is the capability of collecting payments from customers remotely and safely, reducing collection costs, eliminating invoice handling, and transferring delivery costs to the customer.

Although currently limited to bill payment, this system is planned for use in other areas in the future. Indeed, it is expected that the core of wireless payments (including electronic invoices) will be the integration between financial institutions and wireless services providers (Scornavacca & Barnes, 2003).

505i Series

Another dimension of mobile barcode solutions is the development of hardware capable of reading barcodes. DoCoMo's 505i series has a built in camera and barcode-reader software. Figure 4 presents an example of an application that allows a user to transform printed information into digital format. Figure 4a shows a business card with a 2-D barcode. The barcode contains all the information printed on the card. Figure 4b displays the built-in camera on the 505i device. The user takes a picture of the barcode on the business card (Figure 4c) and the information—typically name, title, company, telephone, mobile, and fax numbers—is automatically added to the user's address book (Figure 4d).

The number of possible applications of this technology is limited only by one's imagination. For example, users could retract information from products simply by scanning their barcode. Barcodes could also be used to access Web site addresses, which can be cumbersome using a mobile phone keypad.

Figure 4. NTT DoCoMo's 505i Series barcode reader (NTT DoCoMo, 2003)

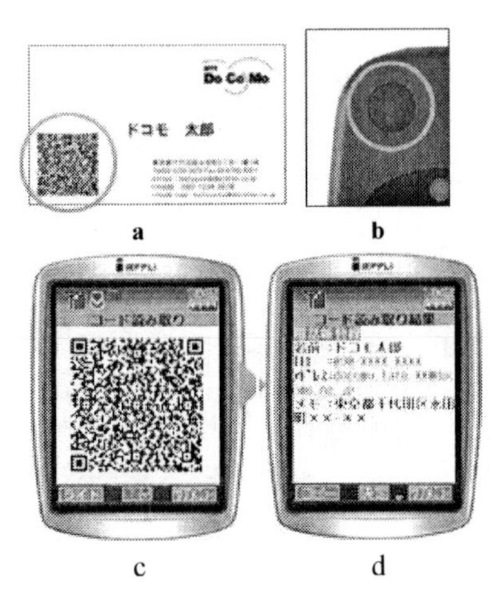

Another possible development that may stem from integration between financial institutions and wireless services providers is using the mobile phone to scan barcodes printed on invoices (as utility bills) and subsequent payment using a mobile phone account. One candidate for this is NTT DoCoMo's DoCommerce payments service. DoCommerce enables both 2G and 3G users to shop using their mobile phone with just a single user ID and password. An account aggregation feature enables DoCommerce users to check—simultaneously and on one single screen—the balances of their various financial accounts held with banks and credit card companies. Some 15 financial institutions, including VISA and MasterCard, are now part of the service (NTT DoCoMo, 2003). One possible development emerging from the barcode-reader capability and future financial service integration could be bill payment.

Vodafone and mTicket (New Zealand)

The Vodafone Group plc acquired its New Zealand business in November 1998 (previously known as BellSouth New Zealand). At the time of purchase, BellSouth had 138,000 customers. Vodafone New Zealand now has 1.3 million customers and a network coverage of 97%. Vodafone provides its GSM (Global System for Mobile) digital communication service on the 900 and 1,800 megahertz parts of the radio spectrum (Vodafone, 2003).

MTicket was the first commercially available mobile ticket system. The company is based in Europe and the Asia Pacific and provides applications for customers such as Vodafone, O2, and EMAP.

Mobile Ticketing

Developed by mTicket, "mobile ticketing" is a box-office system that allows event organizers to market, sell, distribute, and redeem tickets to events using text messages. The service has been available from Vodafone New Zealand since November 2002. In essence, a ticket is a good that can be easily delivered through electronic media. For mTicket, customers are able to send event keywords to a three-digit short code and receive information about events; accordingly, they may then select tickets and make a purchase using short message service (SMS) text messages (MTicket, 2003). Event information, terms, and conditions are all available using system keywords. To purchase tickets, customers are asked to choose seating areas, ticket types, and the quantity of tickets. Ticket costs are added directly to the mobile phone user's monthly bill, or deducted from prepaid credit. The ticket(s) are sent in the form of an SMS or multimedia message service (MMS) message (which contains a unique mTicket number or booking reference) or as a barcode (which may facilitate data input and access to the event).

Events that have previously sold mTickets include the Football Kingz (soccer), Netball New Zealand, the Heineken Open and ASB Classic tennis tournaments, and many music events ranging from dance to live music. The key benefit for consumers is the ability to purchase goods and services anywhere and at anytime using a mobile phone. For Vodafone and the event promoter the main benefits are the capabilities to collect payments from customers and to deliver tickets electronically via the mobile phone. This cuts the costs of collection, handling, delivery, fraud, and call centers (Vodafone, 2003). Alongside, impulse purchases may play a factor in increasing sales.

Clearly this application is less sophisticated than those found in the Japanese market. The principal reason for this is the current technological limitation of most handheld devices in New Zealand. On the other hand, it is extremely creative and it allows users to experience and become familiar with purchases through a mobile phone.

12snap (UK)

12snap is a major player in mobile marketing across Europe (12Snap, 2003). By exploiting the tremendous possibilities offered by the development of the mobile phone as a new media channel, 12snap offers personalized and targeted wireless marketing. The company specializes in "opt-in" advertising schemes. In essence, "opt-in" involves the user agreeing to receive advertising before anything is sent, with the opportunity to change preferences or stop messages at any time (Scornavacca & Barnes, 2003). 12snap currently has over 18 million aggregated permission-based users in the UK, Germany, Scandinavia, and Italy, with a further 3 million across the rest of Europe, making it the mobile media market leader in the these territories.

Figure 5. Business model for mobile barcode coupon campaigns (12Snap, 2003)

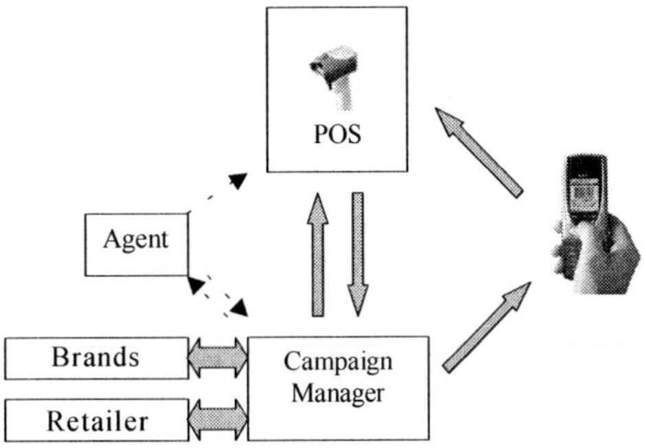

Mobile Barcode Coupons

12snap delivers marketing and customer relationship management (CRM) programmes by combining SMS, voice/sound, the Web, wireless application protocol (WAP), enhanced message service (EMS), and MMS with its own technology. It has built up a reputation for running groundbreaking interactive campaigns and is at the forefront of innovation in the mobile marketing industry. Recently, the company adopted barcodes in its campaigns. Figure 5 illustrates the process of a typical campaign utilizing mobile barcode coupons.

A brand or a retailer contacts the campaign manager in order to deliver coupons directly to targeted customers. An agent can support the development of the campaign strategy. Coupons containing a barcode are delivered via SMS. In order to redeem any relevant offer the customer simply displays the barcode embedded on his/her mobile phone at the POS. The cashier scans the barcode displayed with a barcode reader and the retailer's electronic POS (EPOS) informs the campaign database that a coupon has been redeemed. (Concurrently, the system identifies the client at the point of purchase through its barcode ID and detects other available offers.) Once a coupon is redeemed, it is automatically cancelled. Finally, a "thank-you" message is sent to the customer.

Overall, this business model allows a better measurement and understanding of return on investment (ROI) for mobile marketing (Scornavacca & Barnes, 2003). The solution is considered secure and integrates into EPOS software and hardware providing real-time reporting capabilities, tracking consumer behavior from initial contact through to in-store redemption (12Snap, 2003). A key advantage of the system is that it can be customized to a particular retailer, product, or customer. In addition, because it is based on an electronic format, retailers do not have to manually count, store, and ship coupons via a clearing house. This solution combines the power of couponing with the precision of mobile marketing. However, it is worth bearing in mind that mobile marketing has a more invasive nature than any other media, and a lot of attention must be given to permission issues in order to make the mobile marketing experience pleasant to the users. It must

Figure 6. Barcode-enabled m-commerce business models

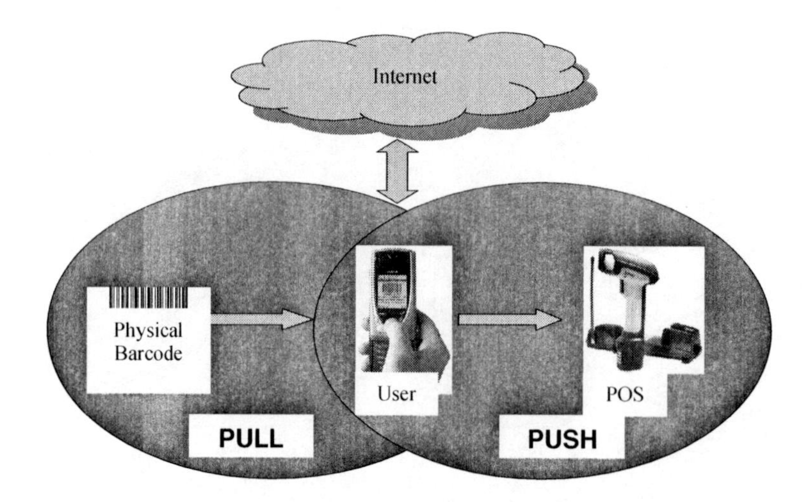

produce a win-win situation between user and advertiser. That said, mobile barcode coupons can certainly help marketers to develop mobile marketing through a better understanding of consumer preferences and behavior.

New Models of M-Commerce Using Barcodes

The examples of mobile barcode applications and the discussion presented above demonstrate the use of two basic barcode-enabled m-commerce business models: "push" and "pull." Figure 6 presents a conceptual model for barcode-enabled m-commerce business models.

Push business models are based in the concept of interaction between a mobile device and a separate OCR device, usually at a POS or via a "gatekeeper" to service provision. This interaction is possible via a barcode displayed on the mobile device screen. Combien?, 12snap's coupons, and mTicket are examples of this kind of business model. This is the most prevalent business model used for m-commerce.

Pull business models are based on the concept of interaction between a mobile device and a physical barcode. This interaction is possible by decoding a printed barcode. This business model is typically dependent on the mobile device being OCR enabled (by camera or barcode reader). The built-in camera and barcode-reader software on DoCoMo's 505i series is a clear example of this application. An interesting example of an OCR-independent pull business model is that of Scan-UK, trialed in 2000. Scan-UK's service enables a user to check if a specific book or CD that they are considering purchasing from

Figure 7. Strategic framework for mobile barcode applications

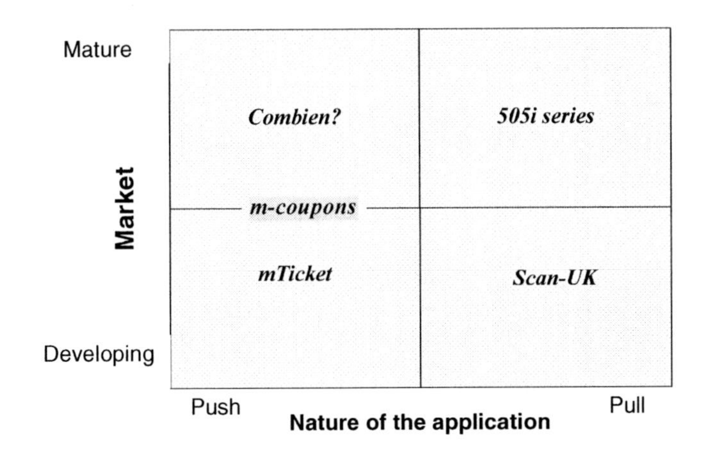

a physical store is cheaper online. The user enters the barcode number of the desired product via SMS, sends the message to the service, and the system sends back a list of how much e-tailers are asking for the same product (BBC News, 2000).

A combination of push and pull business models is also possible. A *push-pull* business model would occur in the situation when a mobile device interacts with a physical barcode and later interacts with an OCR device. An example of this would be when a user scans and stores a physical barcode into his/her mobile phone and later uses the scanned barcode at a POS.

Besides the nature of the application (push and/or pull) another important strategic variable for implementing barcode applications in mobile commerce is the market environment. Figure 7 presents a framework that helps understand the relationship between market maturity and business models. When analyzing barcode applications in mobile commerce, we understand market maturity as the dynamic combination of OCR infrastructure, payment systems integration, m-commerce acceptance, mobile phones market penetration, and hardware development.

The Japanese market is considered a mature market. It has a high penetration of Internet-enabled mobile phones, and most are equipped with a camera and a high-definition colour screen (Scornavacca & Barnes, 2003). Japan also has a solid infrastructure of POS systems, equipped with OCR devices. Combien? can be placed in the upper-left quadrant (mature-push). Typically, this type of application is characterized in a mature market where the advanced mobile technology platform available for customers takes profit of the integration between merchants (e.g., convenience stores) and wireless services providers. DoCoMo's 505i series can be placed in the upper-right quadrant (mature-pull). Notice that this application requires that the user's device is equipped with OCR.

In New Zealand, the mobile market still in a developing stage. Most of the handsets have low definition screens and are without camera or barcode reader. Nevertheless, it also

Table 1. Characteristics of barcode-enabled m-commerce business models

Business model	Characteristics of the model	Benefits	Specific issues
PUSH	A barcode is sent to a mobile device, typically via messaging. Service interaction occurs when the barcode is read from the screen via an external OCR, such as a merchant or another device equipped with OCR capability.	Convenience to the user (time and location) Potential for personalization and CRM Ubiquity of mobile infrastructure (phones) Interaction (passive) Cost reduction to the operator and service provider	Suitable for developing markets where mobile devices are not yet OCR capable. Relies primarily on existing OCR infrastructure of merchants. "Passive" interactivity—the user cannot interact with printed barcodes. Interactivity is driven by the operator or service provider.
PULL	A mobile device is equipped with OCR capability, such as a camera and OCR software. Service interaction occurs when the user pulls information from printed barcodes or other sources, such as screens of mobile devices, computers, or TV.	Convenience to the user (time and location) Customer driven Ubiquity of mobile infrastructure (phones and OCR) Interaction (active) Cost reduction to the operator and service provider	Usually hardware dependent—requires sophisticated mobile devices with OCR capability More suitable for mature markets where advanced devices are in use and commercial activity has begun An "active" model, based on the user's desire for information and interaction

has a high penetration of mobile phones and has a solid infrastructure of POS systems equipped with OCR devices. mTicket occupies the lower-left quadrant due to its push nature and market environment.

The UK market is close to that of New Zealand, but driven by a much larger consumer base, more sophisticated hardware, a greater variety of mobile services, and competition among the larger number of operators. It also has a high penetration of mobile phones and a solid infrastructure of POS systems equipped with OCR devices. The mobile barcode coupons provided by 12snap are positioned in the middle of the left side of the matrix; the UK is supported by a good OCR infrastructure at POS and a reasonable display quality. On the other hand, Scan-UK would be placed in the lower-left quadrant. This application was trialed in 2000. At that point in time, the UK market environment was in a developing stage. Also, this application would be more convenient and accurate if customers could simply scan the product barcode.

The model allows us to understand some strategic implications of barcode-enabled m-commerce business models. Table 1 examines some of the characteristics of barcode-enabled m-commerce business models, as derived from the above discussion.

A key benefit for all the solutions is the *convenience* of the services provided, which can potentially be used by the user on the move at any time and from anywhere. The penetration of mobile telephony means that such services have a high degree of *ubiquity*. The technological platforms enable business models with considerable customer *interaction*, with broad possibilities for the development of services that engage the user. At the same time, by motivating the user to engage in activities enabled by mobile devices, *costs are reduced* significantly via digitization—including such things as collection costs, handling, delivery, fraud, and call centers. Push services, being based on existing OCR infrastructure, are better suited to developing markets. In more mature markets, where devices are more sophisticated, pull services allow the user to interact with printed barcodes, expanding the possibilities for service interaction.

Conclusions

The convergence of traditional OCR technology and mobile telephony has provided some exciting possibilities for new mobile e-commerce business models. In this chapter we have explored the background to barcode technology, and how such technology can be employed in the new mobile environment. In particular, we have examined the case studies of a number of companies leading the field in three countries—the UK, New Zealand, and Japan. Each of these involves real exemplars of commercial use of mobile barcodes. By examining these, we have discovered two distinct business models being used—push and pull—each with their own benefits and problems. Moreover, by combining the choice of business model with an examination the maturity of the market environment, we can distinguish a number of strategic positions for the provision of mobile barcode services.

The development of mobile barcode applications, like m-commerce, is still relatively new and emerging. Technological development may have a strong impact on its future direction. One area of current development that may provide a substitute for barcodes is the use of wireless in personal area networks (PANs) to enable m-commerce. Standards such as Bluetooth and IEEE 802.11 allow a new wave of short-range device interactivity and provide cheap, low-power, high-data-rate connectivity for portable devices in a limited area. Here, for example, the roaming phone user can be provided with information, alerts, or even advertisements based on local interaction with PANs. Customers could also conduct transactions with their mobile phones at the POS—vending machines and ATMs being the best-known examples. In a more complex scenario, the customer looking for a specific product or price could conceivably scan or enter a code into a phone; when the customer walks past a store with the right product or price the phone could send an alert. Stores could even send advertising alerts in an effort to tempt customers inside. Whether short-range wireless will replace barcodes is unclear. Clearly there are also areas of synergy between the technologies, and the use of PANs could enable new types of push and pull application in Figure 7. What is clear, however, is that the infrastructure for barcodes is already in place, and will take some time for POS have short-range-enabled devices of a common standard.

Given the emergent nature of mobile services, this is an area ripe for future research. In particular, we have a number of questions that will provide a future research agenda for mobile barcode research:

- What is the user's acceptance of digital barcode services? Do the user's benefits for these services really outweigh their personal cost? How does this vary among users?
- How can the business use mobile barcodes to build and manage customer relationships?
- How can mobile services become potentially more targeted and measurable? How can businesses measure ROI?
- Which business models and barcode applications are most likely to be successful?
- How will short-range wireless technologies change the nature of interactivity and business models?

Our own research will attempt to examine these questions in more detail utilizing empirical data to provide a firmer base for strategic understanding.

References

12Snap. (2003). Mobile barcode coupons – the marketing revolution for marketeers. Retrieved October 23, 2003, from *www.12snap.com/uk/help/couponsshort.pdf*

Airclic. (2003). Airclic devices. Retrieved October 27, 2003, from *www.airclic.com/devices*

AsiaTech. (2003). PaySmart: Cashless payment solution. Retrieved October 27, 2003, from *www.asiatech.com.au/capter ii/chapterii02 a.htm*

Bai, L., Chou, D.C., Yen, D.C., & Lin, B. (2005). Mobile commerce: Its market analyses. *International Journal of Mobile Communications*, *3*(1), 66–81.

Bango. (2003). Camera-enabled mobiles and PDAs scan barcodes for instant access to WAP. Retrieved June 4, 2003, from *http://corp.bango.net/corporate/media/releases/_31_iwdemo.asp*

Barcode 1. (2003). Barcode 1. Retrieved October 27, 2003, from *www.adams1.com/pub/russadam/isbn.html*

Barnes, S.J. (2003). The mobile commerce value chain in consumer markets. In S.J. Barnes (Ed.), *mBusiness: The strategic implications of wireless communications* (pp. 13–37). Oxford: Elsevier/Butterworth-Heinemann.

Barnes, S.J., & Huff, S.L. (2003). Rising sun: iMode and the wireless Internet. *Communications of the ACM*, *46*(11), 78–84

Barnes, S.J., & Scornavacca, E. (2004). Mobile marketing: The role of permission and acceptance. *International Journal of Mobile Communications*, *2*(2), 128–139.

Barnes, S.J., & Vidgen, R. (2001). Assessing the quality of WAP news sites: The WebQual/m method. *VISION: the Journal of Business Perspective*, *5*, 81–91.

Bayne, K.M. (2002). *Marketing without wires: Targeting promotions and advertising to mobile device users*. London: John Wiley & Sons.

BBC News. (2000). Mobiles bag barcode bargains. Retrieved October 25, 2003, from *http://news.bbc.co.uk/1/hi/sci/tech/782578.stm*

Bergeron, B.P. (2001). *The wireless Web: How to develop and execute a winning wireless strategy*. New York: McGraw-Hill.

Chen, P. (2000). Broadvision delivers new frontier for e-commerce. *M-Commerce, October*, 25.

Clarke, I. (2001). Emerging value propositions for m-commerce. *Journal of Business Strategies, 18*(2), 133–148.

de Haan, A. (2000). The Internet goes wireless. *EAI Journal, April*, 62–63.

EAN. (2003). EAN International. Retrieved October 27, 2003, from *www.ean-int.org/index800.html*

Ecrio. (2003). Mobile commerce. Retrieved October 20, 2003, from *www.ecrio.com/sol mobilecommerce.shtml*

Emarketer. (2002). *One billion mobile users by end of Q2*. Retrieved May 27, 2003, from *www.nua.ie/surveys/index.cgi?f=VS&art_id=905357779&rel=true*

Kalakota, R., & Robinson, M. (2002). *M-business: The race to mobility*. New York: McGraw-Hill.

Kannan, P., Chang, A., & Whinston, A. (2001, January). Wireless commerce: Marketing issues and possibilities. *Proceedings of the 34th Hawaii International Conference on System Sciences*, Maui, HI.

Lau, A.S.M. (2003). A study on direction of development of business to customer m-commerce. *International Journal of Mobile Communications, 1*(1/2), 167–179.

MTicket. (2003). Tickets for a mobile generation. Retrieved October 27, 2003, from *www.mticket.co.uk*

Newell, F., & Lemon, K.N. (2001). *Wireless rules: New marketing strategies for customer relationship management anytime, anywhere*. New York: McGraw-Hill.

NTT DoCoMo. (2003). DoCoMo Net. Retrieved October 27, 2003, from *www.nttdocomo.co.jp* (in Japanese)

NTT DoCoMo. (2005). DoCoMo Net. Retrieved January 27, 2005, from *www.nttdocomo.co.jp* (in Japanese)

Sadeh, M.N. (2002). *M-commerce: Technologies, services, and business models*. London: John Wiley & Sons.

Scornavacca, E., & Barnes, S.J. (2003). Mobile banking in Japan. *International Journal of Mobile Communications, 2*(1), 51–66.

Telecom Trends International. (2003). M-commerce poised for rapid growth. Retrieved October 27, 2003, from *www.telecomtrends.net/pages/932188/index.htm*

Uniform Code Council. (2003). Uniform Code Council. Retrieved October 27, 2003, from *www.uc-council.org*

Vodafone. (2003). What's hot. Retrieved October 27, 2003, from *www.vodafone.co.nz*

Wireless Newsfactor. (2002). Mobile technology gives pointing new meaning. Retrieved September 12, 2003, from *www.wirelessnewsfactor.com/perl/story/16561.html*

Yuan, Y., & Zhang, J.J. (2003). Towards an appropriate business model for m-commerce. *International Journal of Mobile Communications*, *1*(1/2), 35–56.

Section II

Mobile Marketing

Chapter VI

Mobile Advertising: A European Perspective

Tawfik Jelassi, Ecole Nationale des Ponts et Chaussées, France

Albrecht Enders, Friedrich-Alexander-Universität Erlangen-Nürnberg, Germany

Abstract

This chapter is based on research conducted in cooperation with 12Snap, the leading European mobile marketing company, which has implemented large-scale mobile advertising campaigns with companies such as McDonald's, Nestlé, Microsoft, Coca-Cola, Adidas, and Sony. To set the overall stage, we first discuss the advantages and disadvantages of the mobile phone in comparison to other marketing media. Then we propose a framework of different types of advertising campaigns that can be supported through the usage of mobile devices. These campaign types include (1) mobile push campaigns, (2) mobile pull campaigns, and (3) mobile dialogue campaigns. Building on this framework, we analyze different campaigns that 12Snap implemented for different consumer goods and media companies. Drawing from these experiences we then discuss a number of key management issues that need to be considered when implementing mobile marketing campaigns. They include the following themes: (1) the choice of campaign type, (2) the design of a campaign, (3) the targeting of the youth market, and (4) the combination of different media types to create integrated campaigns.

Introduction

The market for mobile phones has expanded rapidly during the past decade and continues to grow quickly. In some European countries such as Finland, Sweden, Norway, and Italy, the mobile phone has reached almost ubiquitous penetration with levels of 80% and higher (*Economist*, 2001). In Germany, mobile phones are more widely used than fixed-line connections (Brechtel, 2002). In addition to voice communications, German users send out 2.2 billion text messages through their mobile phone every month (Brinkhaus, 2002).

The fast spread of mobile phones has created immense profit expectations in the telecommunications industry. Telecommunication companies in many countries have invested large sums of money into acquiring third-generation licenses and building the necessary infrastructure. Yet, as it turns out, it is more difficult to generate revenues than initially anticipated.

In addition to call charges, there are three main revenue sources in mobile communications: (1) **transactions**, (2) **information**, and (3) **advertising**. Transactions are of high interest, yet as of now only to a limited extent, because of the small size of the screen and the clumsy usage of the keypad. With information services (such as weather forecasts or banking services) the crucial issue is the user's willingness to pay for these types of services.

Does mobile advertising have the potential to be a significant source of revenue in the future? First studies on this new advertising medium indicate that mobile advertisement campaigns can be very successful, generating response rates as high as 40%, compared with the 3% response rate generally expected for direct mail and less than 1% for Internet banner ads (Borzo, 2002).

Because of the novelty of the technology, using mobile phones for advertising campaigns presents some challenging questions for marketing departments:

- What are the strategic advantages of the mobile phone in comparison to other advertising media?
- What campaign types can leverage these characteristics?
- What critical issues need to be considered when launching a mobile advertising campaign?

In the remainder of this chapter, we discuss these questions drawing on field research conducted in cooperation with the German mobile marketing company 12Snap.

Advertising through Mobile Phones

With the increasing number of media types, it has become more and more difficult for marketing managers to find appropriate strategies to target potential customers with their messages. First, while it was possible in the past to capture a large segment of society by placing advertisements with the main TV networks, the rise of private channels has led to a high degree of fragmentation, thereby complicating access to consumers. Similar fragmentation can be observed with other mass-media types such as print or radio. As a result, getting time and attention from their audience has turned into a major challenge for advertisers (Davenport & Beck, 2000).

Second, different media types require different approaches because of differences regarding their reach or richness. **Reach** is a function of how easily customers, or in this case, participants in advertising campaigns, can be contacted through a given medium. **Richness**, on the other hand, is defined by (1) bandwidth, that is, the amount of information that can be moved from sender to receiver in a given time, (2) the degree of individual customization of the information, and (3) interactivity, that is, the possibility to communicate bidirectionally (Evans & Wurster, 1997). The communication of rich marketing information, that is, information that ranks high on all three aspects, has traditionally required physical proximity to customers and/or channels specifically dedicated to transmitting the information (see Figure 1).

How does the mobile phone fare within the richness versus reach framework? It can serve as a powerful platform to get in touch with end consumers because it simultaneously

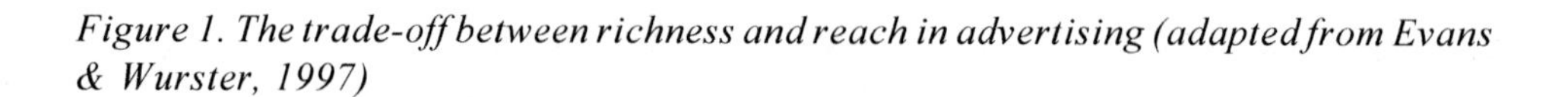

Figure 1. The trade-off between richness and reach in advertising (adapted from Evans & Wurster, 1997)

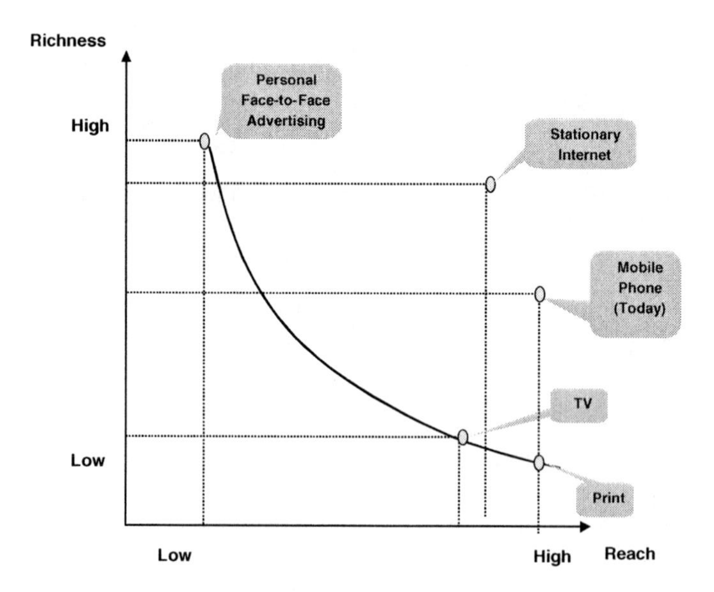

provides expanded reach and a number of richness advantages vis-à-vis most other media types:

- **Ubiquitous Access:** Mobile phone users always have their phone with them and turned on at almost all times (Balasubramanian, Peterson, & Jarvenpaa, 2002; Magura, 2003). This is especially true for teenagers and young users who use the mobile phone to stay in touch with their peers—primarily through SMS (Bughin & Lind, 2001). Ubiquitous access becomes especially important in places like buses, trains and subways, airport lounges, and so forth. The time that people spend traveling is prime time for marketing since it presents a time when people are not occupied with other activities and are thus receptive to other kinds of entertainment. A study by the Boston Consulting Group (2000) found that among private users, the categories "having fun" (71%) and "killing time" (55%) belong to the main motivators for using mobile phones—ranking only behind "keeping in touch with friends" (85%).
- **Detailed user information:** While traditional marketing campaigns only have access to very limited customer information, mobile campaigns can draw on extensive and individual information about each user (such as age, sex, usage profile, etc.). This information helps to launch highly targeted campaigns for specific products and services based on individual preferences of the user.
- **Integrated response channel:** The mobile phone presents the opportunity to interact directly with the user and elicit responses through the same medium. This has two advantages. First, it provides the opportunity for rich interaction. The interactivity and ubiquity of the mobile phone opens up the possibility to turn existing traditional media formats (such as the TV, radio, print, or packaging) interactive. For instance, companies can contact consumers via TV and then subsequently, stay in touch with each one of them through the mobile phone. Second, the integrated response channel also allows mobile marketing companies to measure precisely the impact of their campaigns and then to adapt their strategies accordingly—something that is much more difficult to do with traditional marketing media. For instance, a customer buys a product—with a mobile phone number on the packaging—at a retailer, and as s/he exits the store, s/he completes a quick survey of the shopping experience, which is then transmitted immediately to corporate headquarters. This not only allows the consumer the satisfaction of immediate feedback if they had a positive or negative experience, but it also allows the company to measure quality control in an extremely timely and cost-effective manner (Carat Interactive, 2002).
- **Personal channel:** Unlike other advertising media such as TV, radio, or billboards, the mobile phone belongs to only one person. Therefore, it receives much more attention and, if handled properly (see risks below), can be much more powerful than other, less personal media channels. John Farmer, a cofounder of the SMS application and service provider Carbon Partners, points out that the personal character of the mobile phone is especially important to teenagers (Haig, 2001): "The mobile phone presents the teenage market with the distinct opportunity to

take control of their own communications, free from the previous limitations of the home phone or computer, which were more closely monitored by parents."

Brian Levin, CEO of Mobliss, a U.S. wireless marketing firm, sums up the advantages (Stone, 2001): "When you have a little time to spare, such as in the airport or at the bus stop—then you want to be engaged or entertained. Once you are there, the proximity of this device [the mobile phone] to your face, the intimacy there, is very powerful both in terms of direct response and in terms of branding." At the same time, however, the mobile phone also presents shortcomings and risk factors:

- **Limited media format:** Mobile phones today still have to cope with a very limited set of visual and audio capabilities. In second-generation (2G) phones, screens are typically small, have only low resolution, and are typically not in color. Sound effects are also limited due to the small speakers, and text messages cannot be longer than 160 characters. The challenge is then to ensure at this stage that consumers do not expect an identical experience to what they receive through other devices such as the TV or PC (Carat Interactive, 2002).
- **Private sphere:** The fact that mobile phones belong to only one person does not only present an opportunity but also a challenge for mobile advertisers. Unlike the TV or Internet, the mobile phone is a very personal device to which only family, friends, coworkers, and a selected few others will gain access. Thus, "spamming" is considered much more intrusive than in other media formats (Carat Interactive, 2002).

Developing Effective Mobile Advertising Campaigns

One of the main challenges and opportunities for mobile advertising companies is the personal nature of mobile phones. Advertising campaigns over mobile phones are very sensitive and companies that engage in this type of marketing need to be careful not to offend users. Will Harris, global marketing director for Genie, British Telecom's mobile Internet service, emphasizes (Pesola, 2001): "Sending unsolicited messages is tantamount to brand suicide. Our business is entirely dependent on the goodwill of our customers."

The mobile advertising industry is trying to protect mobile phone users by establishing guidelines for responsible advertising. The main feature of these guidelines is consent, that is, consumers agree or opt-in to receive the advertisements. In addition, they must have a clear understanding of what their personal information is being used for, and if they wish, be removed from the advertiser's databases.

As a result, mobile advertisers have to find ways to entice customers to opt into their campaigns. Cyriac Roeding, 12Snap's marketing director, explains why many companies have difficulties attracting mobile phone users (Pesola, 2001): "A lot of companies make the mistake of coming to this from a technological angle, rather than thinking about what the consumer wants. If advertising is entertaining, if it engages the emotions, it will be accepted."

Although mobile advertising is a relatively recent phenomenon, a number of large corporations, including McDonald's, 20th Century Fox, and Sony, are using this medium in their marketing mix, especially to target young customers. These campaigns differ according to the degree of active involvement of advertiser and recipient (see Figure 2). Level of activity refers here to the involvement both advertiser and consumer show throughout the course of an advertising campaign. Traditional campaigns, which still present the most prominent advertisement type, display low levels of activity on both the advertiser's and the consumer's side since they consist of noninteractive, one-way advertisements in the form of TV spots, radio or print ads, or posters. Mobile campaigns, on the other hand, show high levels of activity either on the side of the advertiser, the consumer, or both. High level of activity on the side of the advertising company implies that the consumer is approached proactively, whereas a high level of activity on the side of the consumer implies that s/he reacts actively to an advertisement or a newspaper ad, for instance, by soliciting further information via the mobile phone.

Through the built-in response channel, mobile phones are suitable both for push and pull campaigns. According to the mobile advertising framework, mobile campaigns can be categorized as follows:

- **Mobile Push Campaigns:** Push advertising is categorized as messages that are proactively sent out to wireless users. Companies use databases with existing

Figure 2. Mobile advertising framework

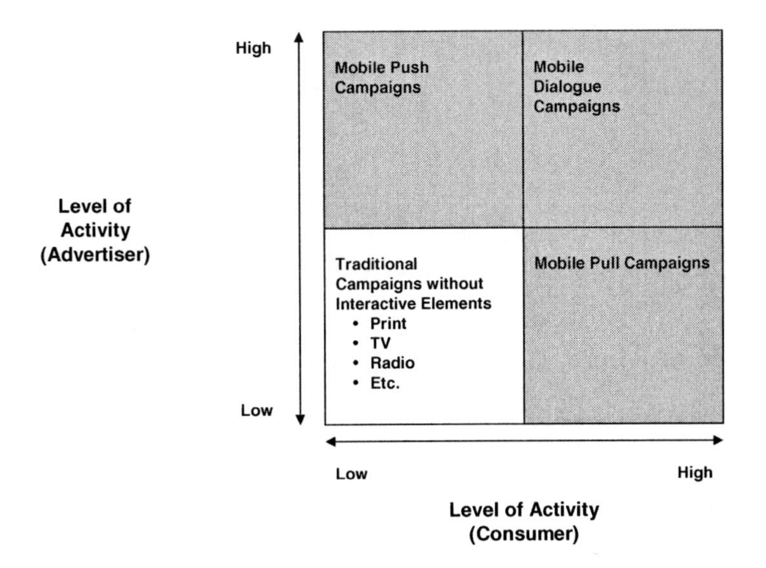

customer profiles, which can be proprietary or purchased externally, to address their target groups (Carat Interactive, 2002). Because of the sensitivity of the mobile phone, it is important to ensure that all members of the database have agreed beforehand (i.e., given their opt-in) to receive mobile advertising. In addition, for the success of a campaign, it is essential to ensure that the selected target group from the database is interested in the specific advertising, which requires extensive profiling of the database (Pearse, 2002). Doing so avoids the alienation of uninterested users, while at the same time maximizing the impact of the advertising budget on those targeted.

- **Mobile Pull Campaigns:** Applying a pull approach, advertisers use their traditional marketing media mix, such as TV, radio, print, or packaging, to promote an interactive mobile campaign. For instance, a telephone number on a French fries box might invite a customer to participate in a raffle for free food through his/her mobile phone. By calling, the consumer gives the "opt-in"—an explicit consent to the campaign—and can then participate.
- **Mobile Dialogue Campaigns:** Dialogue campaigns differ from the above-mentioned campaign types in their duration and the intensity of interaction between advertiser and customer. While simple push and pull campaigns tend to last only 2 to 4 weeks and center around one single theme such as a raffle or a game, dialogue campaigns last several months and include various different themes that build on one another. Their goal is to establish a long-lasting relationship with consumers so as to generate extensive insights into consumers' preferences. A mobile horoscope service, for instance, allows the advertiser to capture the birthday of the consumer, which can then be used for sending out personal birthday greetings later

Figure 3. Taxonomy of mobile advertising campaigns

	Push Campaigns	Pull Campaigns	Dialogue Campaigns
Set-up	Targeted SMS to user in existing database • Company-owned database • Database from external provider	Advertisements for mobile campaigns through other media types, e.g. • Flyer or "on-pack" ad • TV ad	Continuous interaction between advertiser and user
Opt-In	Need to have explicit "opt-in" prior to sending out SMS	Users "opt-in" by calling the phone number	Generation of "Opt-in" either through pull or push
Scope	Single theme • Game • Raffle • Etc.	Single theme • Game • Raffle • Etc.	Multiple themes during the course of a campaign • Different games • Greetings
Length	Short duration (2-4 weeks)	Short duration (2-4 weeks)	Extended duration (several months)
Implementation	• Wella Hair Fashion • Sony • 20th Century Fox • Adidas	• Nestlé KitKat Chunky • Toyota	• McDonald's

on. The in-depth consumer information serves then to distribute mobile coupons—for instance, a free candy bar as a birthday present—to introduce new products or to do market research in a very targeted fashion.

Building on this framework, it is now possible to categorize the actual implementations of mobile advertising campaigns (see Figure 3).

Push Campaigns

Wella, a German manufacturer of hair-care products, developed a push campaign, which featured a "mobile kiss." Wella sent an SMS to members of an externally acquired database offering them to send a kiss message to their friends, who received a voice file with a kiss sound. This was followed by an SMS revealing who had sent the kiss and also providing details on how to return the kiss or send it to someone else. The maximum number of kisses sent by one person was 160. Other components of this mobile campaign included an SMS quiz and a free kissing-lips logo for the mobile phone.

Sony launched a push campaign in the UK which integrated e-mail, Internet, and traditional print media to promote a new PC-compatible MiniDisc Hi-Fi system. The campaign, which was based on MiniDisc Island—an online, interactive adventure playground—had the objective of driving large numbers of users to the Web site. Through an initial SMS 100,000 participants from a database of 14 million permission-based, profiled users were selected and invited to enter a competition to win a stereo system and the PC Link product. Interested participants replied via SMS and were mailed a winning number, with which they could then go to the Web site to see if they had won. Throughout the course of this campaign, 18% of those originally contacted responded to the initial SMS. When contacted with the winning number, over 9% logged on to explore the online adventure game and see if their numbers had come up.

The movie studio **20th Century Fox** launched a push campaign in partnership with the mobile phone operator Vodafone to advertise the UK release of *Planet of the Apes*—a post- apocalyptic movie where apes rule over humans who struggle to survive. The campaign, which started 2 weeks prior to the UK release of the movie, targeted the 2 million 16- to 24-year-old Vodafone customers. They received messages, which invited them to survive a variety of challenging interactive voice response and text games—with names such Ape S-cape and Ape@tak, where callers are asked questions relating to the movie and have to shoot down apes using the keypad when they hear a roar of an ape.

The sports article manufacturer **Adidas** used interactive betting game for the Soccer World Cup 2002 to promote its products in Germany. Users bet on games and received immediate notification after the game about how they did, how they ranked overall within the betting competition, and if they had won a prize such as a shirt from the soccer idol Zinedine Zidane or a personal meeting with the German national soccer team.

Pull Campaigns

Nestlé used a mobile pull campaign to promote the KitKatChunky chocolate bar in Germany. The campaign, which lasted 2 weeks, complemented the overall marketing presence consisting of TV and radio spots and the Web site www.chunky.de. The campaign worked as follows: an SMS offered community members the opportunity to win a 1-year supply of KitKatChunky if they called a specified number. Then, callers were shown the face of a taxi driver who also appears in a KitKatChunky TV commercial, and two other new characters on their screen who were presenting riddles to them. An automated voice then explained: "Each of the three protagonists names a number which makes him shut up. Once you have discovered the number, push the appropriate button on your mobile and a KitKatChunky is stuck in his mouth and you go on to the next round." In the first round, 400,000 users were identified to receive a kick-off SMS at the beginning of the campaign. In the following rounds, only those players who had actively opted-in in the previous round received an SMS. In order to maximize the number of responses, users received alert messages the day before the ad's TV premiere and again 30 minutes before the TV show in which the questions were sent. This illustrates to what extent different media types—here the TV and the mobile phone—can be interlinked, using the respective strengths of each medium, to generate a seamless and entertaining marketing experience for the end user.

The car manufacturer **Toyota** also launched an interactive TV pull campaign during the Soccer World Cup 2002 that displayed a quiz question at the bottom of the TV screen. The question asked viewers to find the license plate number of the Toyota shown in the TV ad and to send this number in via SMS. Within seconds they received notification whether they had made it to the final drawing. In addition to having the chance to win a prize, all callers also received a Toyota ring tone for their mobile phone.

Dialogue Campaigns

Extensive dialogue campaigns are still a rarity because of the novelty of the mobile phone as an advertising medium. In Germany, **McDonald's** launched a mobile dialogue campaign with a focus on interactive mobile games and an evaluation of McDonald's products. The campaign targeted mobile phone users who were informed through in-store flyers placed in McDonald's restaurants. By activating the service (active "opting-in"), participants received automatic messages when music CDs or vouchers were raffled off. In addition, they also received SMS promotions of McDonald's products. The goal of the campaign was (1) to increase in-store traffic, (2) to build a McDonald's customer database of mobile phone numbers, and (3) to increase overall brand awareness. Following this initial pull activity, registered users continued to receive other services such as horoscopes, which in turn allowed McDonald's to capture users' birthdays and to send them personalized birthday greeting subsequently. An additional part of the campaign was a viral activity: McDonald's sent the Christmas greeting "Rockin' Rudi" to users who could then forward it to their friends. The recipient then listened to a taped version of the "Rockin' Rudi" song in combination with a short message from the sender and from McDonald's.

Outlook and Management Issues

There are plenty of opportunities in mobile advertising for companies that thoroughly understand how consumers can benefit from these types of services. The new technology will not be very useful, however, if companies simply use their existing advertising approaches and translate them to the mobile world without addressing the specific characteristics of this new medium (Nohria & Leestma, 2001).

The different innovative types of mobile advertising campaigns mentioned above offer many useful benefits—for instance, highly targeted advertising and interactivity—to those companies that want to add a mobile component to their advertising approach. They also illustrate the difficulties and challenges that are associated with this new approach. Therefore, before embarking on mobile advertising campaigns, managers need to carefully address the following questions.

Which Campaign Type Should We Employ?

For starting a mobile advertising campaign, there are two basic options: push or pull. A push campaign requires an extensive database of customers. Some companies such as telcos or retailers have built up these types of databases in the past through CRM efforts and can now tap into them. However, they always need to keep in mind the personal nature of the mobile phone when doing so. "Spamming" existing customers with unwanted SMS is a sure way to alienate them. Another option is to buy existing profiles from other companies. MTV, for instance, markets its permission-based database through an external mobile advertising company to other companies that want to target the attractive youth market. These companies benefit since they can tap into an extensively profiled, permission-based database of their target group while MTV generates additional revenues.

Setting up a pull campaign is not as sensitive regarding the opt-in, since consumers themselves decide whether they want to participate when they see the advertisement printed on a poster or watch it on TV. Here, the challenge is much more to create compelling advertisements that have the desired pull effect to entice consumers to call in and participate.

How Should We Design Attractive Mobile Advertising Campaigns?

The challenge for any mobile marketing company is to create enough interest within the target group to justify the required investment. Based on the campaigns we have analyzed, four key success factors need to be considered when launching a mobile advertising campaign (Brand & Bonjer, 2001):

- **Interactivity:** just like the Internet, the mobile phone allows advertisers to solicit immediate feedback when contacting recipients. Since the mobile phone is usually always turned on, the inherent interactivity of mobile phones should be integrated in mobile marketing campaigns where possible. The interaction can have many different facets: the number pad can be used to answer riddles or mental agility can be tested through reaction tests. A mobile marketing campaign that does not integrate interactivity would be the equivalent to the broadcasting of a slide show on TV. It would leave a main asset of the medium untapped.
- **Entertainment:** interaction is only fun for users if they find the advertisement exciting. Therefore, mobile campaigns need to combine advertising and entertainment in such a way that users are willing to lend their time to an advertisement. In this respect, the creation of mobile campaigns is similar to more traditional campaigns on TV, for instance. TV viewers watch advertisements mainly because they are entertaining. Ideally, they do not just watch them but they also talk about them to friends thereby creating a viral effect in which the message is passed on by people other than the original sender—as was the case in the Wella and McDonald's campaigns. Therefore, the inclusion of entertaining elements such as a game or a story ought to present an integral part of a mobile marketing campaign.
- **Emotion:** the inclusion of emotional elements—such as visual sequences or music clips in TV ads that aim beneath the conscious understanding of the viewer—has long presented a valuable marketing tool to subliminally reinforce the intended message with consumers. In mobile marketing, however, text, especially if shown on a small mobile phone display, can hardly carry this emotional dimension. Here, just like with TV advertising, it is necessary to leverage the admittedly limited resources of the mobile phone to create "emotion." This can be achieved through the combination of voice and sound. For instance, music jingles such as a short sequence of the soundtrack of the movie *Titanic* can be used as the opening for a partner test or an activity aimed at single people. Again, it is not primarily the technology that drives the quality of any given campaign but instead the creative combination of different effects that ultimately determines its success.
- **Incentive:** the offering of incentives such as product samples increases the willingness of consumers to participate in interactive mobile games. The prospect of winning a prize is especially important due to the above-mentioned opt-in nature of mobile marketing campaigns, as it provides the potential participants with a direct and tangible incentive to participate in a mobile marketing campaign. However, although instant-win competitions are effective in driving volume, they are less suitable to generate a long-term relationship with consumers, since they do not offer incentive to return (Cowlett, 2002).

The overall goal of combining these four factors is to create a game, an image or a jingle that, despite the limitations of the small screen and tiny ring tone of the mobile phone, is so compelling that it is no longer seen as an ad, but takes on a value of its own.

How Should We Target the Difficult-to-Reach Youth Market?

Addressing the lucrative youth market gives marketers a perennial headache, since they do not only vary in their habits, interests, and attitudes and are swayed by rapidly changing fashion trends. They are also hard to pin down, since they do not primarily watch three or four TV stations anymore as was in the past. Instead, their media usage is fragmented between hundreds of TV stations, radio, magazines, newspapers, and the Internet. One thing is generally guaranteed, though—they almost certainly carry a mobile phone and consider SMS an intrinsic part of their lifestyle since it allows them to stay in touch with their peers in a cost-effective and entertaining way (Cowlett, 2002).

Mobile campaigns can effectively leverage the characteristics of this new youth market. Viral effects, used in the McDonald's and Wella campaigns, fulfill the desire to communicate with peers in a fun way. Teens enjoy quizzes or greeting cards they can pass on to friends, because this type of promotion focuses on using the mobile phone for what it was made to do— communicate with other people (Centaur Communications, 2002). In addition, viral elements help to expand significantly the group of recipients beyond the database of the company conducting the campaign and it increases the impact since marketing messages sent from a friend are, because of their personal nature, much more effective than those sent directly from the company itself (Haig, 2001; Kenny & Marshall, 2000). The communication from consumer to consumer helps to generate "buzz"—explosive self-generated demand—where people share their experiences with a product or a service amongst one another (Dye, 2000). At the same time, this approach helps to lower costs since users themselves target new consumers.

How Should We Combine the Mobile Phone with Other Media Types to Create Integrated Campaigns?

Because of its limitations regarding screen size, sound, and handling, the mobile phone is not suitable for stand-alone campaigns. Instead, it should be used to extend the presence of a company into an additional channel (Carat Interactive, 2002). Doing so, the mobile phone plays the role of the natural glue between other media types because of its ubiquitous nature: it is handy and turned on when watching TV, looking at a billboard on the subway, buying groceries at the supermarket, or listening to the radio. All the campaigns mentioned above make extensive usage of this cross-linking of different media types, leveraging the unique strengths of each. It is not only other media types that benefit from the integration of the mobile in multichannel advertising campaigns: tangible support mechanism from other media types that have been around for years—such as a flyer or an in-store promotion—give mobile campaigns higher legitimacy because they have a physical component (Enders & Jelassi, 2000).

From a market research perspective, the inclusion of mobile components in advertising campaigns has the added benefit that it allows to measure directly the effect of different advertising approaches. Take, for instance, a TV advertisement that is aired on different channels and broadcasting times, or a billboard advertisement placed in different locations that asks viewers to participate in an SMS contest. Based on the measurement of actual response rates in different channels or locations, it becomes possible to steer placement more effectively than via traditional indirect measurements.

References

Balasubramanian, S., Peterson, R., & Jarvenpaa, S. (2002). Exploring the implications of m-commerce for markets and marketing. *Journal of the Academy of Marketing Sciences, 30*(4), 348–361.

Borzo, J. (2002). Advertisers begin dialing for dollars. *Asian Wall Street Journal*, February 18.

Boston Consulting Group. (2000). Mobile commerce—winning the on-air consumer. November.

Brand, A., & Bonjer, M. (2001, November). *12Snap: Mobiles Marketing im Kommunikations-Mix innovativer Kampagnenplanung* (White paper). Munich: 12Snap AG.

Brechtel, D. (2002). Bei Anruf Werbung. *Horizont, September 12*, 80–81.

Brinkhaus, G.B. (2002). Keine Massenmailings: Wie Mobile Marketing Funktioniert. *FAZ-online, September 29*. Retrieved from *www.faz.net*

Bughin, J., Lind, F. et. al. (2001). Mobile portals mobilize for scale. *McKinsey Quarterly, March*, 118–125.

Carat Interactive. (2002). *The future of wireless marketing* (White paper).

Centaur Communications. (2002). Good text guide. *In-Store Marketing, October 7*, 23–27.

Cowlett, M. (2002). Mobile marketing—"text messaging to build youth loyalty." *Marketing, October 31*, 29–34.

Davenport, T., & Beck, J. (2000). Getting the attention you need. *Harvard Business Review, September*, 118–125.

Dye, R. (2000). The buzz on buzz. *Harvard Business Review, November*, 139–144.

The Economist. (2001). The Internet untethered. *The Economist, October 13*, pp. 3–26.

Enders, A., & Jelassi, T. (2000). The converging business models of Internet and bricks-and-mortar retailers. *European Management Journal, 18*(5), 542–550.

Evans, P., & Wurster, W. (1997). Strategy and the new economics of information. *Harvard Business Review, September–October*, 71–82.

Haig, M. (2001). KIDS—talking to the teen generation. *Brand Strategy, December*.

Jelassi, T., & Enders, A. (2005). *Strategies for e-business: Creating value through electronic and mobile commerce*. Essex, UK: Financial Times/Prentice Hall.

Kenny, D., & Marshall, J. (2000). Contextual marketing: The real business of the Internet. *Harvard Business Review, November*, 119–124.

Magura, B. (2003). What hooks m-commerce customers? *MIT Sloan Management Review, Spring*, 9.

Nohria, N., & Leestma, M. (2001). A moving target: The mobile-commerce customer. *Sloan Management Review, 42*, 104.

Pearse, J. (2002). NMA wireless—mobile conversations. *New Media Age, October 31*, pp. 37–43.

Pesola, M. (2001). The novelty could quickly wear off. *Financial Times.com*, July 17.

Stone, A. (2001, January 3). Mobile marketing strategies Q & A. Retrieved from *www.mcommercetimes.com/Marketing/200*

Endnote

- This chapter is based on a teaching case study that features the German wireless advertising company 12Snap (see Jelassi & Enders, 2005).

Chapter VII

Key Issues in Mobile Marketing: Permission and Acceptance

Stuart J. Barnes, University of East Anglia, UK

Eusebio Scornavacca, Victoria University of Wellington, New Zealand

Abstract

The growth and convergence of wireless telecommunications and ubiquitous networks has created a tremendous potential platform for providing business services. In consumer markets, mobile marketing is likely to be a key growth area. The immediacy, interactivity, and mobility of wireless devices provide a novel platform for marketing. The personal and ubiquitous nature of devices means that interactivity can, ideally, be provided anytime and anywhere. However, as experience has shown, it is important to keep the consumer in mind. Mobile marketing permission and acceptance are core issues that marketers have yet to fully explain or resolve. This chapter provides direction in this area. After briefly discussing some background on mobile marketing, the chapter conceptualises key characteristics for mobile marketing permission and acceptance. The chapter concludes with predictions on the future of mobile marketing and some core areas of further research.

Introduction

The proliferation of mobile Internet devices is creating an extraordinary opportunity for e-commerce to leverage the benefits of mobility (Chen, 2000; Clarke, 2001; de Haan, 2000; Durlacher Research, 2002; Evans & Wurster, 1997; Kalakota & Robinson, 2002; Siau & Shen, 2003; Yuan & Zhang, 2003). Mobile e-commerce, commonly known as m-commerce, is allowing e-commerce businesses to expand beyond the traditional limitations of the fixed-line personal computer (Barnes, 2002a; Bayne, 2002; Clarke, 2001; Lau, 2003; Siau & Shen, 2003; Sigurdson & Ericsson, 2003). According to a study by Telecom Trends International (2003), global revenues from m-commerce could grow from $6.8 billion in 2003 to over $554 billion in 2008.

Mobile commerce has a unique value proposition of providing easily personalized, local goods and services, ideally, at anytime and anywhere (Durlacher Research, 2002; Newell & Lemon, 2001). Due to current technological limitations, some problems, such as uniform standards, ease of operation, security for transactions, minimum screen size, display type, and the relatively impoverished web sites, are yet to be overcome (Barnes, 2002b; Clarke, 2001).

As each mobile device is typically used by a sole individual, it provides a suitable platform for delivering individual-based target marketing. This potential can improve the development of a range of customer relationship management (CRM) tools and techniques (Seita, Yamamoto, & Ohta, 2002). It is believed that in the near future marketing through the mobile phone will be as common a medium as the newspaper or TV. However, mobile marketing is unlikely to flourish if the industry attempts to apply only basic online marketing paradigms to its use; the medium has some special characteristics that provide quite a different environment for ad delivery, including time sensitivity, interactivity, and advanced personalization. Moreover, a key tenet is likely to be that consumers receive only information and promotions about products and services that they want or need; one of the most important aspects to consider is that wireless users demand packets of hyperpersonalized information, not scaled-down versions of generic information (Barnes, 2002c). Sending millions of messages to unknown users (known as spam) or banner ads condensed to fit small screens (Forrester Research, 2001) are doubtless unlikely to prove ideal modes of ad delivery to a captive mobile audience.

This chapter aims to explore the peculiarities of mobile-oriented marketing, focusing on issues of permission and acceptance, and some of the possible business models. The following two sections provide a basic review of the technological platform for mobile marketing and an introduction to marketing on the mobile Internet (focusing on advertising), respectively. The fourth section presents a conceptual definition and model for permission on mobile marketing applications, while section five provides a model for mobile marketing acceptance and examines a number of possible scenarios for mobile marketing, based on the previous analysis. Finally, the chapter rounds off with some conclusions, and further research questions, and provides some predictions on the future of wireless marketing.

The Technological Platform for Mobile Marketing

Kalakota and Robinson (2002) define mobile marketing as the distribution of any kind of message or promotion delivered via a mobile handset that adds value to the customer while enhancing revenue for the firm. It is a comprehensive process that supports each phase of the customer life cycle: acquisition, relationship enhancement, and retention. A variety of technological platforms are available to support mobile marketing. Here we describe briefly some of the principal components. (For a more detailed discussion, see Barnes [2002b, 2002c].) The m-commerce value chain involves three key aspects of technology infrastructure:

- **Mobile transport.** Current networks have limited speeds for data transmission and are largely based on second-generation (2G) technology. These "circuit-switched" networks require the user to dial up for a data connection. The current wave of network investment will see faster, "packet-switched" networks, such as General Packet Radio Service (GPRS), which deliver data directly to handsets, and are, in essence, always connected. In the near future, third-generation (3G) networks promise yet higher transmission speeds and high-quality multimedia.
- **Mobile services and delivery support.** For marketing purposes, SMS (a text-messaging service) and WAP (a proprietary format for Web pages on small devices) are considered the key platforms in Europe and the United States, with iMode (based on compact hypertext markup language or cHTML) and iAppli (a more sophisticated version of iMode based on Java) taking precedence in Japan (WindWire, 2000). For PDAs, "Webclipping" is often used to format Web output for Palm or Pocket PC devices.
- **Mobile interface and applications.** At the level of the handset and interface, the brand and model of the phone or PDA are the most important part of the purchase decision, with "image" and "personality" being particularly important to young customers (Hart, 2000).

The next section explores the possibilities and experiences of using wireless marketing on these technology platforms.

Marketing on the Wireless Medium

The wireless Internet presents an entirely new marketing medium that must address traditional marketing challenges in an unprecedented way (WindWire, 2000). Key industry players in the value chain providing wireless marketing to the consumer are agencies, advertisers, wireless service providers (WSPs), and wireless publishers. For agencies and advertisers, the wireless medium offers advanced targeting and tailoring of messages for more effective one-to-one marketing. For the WSP, the gateway to the wireless Internet (e.g., British Telecom, AT&T, and TeliaSonera), wireless marketing presents new revenue streams and the possibility of subsidizing access. Similarly, wireless publishers (e.g., the *Financial Times, New York Times*, and CBS Sportsline), as a natural extension of their wired presence, have the opportunity for additional revenue and subsidizing access to content. At the end of the value chain, there is potential for consumers to experience convenient access and content value, sponsored by advertising (Kalakota & Robinson, 2002; WindWire, 2000).

Like the wired medium, marketing on the wireless medium can be categorized into two basic types: push and pull, which are illustrated in Figure 1. *Push* marketing involves sending or "pushing" advertising messages to consumers, usually via an alert or SMS (short message service) text message. It is currently the biggest market for wireless advertising, driven by the phenomenal usage of SMS—in December 2001, 30 billion SMS messages were sent worldwide (Xu, Teo, & Wang, 2003). An analysis of SMS usage has shown unrivalled access to the 15 to 24 age group—a group that has proved extremely difficult to reach with other media (Puca, 2001).

Pull marketing involves placing advertisements on browsed wireless content, usually promoting free content. Any wireless platform with the capacity for browsing content can be used for pull advertising. WAP and HTML-type platforms are the most widely used. Japan has experienced positive responses to wireless pull marketing, using iMode. Interestingly, wireless marketing in Japan has more consumer appeal than marketing on the conventional Internet. Click-through rates for mobile banner ads during the summer of 2000 averaged 3.6%, whilst those for wireless e-mail on iMode averaged 24.3%. Click-through rates for online banner ads on desktop PCs in Japan often average no more than 0.5 or 0.6% (Nakada, 2001).

Overall, current push services are very much in the lower left-hand quadrant of Figure 1. Until the availability of better hardware, software, and network infrastructure, services

Figure 1. Categorization of wireless marketing—with examples (Barnes, 2002c)

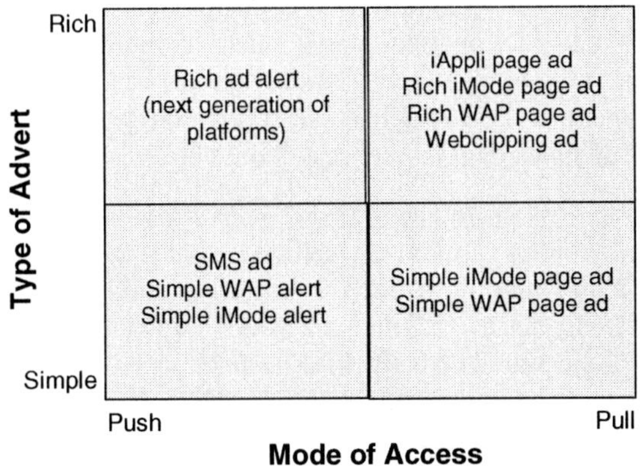

will remain basic. With faster, packet-based networks and more sophisticated devices, protocols and software, richer push-based marketing is likely to emerge, pushing the possibilities into the top left-hand quadrant.

Permission Issues for Mobile Marketing Applications

The discussion above has provided some insights about mobile marketing, particularly in terms of the wireless technological platform and basic applications of the medium. However, as yet, we have provided little conceptual discussion. The purpose of this section is to discuss the key variables of mobile marketing and present a conceptual model of permission for applications on this field.

In order for mobile marketing to reach its full potential of personalized information available anytime, anyplace, and on any device, it is necessary to understand the key characteristics of the mobile medium involved. We believe that any mobile marketing application should contemplate the following aspects:

- **Time and Location.** Although two different aspects, we consider them strongly related. An individual's behavior and receptiveness to advertisement is likely to be influenced by their location, time of day, day of week, week of year, and so on. Individuals may have a routine that takes them to certain places at certain times, which may be pertinent for mobile marketing. If so, marketers can pinpoint location and attempt to provide content at the right time and point of need, which may, for example, influence impulse purchases (Kannan, Chang, & Whinston, 2001). Feedback at the point of usage or purchase is also likely to be valuable in building a picture of time-space consumer behavior.
- **Information.** In particular, data given a context by the user. By itself, data do not contain an intrinsic meaning. It must be manipulated appropriately to become useful. Therefore, information can be defined as the result of data processing, which possesses a meaning for its receiver. Murdick and Munson (1988) point out that quantity of data does not necessarily result in quality of the information. The most important thing is what people and organizations do with the information obtained and its ability of extraction, selection, and presentation of information pertinent to the decision-making process should be considered as a decisive factor.
- **Personalization.** One of the most important aspects to consider is that wireless users demand packets of hyperpersonalized information, not scaled-down versions of generic information (Barnes, 2002c). The nature of the user, in terms of a plethora of personal characteristics such as age, education, socioeconomic group, cultural background and so on is likely to be an important influence on how ads are processed. These aspects have already proven to be important influences on Internet use (OECD, 2001), and as indicative evidence has shown above, elements

> such as user age are proving an important influence on mobile phone usage. The wireless medium has a number of useful means for building customer relationships. Ubiquitous interactivity can give the customer ever more control over what they see, read, and hear. Personalization of content is possible by tracking personal identity and capturing customer data; the ultimate goal is for the user to feel understood and simulating a one-to-one personal relationship. Through relational links of personal preferences, habits, mobile usage, and geographic positioning data the process of tailoring messages to individual consumers can become practical and cost effective.

The combination of the variables mentioned above allows us to understand one of the most important issues in mobile marketing: permission. Godin and Peppers (1999) refer to the traditional way of delivering marketing to customers as "interruption marketing." The authors suggest that instead of interrupting and annoying people with undesired information, companies should develop long-term relationships with customers and create trust through "permission marketing." The concept of permission marketing is based on approaching customers to ask for their permission to receive different types of communication in a personal and intimate way. It is well known among marketers that asking for a customer's permission is better and easier than asking for forgiveness (Bayne, 2002). In the wireless world, there is evidence to suggest that customers do not want to be interrupted—unless they ask to be interrupted (Newell & Lemon, 2001).

A mobile phone is a more personal environment than a mailbox or an e-mail inbox, and an undesired message has a very negative impact on the consumer (Enpocket, 2003; Godin & Peppers, 1999; Newell & Lemon, 2001). As mobile marketing has a more invasive nature than any other media, much attention must be given to permission issues in order to make the mobile marketing experience pleasant to the users. The information received must be of high value to gain the user's permission. It must produce a win–win situation between user and advertiser.

We understand permission as the dynamic boundary produced by the combination of one's personal preferences, that is, personalization, of time, location, and information. The user should be able to indicate when, where, and what information he/she would like to receive. Here are a couple examples of how mobile marketing can help consumers and businesses:

- You are getting ready to go to the airport and you receive a sponsored message saying that your flight is delayed for 4 hours. Because of this information, instead of spending 4 long and boring hours waiting at an airport lounge, you manage to have an enjoyable dinner with your friends.
- You let your wireless service provider know that you would like to receive during weekdays, from 12 p.m. to 1 p.m., information about the menu specials of all Italian restaurants costing less than $20 and within a 1-mile radius of where you are located.

Figure 2. Concept of permission for mobile marketing

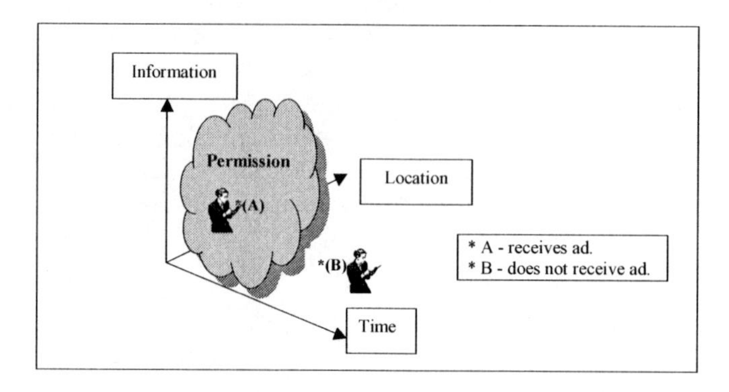

Now, let us consider the situation if this information was not customer relevant, or time and location sensitive. For example, imagine the following scenario. You are on a business trip, it is 3:30 p.m. and you had to forgo lunch due to an important meeting. Next, your cell phone beeps and you receive an offer of a menu special of an unknown restaurant in your hometown. The value to the recipient of this information is zero; moreover, it is more likely to have a negative impact. Figure 2 helps us visualize the concept of permission on mobile marketing.

The idea of a message being sent directly to an individual's phone is not without legislative concerns. Indeed, all over the world, privacy and consumer rights issues lead to the promotion of "opt-in" schemes. In essence, "opt-in" involves the user agreeing to receive marketing before anything is sent, with the opportunity to change preferences or stop messages at any time. Several current initiatives and industry groups, such as the Mobile Data Association, are helping to build standards of best practice for the mobile data industry (MDA, 2003).

As permission for mobile marketing applications should be dynamic, it is important to be able to identify customer responses to events. Stemming from the technological capabilities of mobile Internet-enabled devices, the measurement of reaction marketing is facilitated. As a consequence, the planning and justification of marketing expenditure becomes more precise. It also will help the identification of which mobile marketing strategies work and which do not. The constant feedback permits marketing strategies to be dynamically adjusted to produce better results for marketers.

Acceptance of Mobile Marketing

Now we have discussed the technological and conceptual factors surrounding mobile marketing, let us examine the variables that influence customer acceptance. Specifically, this section aims to explore the few studies already accomplished on mobile marketing acceptance and provide a model that summarizes the main variables concerning this issue.

There is no doubt that mobile marketing is still at an embryonic stage. However, several recent studies help us to understand some key factors contributing to the penetration and acceptance of mobile marketing among consumers (Enpocket, 2002a, 2002b; Ericsson, 2000; Godin & Peppers, 1999; Quios, 2000). The study by Ericsson (2000) had a sample of approximately 5,000 users and 100,000 SMS ad impressions in Sweden; the Quios study (2000) examined 35,000 users and 2.5 million SMS ad impressions in the UK; and the Enpocket study (Enpocket, 2002a, 2002b, 2003) researched over 200 SMS campaigns in the UK, surveying over 5,200 consumers—after they had been exposed to some of the SMS campaigns—from October 2001 to January 2003. The results of the three studies tend to converge, each pointing out that more than 60% of users liked receiving wireless marketing. The reasons cited for the favorable attitudes to mobile marketing include content value, immersive content, ad pertinence, surprise factor, and personal context.

The Enpocket study (2002a, 2002b, 2003) found that consumers read 94% of marketing messages sent to their mobile phones. It is important to point out that all these customers had given permission to receive third-party marketing. Moreover, the viral marketing capability of mobile marketing was identified by the fact that 23% of the customers surveyed by Enpocket showed or forwarded a marketing message to a friend. Another interesting finding is that the average response rate for SMS campaigns (15%) was almost three times higher than regular e-mail campaigns (6.1%). If delivered by a trusted source such as a wireless service provider (WSP) or major m-portal, acceptance of SMS marketing (63%) was considered comparable to that of TV (68%) or radio (65%). Notwithstanding, SMS marketing delivered by another source was far less acceptable—at just 35% of respondents. Similarly, the rejection level of SMS marketing from a WSP or portal was just 9%, while SMS from other sources was rejected by 31% of those surveyed. Telesales was rejected by 81% of respondents.

The indicative evidence about customer trust was further strengthened by other findings from the surveys. For example, 74% of customers indicated that WSPs were the most trusted organisation to control and deliver SMS marketing to their mobile devices. Major

Figure 3. Model for mobile marketing acceptance

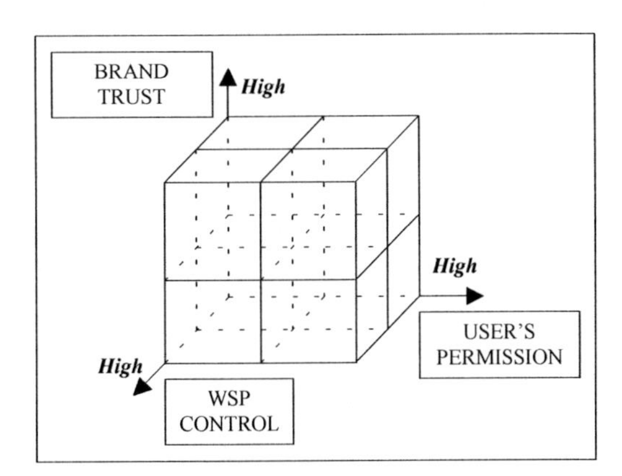

brands such as Coca-Cola and McDonald's were preferred by only by 20% of respondents (Enpocket, 2002a). As a result of the close relationship with the user, SMS marketing typically helps to build stronger brand awareness than other medias (Enpocket, 2002b).

It is important to highlight that the statistics presented above are being materialized in the form of profits mainly by mobile marketing and content sponsorship. Some marketers are using the sponsorship revenue model by conveying brand values through association with mobile content that fits the company's product or corporate image (Kalakota & Robinson, 2002). Features such as mobile barcode coupons are allowing a better measurement and understanding of return on investment (ROI) for mobile marketing (12Snap, 2003).

The indicative evidence and discussion above provide strong hints towards three main variables that influence a consumer's acceptance of mobile marketing: user's permission, WSP control, and brand trust. Figure 3 presents a conceptual model for mobile marketing acceptance based on these factors. Note that user permission is weighted in the model (see below).

The model allows us to forecast eight scenarios for mobile marketing acceptance. Table 1 summarizes the different scenarios. An example for scenario 1 would be if a trusted brand such as Coca-Cola sent a marketing message through the user's WSP (e.g., Vodafone) with his/her permission. In this situation, all the variables have a high level and the message should be highly acceptable to the customer. At the opposite end of the spectrum, in scenario 8, an unknown company (brand) sends a message without WSP control and without the user's permission. Here, the probability of rejection is very high.

Scenarios 4 and 5 point out an element that requires further detailed investigation. We believe that the most important variable in this model is "user permission." For example, if Coca-Cola sends a message via an operator to a user who has not granted permission (scenario 4), it should have a lower acceptance than a brand with low trust that sends

Table 1. Scenarios for mobile marketing acceptance

Scenario	Brand Trust	WSP Control	User's Permission	Acceptance
1	High	High	High	High acceptance
2	Low	High	High	Acceptable
3	High	Low	High	Acceptable
4	High	High	Low	Low acceptance
5	Low	Low	High	Acceptable
6	Low	High	Low	Low acceptance
7	High	Low	Low	Low acceptance
8	Low	Low	Low	Unacceptable

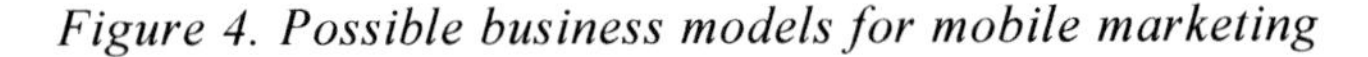

Figure 4. Possible business models for mobile marketing

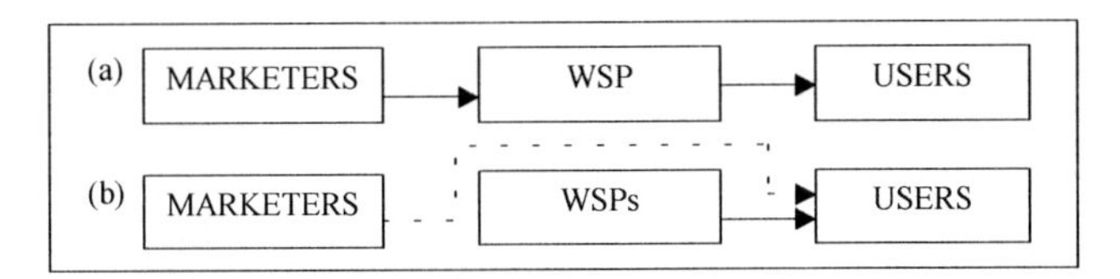

a message without WSP control to a customer who granted permission. This assumption is supported by the fact that the great majority of the consumers interviewed by Enpocket (2002a) are fearful that SMS marketing will become comparable to e-mail marketing with high levels of unsolicited messages.

The scenarios presented above are based on literature and on secondary data from the three studies previously approached. It would be interesting in the near future to substantiate this conceptual grid with primary data.

WSP control can directly affect how mobile marketing business models are configured. Based on the findings from the above analysis, we present two basic business models in which WSP control is the main differentiator (Figure 4). Figure 4a presents a model where the WSP has full control of the marketing delivery. On the other hand, Figure 4b shows a model where marketers can send messages directly to users without the control of the WSP.

The results of the studies presented by Ericsson (2000), Quios (2000), and Enpocket (2002a, 2002b, 2003) allow us to presume that the model presented by Figure 4a should be more successful than the one in Figure 4b. This assumption can also be supported by the fact that a WSP is usually more highly trusted by the consumers and possesses the technological capabilities to limit the delivery of messages. In addition, consumers interviewed by Enpocket (2002a) expressed a strong preference for the WSPs to become the definitive media owners and permission holders—possibly as a consequence of bad experiences with Internet marketing using nontargeted spam mail. Another issue to be taken into consideration is how WSP control can affect the revenue model for mobile marketing. In Figure 4a, the WSP can easily charge marketers for using its services, but in Figure 4b, this becomes a difficult task.

Conclusions

The immediacy, interactivity, and mobility of wireless devices provide a novel platform for marketing. The personal and ubiquitous nature of devices means that interactivity can be provided anytime and anywhere. Marketing efforts are potentially more measurable and traceable. Furthermore, technologies that are aware of the circumstances of the user can provide services in a productive, context-relevant way, deepening customer relationships. The convergence between marketing, CRM, and m-commerce represents a potentially powerful platform for wireless marketing.

Notwithstanding, it is important to keep the consumer in mind; the key to success is the management of and delivery upon user expectations. A key aspect of mobile marketing is likely to be obtaining permission from the users to send information to their mobile devices. Already, the wireless Internet has demonstrated the need for temperance; the wireless Internet is not an emulator of or replacement for the wired Internet, it is merely an additional, complementary channel for services. Further, aside from initial pilot investigations, it is not abundantly clear how consumers will respond to the idea of mobile marketing. Clearly, the issues concerning mobile marketing acceptance need to be further investigated. Alongside, a deeper investigation into business and revenue models is needed; for example, how can companies, marketers, WSPs, and consumers create a win–win environment? In addition, although it is expected that consumers will not tolerate receiving messages without permission, more work is still needed to explain how consumers give permission to receive mobile marketing.

Currently, wireless marketing is embryonic and experimental—the majority of wireless marketing is SMS based (simple push services—lower left-hand quadrant of Figure 1). The next generation of devices and networks will be important in the evolution of wireless marketing; higher bandwidth will allow rich and integrated video, audio and text. In addition, considerable effort is needed in building consumer acceptance, legislation for privacy and data protection, standardizing wireless ads, and creating pricing structures. If these conditions hold, wireless could provide the unprecedented platform for marketing that has been promised. Clearly, it is too early to tell, but future research aimed at examining these fundamental issues will help to further understand the implications of permission-based mobile marketing.

References

12Snap. (2003). Mobile barcode coupons—The marketing revolution for marketeers. Retrieved May 18, 2003, from *www.12snap.com/uk/help/couponsshort.pdf*

Barnes, S.J. (2002a). Under the skin: Short-range embedded wireless technology. *International Journal of Information Management, 22*(3), 165–179.

Barnes, S.J. (2002b). The mobile commerce value chain: Analysis and future developments. *International Journal of Information Management, 22(2)*, 91–108.

Barnes, S.J. (2002c). Wireless digital advertising: Nature and implications. *International Journal of Advertising, 21*(3), 399–420.

Bayne, K.M. (2002). *Marketing without wires: Targeting promotions and advertising to mobile device users*. London: John Wiley & Sons.

Chen, P. (2000). Broadvision delivers new frontier for e-commerce. *M-commerce, October*, 25.

Clarke, I. (2001). Emerging value propositions for m-commerce. *Journal of Business Strategies, 18*(2), 133–148.

de Haan, A. (2000). The Internet goes wireless. *EAI Journal, April*, 62–63.

Durlacher Research. (2002). *Mobile commerce report.* Retrieved July 10, 2002, from *www.durlacher.com*

Enpocket. (2002a). Consumer preferences for SMS marketing in the UK. Retrieved March 13, 2003, from *www.enpocket.co.uk*

Enpocket. (2002b). The branding performance in SMS advertising. Retrieved March 13, 2003, from *www.enpocket.co.uk*

Enpocket. (2003). The response performance of SMS advertising. Retrieved March 13, 2003, from *www.enpocket.co.uk*

Ericsson. (2000). *Wireless advertising*. Stockholm: Ericsson Ltd.

Evans, P.B., & Wurster, T.S. (1997). Strategy and the new economics of information. *Harvard Business Review, 75*(5), 70–82.

Forrester Research. (2001). Making marketing measurable. Retrieved February 10, 2002, from *www.forrester.com*

Godin, S., & Peppers, D. (1999). *Permission marketing: Turning strangers into friends, and friends into customers*. New York: Simon & Schuster.

Hart, Peter D. (2000). *The wireless marketplace in 2000*. Washington, DC: Peter D. Hart Research Associates.

Kalakota, R., & Robinson, M. (2002). *M-business: The race to mobility*. New York: McGraw-Hill.

Kannan, P., Chang, A., & Whinston, A. (2001,). Wireless commerce: Marketing issues and possibilities. In *Proceedings of the 34th Hawaii International Conference on System Sciences*, Maui, HI.

Lau, A.S.M. (2003). A study on direction of development of business to customer m-commerce. *International Journal of Mobile Communications, 1*(1/2), 167–179.

Mobile Data Association (MDA). (2003). *Mobile Data Association.* Retrieved May 1, 2003, from www.mda-mobiledata.org/

Murdick, R.G., & Munson, J.C. (1988). *Sistemas de Informacíon Administrativa*. Mexico: Prentice-Hall Hispano Americana.

Nakada, G. (2001). *I-Mode romps*. Retrieved March 5, 2001, from *www2.marketwatch.com/news/*

Newell, F., & Lemon, K.N. (2001). *Wireless rules: New marketing strategies for customer relationship management anytime, anywhere*. New York: McGraw-Hill.

NTT DoCoMo. (2003). Sehin Rain-Apu. Retrieved March 13, from *http://foma.nttdocomo.co.jp/term/index.html* (in Japanese)

Organisation for Economic Co-operation and Development (OEC). (2001). *Understanding the digital divide*. Paris: OECD Publications.

Puca. (2001). Booty call: How marketers can cross into wireless space. Retrieved May 28 2001, from *www.puca.ie/puc_0305.html*

Quios. (2000). *The efficacy of wireless advertising: Industry overview and case study*. London: Quios Inc./Engage Inc.

Sadeh, M.N. (2002). *M commerce: Technologies, services, and business models*. London: John Wiley & Sons.

Seita, Y., Yamamoto, H., & Ohta, T. (2002). Mobairu wo Riyoushitari Aiaru Taimu Maaketingu ni Kansuru Kenkyu. In *Proceedings of the 8th Symposium of Information Systems for Society*, Tokyo, Japan.

Siau, K., & Shen, Z. (2003). Mobile communications and mobile services. *International Journal of Mobile Communications*, *1*(1/2), 3–14.

Sigurdson, J., & Ericsson, P. (2003). New services in 3G—new business models for strumming and video. *International Journal of Mobile Communications*, *1*(1/2), 15–34.

Telecom Trends International. (2003). M-commerce poised for rapid growth, says Telecom Trends International. Retrieved October 27, 2003, from *www.telecomtrends.net/pages/932188/index.htm*

WindWire. (2000). *First-to-wireless: Capabilities and benefits of wireless marketing and advertising based on the first national mobile marketing trial*. Morrisville, NC; WindWire Inc.

Xu, H., Teo, H.H., & Wang, H. (2003,). Foundations of SMS commerce success: Lessons from SMS messaging and co-opetition. In *Proceedings of the 36th Hawaii International Conference on Systems Sciences*, Big Island, HI.

Yuan, Y., & J.J. Zhang (2003). Towards an appropriate business model for m-commerce. *International Journal of Mobile Communications*, *1*(1/2), 35–56.

Note

An earlier and shorter version of this paper appeared as Barnes, S. J., & Scornavacca, E. (2004). Mobile marketing: The role of permission and acceptance. *International Journal of Mobile Communications*, *2*(2), 128–139.

Chapter VIII

Consumer Perceptions and Attitudes Towards Mobile Marketing

Amy Carroll, Victoria University of Wellington, New Zealand

Stuart J. Barnes, University of East Anglia, UK

Eusebio Scornavacca, Victoria University of Wellington, New Zealand

Abstract

Mobile marketing is an area of m-commerce expected to experience tremendous growth in the next 5 years. This chapter explores consumers' perceptions and attitudes towards mobile marketing via SMS through a sequential, mixed-methods investigation. Four factors were identified and proven as all having a significant impact on mobile marketing acceptance—permission, content, wireless service provider (WSP) control, and the delivery of the message, which guided the development of a revised and empirically tested model of m-marketing consumer acceptance. The findings also suggest that marketers should be optimistic about choosing to deploy mobile marketing, but exercise caution around the factors that will determine consumer acceptance. The chapter concludes with a discussion about directions for future research.

Introduction

One area of m-commerce that is expected to experience tremendous growth is global wireless advertising. It has been predicted that the mobile marketing industry will grow from $4 billion to $16 billion from 2003 to 2005 (Ververidis & Polyzos, 2002). Mobile marketing provides new revenue streams and the opportunities for subsidized access, along with the potential for customers to experience more convenient and relevant content value, sponsored by advertising (Barnes & Scornavacca, 2004). It is expected that 33% of cellular service provider's revenue will be coming from advertising and from payments and commissions from mobile commerce activities (Ververidis & Polyzos, 2002).

Wireless marketing allows effective targeting and tailoring of messages to customers to enhance the customer-business relationship (Barnes & Scornavacca, 2004). Studies on this new advertising medium indicate that mobile advertising campaigns can generate responses, which are as high as 40% compared with a 3% response rate through direct mail and 1% with Internet banner ads (Jelassi & Enders, 2004). Despite this phenomenal marketing potential, there has been very little research on mobile marketing and particularly through its most successful application, short message service (SMS) (Barnes & Scornavacca, 2004). According to GSM Association, cell phone users send more than 10 billion SMS messages each month, making SMS the most popular data service (Dickinger, Haghirian, Murphy, & Scharl, 2004). Conceptual frameworks and models identified in the literature provide insight into the critical success factors of m-commerce marketing; however, very few of these studies have empirically tested or generated models from a consumer's perspective (Barnes & Scornavacca, 2004; Dickinger et al., 2004; Scornavacca & Barnes, 2004).

The aim of this chapter is to explore consumers' perceptions and attitudes towards mobile marketing via SMS, and to empirically test Barnes and Scornavacca's (2004) m-marketing acceptance model. The following section provides a background to mobile marketing and identifies some of the prominent models in the m-business literature. It also examines the factors believed to influence consumer acceptance of mobile marketing. The third section discusses the methodology, while the fourth and fifth sections provide the results of the study and a revised model for mobile marketing acceptance. The chapter concludes with a discussion about the future for SMS mobile marketing, and directions for further research.

Background on Mobile Marketing

Mobile marketing can be defined as "Using interactive wireless media to provide customers with time and location sensitive, personalized information that promotes goods, services and ideas, thereby generating value for all stakeholders" (Dickinger et al., 2004). This definition includes an important concept of adding value not just for the marketing party, but also for the consumer. The literature shows a variety of technological platforms such as wireless application protocol (WAP), SMS, and multimedia message service (MMS) that are available to support mobile marketing applications (Barnes & Scornavacca, 2004; Dickinger et al., 2004).

SMS is the most popular mobile data application to date, showing phenomenal usage with 580 million mobile messaging users sending over 430 billion messages worldwide in 2002 (TTI, 2003). Text message services have been hugely popular for interpersonal communication, allowing users of all ages to exchange messages with both social and business contacts (Dickinger et al., 2004; Xu, Teo, & Wang, 2003). Xu, Teo, and Wang (2003) identified three consistent success indicators for SMS messaging. The first factor is the cost effectiveness and interoperability of the wireless infrastructure, the second is the high penetration of mobile phones (ubiquitous penetration levels of over 80% in some countries), and the third is the relatively low cost of the SMS messaging service.

Countries such as Japan, New Zealand, Germany, and the UK have cost-effective and interoperable wireless structures, a high penetration of mobile phones, and a relatively low cost for the SMS messaging service have experienced remarkable success with the SMS application (Barnes & Scornavacca, 2004). The success that SMS has had as a messaging service provides a potentially huge SMS messaging customer base which could lend itself as a SMS mobile marketing customer base, making it an attractive opportunity for marketers (Kellet & Linde, 2001).

One of the main challenges and opportunities for mobile advertising companies is to understand and respect the personal nature of the usage of mobile phones (Barnes & Scornavacca, 2004; Barwise & Strong, 2002; Jelassi & Enders, 2004; Heinonen & Strandvik, 2003).

Consumer Acceptance of Mobile Marketing

The acceptance of a mobile marketing message is likely to be influenced by the consumer's acceptance of the mobile medium, the relevance of the content, and the context of the marketing message (Barnes & Scornavacca, 2004; Dickinger et al., 2004; Enpocket, 2003; Heinonen & Strandvik, 2003). Messages that are concise, funny, interactive, entertaining, and relevant to the target group usually achieve higher levels of success (Dickinger et al., 2004; Jelassi & Enders, 2004). The recent m-business literature offers a couple of frameworks that investigate user acceptance of SMS based mobile marketing (Barnes & Scornavacca, 2003; Dickinger et al., 2004).

The guiding model used for this research is the conceptual model of permission and acceptance developed by Barnes and Scornavacca (2004). This model was selected as it looks at a small subset of factors identified in the literature, which are believed to be the *most* important variables influencing consumer acceptance.

Barnes and Scornavacca (2004) believed that *user permission, wireless service provider control (WSP),* and *brand recognition* are the three most important variables that could influence consumers' acceptance of mobile marketing.

Among those, user permission was believed to be the most important variable, the main reason for this being that most consumers are fearful of SMS mobile marketing becoming like e-mail marketing, that is, with high levels of spam. WSP control is found to increase

Table 1. Scenarios for m-marketing acceptance (Barnes & Scornavacca, 2004)

User's Permission	WSP Control	Brand Trust	Acceptance
High	High	High	High Acceptance
High	High	Low	Acceptable
High	Low	High	Acceptable
High	Low	Low	Acceptable
Low	High	High	Low Acceptance
Low	High	Low	Low Acceptance
Low	Low	High	Low Acceptance
Low	Low	Low	Not Acceptable

the probability of user acceptance to mobile marketing. This was supported by the fact that users are likely to have high levels of trust with their WSP (Enpocket, 2002b; Ericsson, 2000).

The model also puts forward eight propositions of varying levels of acceptance according to the different combinations of factors. Table 1 presents Barnes and Scornavacca's (2004) hypothesized acceptability of SMS marketing messages based on high and low levels of permission, WSP control, and brand trust. This model is yet to be empirically tested with primary data.

These propositions provide a starting point in further exploring the factors that could contribute to consumer acceptance of mobile marketing.

Methodology

The chosen strategy of inquiry for this research is sequential exploratory mixed methods. Sequential procedures are ones in which the researcher uses the findings of one method to elaborate on or expand with another method (Creswell, 2003; Green, Caracelli, & Graham, 1989). The objectives of the sequential exploratory approach for the purpose of this study is to use two qualitative focus groups to explore the perceptions of mobile marketing, focusing on the main variables believed to influence mobile marketing acceptance, and then elaborate on this through experimental research in which the findings of the initial phase will be used. The empirical data will hopefully confirm what has been identified from the literature and the findings from the focus groups.

Focus Groups

The samples for the focus groups were purposely selected based on convenience sampling, availability, and profiling. Participants for both groups were in the age range

20–28 reflective of one of the major target groups for SMS mobile marketing. Four participants were selected for focus group A and five participants for focus group B. The participants in focus group A had a greater knowledge of mobile commerce technologies and applications than the participants in focus group B, which was purposely achieved in order to canvas a range of experiences and provide differing viewpoints. The participants in this study were students of a university in New Zealand as well as professionals working in the local central business district.

Interviews were based on open-ended questions and triggers. Video recording was used to tape the focus group discussions, with additional notes being taken by the facilitator. The advantages of using a focus group was that a range of ideas and perceptions were derived and the dynamics of the group provided a rich understanding of the research problem. These focus groups generated new propositions that were tested in the survey questionnaire phase.

Data analysis for the focus groups involved initially transcribing interviews and sorting the data into groups of information based on various topics. The transcriptions were then read over to look for ideas, depth, and credibility of the information from participants; thoughts were noted down in the margins of the transcript (Creswell, 2003). A coding process was then carried out where the data was organized into clusters before any meaning was derived from it (Rossman & Rallis, 1998). The themes and categories identified from the analysis are the major findings of the qualitative phase, and have been shaped into a general description of the phenomenon of mobile marketing acceptance (see the results section for details). Reliability measures were used to check for consistency of themes and patterns, while validity measures (triangulation, member checking, bias discussion, and peer debriefing) were used to determine the accuracy of the findings (Creswell, 2003).

Survey Questionnaire

This phase involved the use of a cross-sectional survey questionnaire to test the acceptance of mobile marketing messages against 16 various propositions that were formulated from the results of the focus groups. The advantage of using a survey in this study was the economy and rapid turnaround of data collection that a survey provides. Surveys are also advantageous in their ability to make inferences about consumer behaviour for given populations based on a sample (Babbie, 1990).

A survey questionnaire was chosen due to its cost effectiveness, data availability, and convenience. Seventy-eight participants for the quantitative phase of the research were selected using random convenience sampling with eight members of the sample being nonrespondents.

The instrument used in the survey was a modified version of the permission and acceptance model of mobile marketing developed by Barnes and Scornavacca (2004) with four variables: permission, WSP control, content, and delivery of the message. Sixteen propositions were formulated around these variables that were tested with a 4-point Likert scale ranging from "unacceptable" to "accept enthusiastically."

The data that was collected from the surveys was entered into an Excel spreadsheet, and statistical calculations were carried out. The 16 propositions were then placed in a table with the expected and actual levels of acceptance that were found for each proposition (see Tables 2 and 3). Tabular analysis was conducted in order to analyze the change in SMS mobile marketing acceptance through the various combinations of the set of variables (permission, WSP control, content, and delivery). The results from the quantitative phase were then compared against previous literature in order to provide further insight of the findings.

To avoid possible threats to validity, caution was taken when the results of this experiment were generalized to other populations and environments, when conducting statistical analysis on the data, and when the definitions and boundaries of the terms were defined.

Results from the Focus Groups

While focus group A was more knowledgeable in the area of mobile commerce, mobile technologies, and the potential of mobile marketing; both focus groups had only ever experienced mobile marketing through their wireless service providers. To some extent the participants' experience of receiving marketing messages from their service provider influenced their individual perceptions and perceived importance of varying factors contributing to consumer acceptance. The results of both focus groups were consistent with little disparity between the two.

Factors identified in the focus groups as having a significant impact on consumer acceptance of mobile marketing were permission to receive mobile marketing messages, control of the wireless service provider, relevance of the content, timeliness and frequency of the messages, simplicity and convenience of the messages, the brand or company sending the message, the control of the marketing from the consumer, and the privacy of the consumer. Consistent with Barnes and Scornavacca's (2004) model, permission and WSP control were perceived to have a heavy bearing on the acceptance of a mobile marketing message; however, brand was found to have little or no impact on acceptance than the likes of content, and time and frequency of the messages. The emerging there are classified as follows:

- **Permission:** Permission raised the most discussion in each focus group, and it was concluded by the participants as the most important success factor. Participants stated that consumers should have to "opt in" before they receive mobile marketing messages of any kind, and have the option to "opt out" at any stage.
- **Wireless service provider (WSP) control:** Although there was great emphasis on permission, it was also strongly felt that there needed to be a degree of filtering from the service provider. As participant A stated, "there has to be some sort of protection; they can't just open it up to anyone—if companies want to market to customers they should have to go through Vodafone." The idea was raised that if

participants had just one company to go to which was linked to their service provider, then there would be just one point of contact allowing consumers to easily "opt in" and "opt out" rather than tracking down several different companies. Participants agreed that it should be evident in the message that it is being filtered by the service provider and legitimate.

- **Personalization and content:** It was agreed that permission regarding time of day, frequency, and content would also be critical to the acceptance of mobile marketing. Both focus groups agreed that content and its relevance would play a key role in the acceptance of a mobile marketing message, with some participants arguing this as the most important factor. It was believed that marketers should make use of the technology and the advantages it provides over traditional forms of marketing and the Internet, looking to add value other than just advertising. Other ideas discussed in the focus groups were to tie content with location, timing, and ensure that the format of the message works with the limitations of the phone.
- **Frequency:** Participants agreed that there would be a limit to the number of mobile marketing messages they wished to receive, and there should be some control over the number of messages they are receiving depending on what good or service was being marketed or the industry (e.g., food/flowers). Both focus groups agreed that if consumers were to be hounded by marketing messages, it may result in switching providers, or deleting messages without reading them.
- **Time:** Participants raised the issue of time playing an important role in the acceptance of mobile marketing messages. It was believed that it is important for consumers to receive marketing messages at times suitable for them, and consumers are able to not only give permission to receive messages but also choose the times they wish to receive them.
- **Brand:** As far as the brand or company that was marketing was concerned, the general feeling among both focus groups was that as long as the marketing messages were being filtered by the service provider it would not matter too much who it was from; however, if it was third party, they would be annoyed right away. The majority of participants argued that it would be the more well-known brands or brands that the individual consumer recognizes. However, some consumers may prefer to receive messages from a little boutique shop down the road and there should be a way smaller companies can afford mobile marketing. Again if the brand or company doing the marketing was to go through the wireless service provider, this would result in an even higher level of trust. Focus group B believed that consumers should be able to select which companies and brands they receive messages from to a very specific point.
- **Technology/Ease of use:** A number of important issues were raised with regard to the mobile technology and convenience of the marketing message, some of which have already been pointed out in the previous sections. The main point raised that falls under this section is that marketing messages should not be a hassle for consumers to receive, they should work with the limitations of the phone, and there should be a manageable way to deal with them.

Table 2. Revised model with the 16 scenarios for marketing acceptance

Proposition	Permission	WSP Control	Content	Delivery	Expected Acceptance Level
1	High	High	High	High	Accept Enthusiastically (4)
2	High	High	High	Low	Acceptable (3)
3	High	High	Low	High	Acceptable (3)
4	High	Low	High	High	Acceptable (3)
5	Low	High	High	High	Acceptable (3)
6	High	High	Low	Low	Accept reluctantly (2)
7	High	Low	High	Low	Accept reluctantly (2)
8	High	Low	Low	High	Accept reluctantly (2)
9	Low	High	High	Low	Accept reluctantly (2)
10	Low	High	Low	High	Accept reluctantly (2)
11	Low	Low	High	High	Accept reluctantly (2)
12	High	Low	Low	Low	Unacceptable (1)
13	Low	High	Low	Low	Unacceptable (1)
14	Low	Low	High	Low	Unacceptable (1)
15	Low	Low	Low	High	Unacceptable (1)
16	Low	Low	Low	Low	Unacceptable (1)

Revised Model and Survey Results

Four conceptual factors emerged as having the most influence on consumer acceptance based on the tabular analysis and findings of the focus groups. Similar topics were merged as conceptualized themes and then these themes were analyzed according to the number of times they were mentioned in the focus groups, whether these comments were implying that they were important factors and whether the participants explicitly stated them as being one of the *most* important factors.

Table 2 presents 16 new propositions based on varying combinations of the identified factors, ranked according to the importance of factors: (1) permission, (2) WSP control, (3) content, (4) delivery, and also the number of factors which are low (0, 1, 2, 3, or 4).

The results obtained in the survey demonstrated that propositions 6, 7, 8, 11, 12, 13, 14, 15, and 16 were supported, while propositions 1, 2, 3, 4, 5, 9, and 10 were not found to be supported by the data collected. Tables 3 and 4 show the revised propositions with the expected and actual levels of acceptance for mobile marketing. Notice that the second table actually shows the propositions reshuffled in order to demonstrate their rank of acceptance according to the results.

Overall, consumer acceptance of mobile marketing messages was much lower than expected. Over 50% of respondents answered unacceptable to more than 10 out of the 16 scenarios put forward to them, with the average number of scenarios answered as unacceptable being 9. On the other hand, nearly 70% of the respondents did not answer

Table 3. Revised model ranked according to expected results

Proposition	Permission	WSP Control	Content	Delivery	Expected Acceptance Level	Average	Rank	Actual level
1	High	High	High	High	Accept Enthusiastically (4)	3.16	1	Acceptable
2	High	High	High	Low	Acceptable (3)	1.60	8	Accept reluctantly
3	High	High	Low	High	Acceptable (3)	1.99	3	Accept reluctantly
4	High	Low	High	High	Acceptable (3)	2.29	2	Accept reluctantly
5	Low	High	High	High	Acceptable (3)	1.91	4	Accept reluctantly
6	High	High	Low	Low	Accept reluctantly (2)	1.50	9	Accept reluctantly
7	High	Low	High	Low	Accept reluctantly (2)	1.70	5	Accept reluctantly
8	High	Low	Low	High	Accept reluctantly (2)	1.63	7	Accept reluctantly
9	Low	High	High	Low	Accept reluctantly (2)	1.43	10	Unacceptable
10	Low	High	Low	High	Accept reluctantly (2)	1.41	11	Unacceptable
11	Low	Low	High	High	Accept reluctantly (2)	1.66	6	Accept reluctantly
12	High	Low	Low	Low	Unacceptable (1)	1.41	12	Unacceptable
13	Low	High	Low	Low	Unacceptable (1)	1.30	14	Unacceptable
14	Low	Low	High	Low	Unacceptable (1)	1.30	15	Unacceptable
15	Low	Low	Low	High	Unacceptable (1)	1.39	13	Unacceptable
16	Low	Low	Low	Low	Unacceptable (1)	1.19	16	Unacceptable

"accept enthusiastically" to anything, and of the 30% who did give this response for at least one scenario, more than 80% only gave this response for one or two of the questions (Tables 3 and 4).

Of all the propositions the highest level of acceptance for mobile marketing was as expected for proposition 1. However, it can be seen that even where consumers have given permission, the content of the message was relevant, the delivery appropriate, and the message had come through the WSP, it was found on average to be only acceptable, with just 31% of respondents accepting this message enthusiastically. Thus disproving proposition 1. Alternatively on average the lowest level of acceptance (unacceptable) was found where there was a low level of all these factors. Only 9 out of the 70 participants answered anything other than unacceptable for this question. This result was expected and consistent in proving proposition 16.

Permission and delivery of the message were the two variables that were found to equally have the most influence on the participant's level of acceptance, while content was found to be the next most important factor with control of the WSP having the least amount of impact on the level of acceptance. Participants were more likely to accept messages that had a lower level of WSP control or irrelevant content than messages that they had not given permission for or that came at an inappropriate time or frequency. This was shown again in Table 4, rows 12–15, where participants found scenarios 13 and 14 more unacceptable, despite having high levels of WSP control and content, respectively, than scenarios 12 and 15 where there were higher levels of permission and appropriate delivery, respectively.

Table 4. Revised model ranked according to actual results

Proposition	Permission	WSP Control	Content	Delivery	Expected Acceptance Level	Average	Rank	Actual level
1	High	High	High	High	Accept Enthusiastically (4)	3.16	1	Acceptable
4	High	Low	High	High	Acceptable (3)	2.29	2	Accept reluctantly
3	High	High	Low	High	Acceptable (3)	1.99	3	Accept reluctantly
5	Low	High	High	High	Acceptable (3)	1.91	4	Accept reluctantly
7	High	Low	High	Low	Accept reluctantly (2)	1.70	5	Accept reluctantly
11	Low	Low	High	High	Accept reluctantly (2)	1.66	6	Accept reluctantly
8	High	Low	Low	High	Accept reluctantly (2)	1.63	7	Accept reluctantly
2	High	High	High	Low	Acceptable (3)	1.60	8	Accept reluctantly
6	High	High	Low	Low	Accept reluctantly (2)	1.50	9	Accept reluctantly
9	Low	High	High	Low	Accept reluctantly (2)	1.43	10	Unacceptable
10	Low	High	Low	High	Accept reluctantly (2)	1.41	11	Unacceptable
12	High	Low	Low	Low	Unacceptable (1)	1.41	12	Unacceptable
15	Low	Low	Low	High	Unacceptable (1)	1.39	13	Unacceptable
13	Low	High	Low	Low	Unacceptable (1)	1.30	14	Unacceptable
14	Low	Low	High	Low	Unacceptable (1)	1.30	15	Unacceptable
16	Low	Low	Low	Low	Unacceptable (1)	1.19	16	Unacceptable

It is interesting to note that consistent with the propositions, the level of acceptance declined with the number of factors that had low levels, except in the case of proposition 2, which was expected to generate the second highest level of acceptance and in actual fact dropped down to position 8. Where all other factors were high, yet the delivery of the message was inappropriate, more than 50% of respondents found this message unacceptable, compared to just 26% of respondents who considered a message with low levels of WSP control unacceptable.

Looking at the other rankings of propositions from their expected to actual perceived influence on acceptance, just three propositions stayed in the same ranked position. However, of the propositions that did get shuffled in rank, nine of these moved only within one or two ranks, with just three propositions moving three places or more. Participants found all messages that had three or more factors with low levels to be completely unacceptable. This was consistent with the expected results, and supported the propositions 12, 13, 14, 15, and 16. Messages that had only high levels of WSP control or relevant content were found to be 10% less unacceptable than messages with only high levels of permission or appropriate delivery—thus supporting the theory that permission and delivery of the messages are perceived to be the most important factors.

Discussion

The findings indicated a number of factors that are critical to the acceptance of mobile marketing by consumers. While the empirical testing showed that some factors are more important than others in influencing the overall level of acceptance, it was found that all factors played a significant role.

Consistent with the literature explicit permission was found to be essential (Barnes & Scornavacca, 2004; Enpocket, 2003; Godin et al., 1999). The wireless channel is relatively protected and spam free with consumers having little experience with mobile marketing. Due to the personal nature of the phone, and experiences with unsolicited spam via e-mail users were weary of receiving marketing to their cell phones, and a number of privacy issues were raised in the focus groups. Another finding that emerged from the study was the importance of delivery with the marketing message. Literature has suggested that frequency and time are linked to targeting, where users are happy to receive messages at a higher frequency so long as the relevance to them is maintained (Enpocket, 2002b). This was supported by the empirical testing where it shows messages with a low level of relevant content yet appropriate delivery were found to be much more acceptable than messages with a low level of relevant content and inappropriate delivery (a higher frequency). While participants in the focus groups made a point of saying that it is useless receiving any messages containing content that is irrelevant, there are a number of possible reasons why the respondents may have found delivery to be more important. If a consumer receives a message that is irrelevant to them once in a blue moon, and it does not come at a disturbing time, they may not be that bothered by it. On the other hand, if they were messages on something that was relevant to them but were receiving these messages continuously and at interruptive times, it is likely to be more unacceptable.

It was interesting to see that the control of the WSP had the least impact on consumer acceptance in the survey results, conflicting with the results of the focus groups where participants expressed their strong opinions towards the importance of WSP control. The results may in fact indicate that where consumers receive messages they find disturbing or intrusive, they would rather it had not come from the service provider they trust. The focus groups indicated this, stating that they trust their service provider's judgment and would expect them to behave responsibly. Consumer attention seems more likely to divert to the filter when they are receiving unsolicited messages that they find disturbing.

Despite literature showing mobile marketing to be a successful tool in building brand awareness, and an important factor in consumer acceptance (Dickinger et al., 2004; Enpocket, 2002a), the study revealed that the brand being marketed may have very little impact. Consumers are more likely to care whether a brand has been accepted by their service provider and has come through a filter, than about their level of trust between two different brands. Despite having a high trust in a brand, consumers are still doubtful of the bona fide of these messages when they have come direct. They are also less likely to care about the brand that is being marketed to them than whether the content is relevant. The importance that is placed on brand is likely to increase when all other factors are high, and there is more choice in the market. Currently there are a limited number of brands being marketed through the mobile phone in New Zealand and more attention to brand is likely to arise in the future where consumers receive similar messages, with all other factors being equal, from competing brands.

Conclusions

This research highlights the importance of consumer perceptions and acceptance levels of mobile marketing. The literature showed the powerful marketing potential that mobile marketing can offer companies through its anytime and anywhere nature, yet limited research looking at consumers' perceptions and acceptance of mobile marketing has been carried out. This study set out to overcome the apparent gap in the literature, and through the use of both qualitative and quantitative methodology, a model has been adopted, explored, developed, and empirically tested and validated.

This study suggests that marketers should be optimistic about choosing to deploy mobile marketing; however, exercise caution around the factors that will determine consumer acceptance. While consumers can see the potential in the mobile medium, they are weary of receiving unsolicited messages they do not want. Obtaining user trust and permission will be the main challenge faced by marketers, and future research should focus on ways to overcome these challenges. Consumers are more likely to trust messages coming from their service providers than anywhere else, so it is important that service providers provide a high level of filtering and protection as reassurance for their users. Trust and permission are necessary factors of consumer acceptance; however, they should not be seen as the only objectives. Attention needs to be focused around the relevance of the content and the timeliness and frequency of the delivery of marketing messages.

The research showed that simply focusing on contextual, content, or permission/control factors in isolation is unlikely to result in a high or even moderate level of acceptance. Instead, marketers need to take into account all these factors and how varying combinations of these factors will impact consumer acceptance.

The permission and acceptance model, which has been developed and tested in this research, provides a foundation for further SMS mobile marketing research to be built upon. Academics can refer to this model as a guide for further understanding of consumer acceptance to mobile marketing, while practitioners may find this model useful in providing direction for mobile marketing strategies. The device media aspects discussed in the focus groups may also provide an indication as to what new technologies and mobile devices will be of significance in meeting consumers' needs for the future.

The generalizability of this study is limited by it being conducted only in New Zealand as well as the lack of further qualitative interviews to further elaborate on the initial quantitative analysis. This cross-sectional study only looked at consumer acceptance at one point in time, and little is known about the sample frame that was used for the survey questionnaire. Furthermore the sample of the participants for the quantitative phases was only a small number which leaves possibility for self-selection bias. Longitudinal research testing consumer perceptions and acceptance over a set amount of time, and taking into account demographics when testing consumer acceptance levels would provide some deeper insight into these areas.

References

Babbie, E. (1990). *Survey research methods* (2nd ed.). Belmont, CA: Wadsworth.

Barnes, S.J., & Scornavacca, E. (2004). Mobile marketing: The role of permission and acceptance. *International Journal of Mobile Communications, 2*(2), 128–139.

Barwise, P., & Strong, C. (2002). Permission-based mobile advertising. *Journal of Interactive Marketing, 16*(1), 14–24.

Creswell, J. (2003). *Research design qualitative, quantitative, and mixed methods approaches* (2nd ed.). Thousand Oaks, CA: Sage Publications.

Dickinger, A., Haghirian, P., Murphy, J., & Scharl, A. (2004). *An investigation and conceptual model of SMS marketing.* Paper presented at the 37th Hawaii International Conference on System Sciences, HI.

Enpocket. (2002a). The branding performance of SMS advertising. Retrieved March 13, 2003, from *www.enpocket.co.uk*

Enpocket. (2002b). Consumer preferences for SMS marketing in the UK. Retrieved March 13, 2003, from *www.enpocket.co.uk*

Enpocket. (2003). The response performance of SMS advertising. Retrieved March 12, 2003, from *www.mda-mobiledata.org*

Ericsson. (2000). *Wireless advertising.* Stockholm: Ericsson Ltd.

Godin, S., Hardcover, p., 1 edition (May 1, & 0684856360., S. S. I. (1999). *Permission Marketing: Turning strangers into friends, and friends into customers.*

Green, J.C., Caracelli, V.J., & Graham, W.F. (1989). Toward a conceptual framework for mixed method evaluation designs. *Educational Evaluation and Policy Analysis, 11*(3), 255–274.

Heinonen, K., & Strandvik, T. (2003, May 22–23). *Consumer responsiveness to mobile marketing.* Paper presented at the Stockholm Mobility Roundtable, Stockholm, Sweden.

Jelassi, T., & Enders, A. (2004, June 14–16). *Leveraging wireless technology for mobile advertising.* Paper presented at the 12th European Conference on Information Systems, Turku, Finland.

Kellet, K., & Linde, A. (2001). EMS, MMS, & the future of mobile messaging, white paper. Retrieved , from, *www.magic4.com.*

Rossman, G.B., & Rallis, S.F. (1998). *Learning in the field: An introduction to qualitative research.* Thousand Oaks, CA: Sage.

Scornavacca, E., & Barnes, S.J. (2004, March). *Raising the bar: Barcode-enabled m-commerce solutions.* Paper presented at the Austin Mobility Roundtable, Austin, TX.

TTI. (2003). *Mobile messaging: Which technologies and applications will succeed?* Retrieved July 5, 2004, from *www.telecomtrends.net*

Ververidis, C., & Polyzos, G. (2002). *Mobile marketing using location based services.* Paper presented at the First International Conference on Mobile Business, Athens, Greece.

Xu, H., Teo, H.H., & Wang, H. (2003, January 7–10). *Foundations of SMS commerce success: Lessons from SMS messaging and co-opetition.* Paper presented at the 36th Hawaii International Conference on System Sciences, Big Island, HI.

Section III

Organizational Applications of M-Business

Chapter IX

How Mobile Technologies Enable Best Business Practice: A Case in the Fine-Paper Industry

Vaida Kadytë, Åbo Akademi University, Finland

Abstract

Within the last 10 years, a number of sophisticated mobile devices have become available to assist not only in managing appointments and contacts but also to provide a tool for enhancing user experience and introducing new collaborative ways of doing business. However, little conceptual thought and empirical illustration has been given to how industrial organisations are using the potential offered by mobile technology. This chapter is based on putting into practice the main conceptual ideas of the Freedom Economy in the domain of mobile business applications through action research methodology. We present an in-depth case study on implementing mobile solutions in the fine-paper industry and aim to investigate to what extent mobile technologies pose a challenge to contemporary industrial life and how they will eventually enable firms to achieve the best business practice.

Introduction

With the increasing interest in mobile business applications, companies across industries are promised huge productivity benefits, faster business reporting for decision making, reduced operational costs, and increased customer satisfaction. As business always seeks ways to enhance productivity and profits, mobile technology seems to be the next logical step forward when considering IT investment decisions. Intelligent firms are considering the possible implications of m-commerce to boost productivity across many areas of their business by deploying both internal (m-workforce) and external (m-CRM, m-supply chain management) applications for enterprise systems. However, most predictions about the impact of mobile commerce in business look either uncertain or are overhyped by consulting bodies and most companies are holding back on investing in mobile technology until it is stable and the benefits are obvious. A clear expectation that the technology will be of great importance in the coming years is not enough. The inadequate knowledge of managers who do not have a clear idea of how to use the opportunities presented by mobile technology is one major factor inhibiting mobile business (Lehmann, Kuhn, & Lehner, 2004). Designing mobile enterprise solutions is a complex undertaking and requires maintaining a balance between a variety of existing corporate standards in industry and the innovativeness offered by modern technology. The ability to successfully implement modern technologies in organisations also needs a continuous upgrade of software and the adoption of new hardware where required skills and knowledge are not easily available (Bendeck, Kötting, Schaaf, Maurer, Valenti, & Robert, 2001). It is difficult to minimise the uncertainty and manage the risks involved in implementing mobile business solutions because the technology is immature and developing rapidly and the user base is inexperienced. The research reported in this paper was designed in help industrial organisations realise the locked up value of mobile technology. To facilitate an understanding of the potential of mobile technology for industry, the main conceptual ideas were adapted from the Freedom Economy developed by Keen and Mackintosh (2001) and were used to study and evaluate the business impact of the implementation of mobile technologies in three organisations.

After positioning our work with respect to mobile commerce, we outline our research framework powered by three essential rules of the Freedom Economy and describe the idea of business value with respect to mobile technology for industrial users. This will be followed by an in-depth analysis of the industrial case study, which involved three organisations where actual implementations of mobile technologies were designed to enhance customer service and raise the industry to another business level, before we draw conclusions.

Three Freedoms in Mobile Business

New mobility-related opportunities are open for business transformation, as a result of technology convergence in several areas including IP (Internet Protocol) networks and mobile devices. Mobile technology-related applications are likely to follow the revolutionary pathway of electronic marketplaces, where state-of-the-art technologies have been primarily applied in consumer-oriented areas but the largest share of revenues was generated in the business-to-business (B2B) segment. Before we present the conceptual framework and case results, it is important to define several terms and concepts related to the mobile business domain that are often used in an ambiguous way in practice. Many definitions characterizing the business application area of mobile commerce focus on enabling business transaction through wireless devices, confusing mobile commerce and mobile business. A commonly adopted definition, by Durlacher, defines mobile commerce as "any transaction with a monetary value that is conducted via a mobile telecommunication network" (Müller-Veerse, 1999). Rather than looking for similarities between e-commerce and m-commerce, it is more useful to identify existing differences and avoid a misleading interpretation of definitions. The fact is that m-commerce also has the potential to be transactional, but not necessarily. According to Stafford and Gillenson (2003), a primary distinction between m-commerce and e-commerce lies in the differences between transaction and the facilitation of enhanced information network access. This seems to be the case with business applications, where most of the m-commerce-related applications are expected to be data driven. We are already experiencing growth in two major application areas in the m-commerce business domain. One is B2B applications, which aim to improve the effectiveness and productivity of interbusiness interaction between corporations, companies, business units, and so forth, and create new solutions for production processes with mobile technologies (Carlsson, Walden, & Veijalainen, 2004). The second is defined by Carlsson, Walden, and Veijalainen as the business-to-employee (B2E) applications area. Here the same effectiveness and productivity-related value propositions can be applied to groups/teams and individuals and mobile technologies can be realised by boosting distance working, simplifying administrative procedures, and enhancing the effectiveness of teamwork. Consequently, we prefer to adopt a broader definition of mobile business (Ritz & Stender, 2003), which is to be understood as "the general use of mobile applications by employees at the organisational interface to the customer for providing value added service" (p. 885).

But what are the core issues involved in developing value-added services in the business domain? In order to understand this, it is important to note that an application is a collection of software components that implement business tasks or processes. The value added of mobile applications depends, of course, on the particular context of business use. We assume that a particular business environment is the subject of an evaluation regarding the value of mobile business applications. An application is enhanced with a context-dependent content, which inherits its own distinguishing value and brings it into the form of a service. Consequently, we define mobile service as a value-added service that is based on a mobile network and consists of a user-specific content and applications that enable access to the content on a mobile device. The common wisdom that mobile applications and services should bring value-adding solutions for the users that really outweigh conventional alternatives has been widely acknowledged but little elaboration has been done. Therefore we use Keen's ideas that can help explain the uniqueness of mobile applications and services. The Braudel Rule is the key to explaining the Freedom Economy or so-called future of mobile business (Keen & Mackintosh, 2001, pp. 31–56): a freedom attains value when it changes the limits of the possible in the structure of everyday life. In other words, the value of mobile technologies lies not in their convenience, simplicity or immediacy within existing routines but in the freedom they create through the set of rules of the Freedom Economy (Keen & Mackintosh, 2001, p. 20):

- 1st Rule means adding value to the customer relationship by exploiting mobility, personalisation, and fusion of telephony and the Internet, and this creates *relationship freedoms*.
- 2nd Rule creates *process freedoms* if value is added along the entire supply chain and in related logistics operations and business partner relationships by making fully mobile as many as possible of the steps, people, information items, and communication needed in order to design an effective business process.
- 3rd Rule creates *knowledge freedoms*, which means adding value to the workers of organisation through knowledge mobilisation by bringing information, communication, and collaboration to them instead of their having to go to the sources themselves.

When applied specifically to the business domain, the value of mobile technologies may be understood through the same set of rules of the Freedom Economy. It may be realised through the creation of three freedoms within existing employee work routines, processes, and interorganisational interactions. In fact, it is argued here that to successfully transform business it is necessary to address the economic peculiarities of an industry and gain access to its operating business environment. Though there are several issues common to all industries that impact on the daily life of large organisations, such as globalisation and the fact that the workforce is becoming increasingly mobile, speed is becoming a survival factor in a dynamic business environment, and more demanding customers create more vulnerable relationships. These changes in the business environment, or so-called value drivers, require radical transformation to enable organisations to survive. The management of such a transformation is a delicate art of aligning the various components that comprise the organisation with its industrial environment. The key requirements related to the transformation that enables companies to move to next generation stages are related to mobile technology alignment at several organisational levels. We suggest three levels of alignment that correspond to the rules of the Freedom Economy. The first level constitutes a *micro-alignment* between the mobile technology and employee fit; it deals with mobile technology adoption over time and emphasises the need to provide usable tools that support the employee's choices and working needs. The second level corresponds to the classical *intra-alignment* of mobile technology with the core business processes. It is expected that mobile technology investments will drive critical business processes towards achieving excellence. Finally, the third level considers an *inter-alignment* between mobile technology and organisational capabilities at the interfaces with customers and partners. Even though the second level of alignment remains essential for transforming organisations to the next generation stage, we estimate that two other levels of alignment are necessary first steps in achieving best business practice in the dynamic, complex, and uncertain business environment of many contemporary industries. When discussing mobile technology and its applications, it is also crucial to understand the strategic business goals and objectives that determine the requirements for creating the additional value that will improve the overall performance of the company (Paavilainen, 2001). Moreover, we believe that the ability to divert the business value of mobile technologies to the overall performance of the company depends to a large extent on and varies within different industries. What interests us in this paper is improved organisational performance via mobile technology investment in the fine-paper industry, where the framework is used to analyse the data collected in a research project. Value drivers and strategic business goals related to our study will be outlined in the next section.

The Industrial Case Study and Research Method

The industrial case study contributes to the research into mobile applications in business that has arisen out of the author's work in the Mobile Commerce Research project at the Institute of Advanced Management Systems Research, sponsored by TEKES (National Technology Agency of Finland) for about 2 years. It involved the third largest company manufacturing fine paper in Europe, one of its key customer—the largest printing firm in Finland—and its business partner organisation—one of the largest paper merchants in Europe. The fine-paper business was selected as a vertical market for mobile business applications for several reasons. First, it is considered one of the most complex and strong industrial clusters in Finland and currently faces economic recession. The fine-paper production industry has its own distinctive features, for example, a multitude of actors involved in the value chain, enduring business relationships, complex business influenced by globalisation, and fierce competition within the industry and from other emerging e-industries. Second, while in most cases information and communication technologies (ICT) is seen as a substitute for paper, it also offers enhanced prospects for the paper industry (Kuuluvainen, 2002). For many successful companies in the industry customer service seems to be a decisive factor, since prices are about the same and the quality of the paper products marketed today is generally perceived as quite adequate. Therefore the success of the companies operating within this industry is likely to depend on their ability to exploit ICT to ensure the best levels of service to their business customers. Mobile technology, the most modern form of ICT and most efficient communication medium so far, can probably offer better than anything else the motive for coping with the challenges posed by the network economy. The main research objective of the project was the development of novel mobile technology applications for the fine-paper industry with the aim of improving B2E communications and B2B relationships with industrial customers.

The case research strategy was chosen for its suitability when studying new topic areas (Eisenhardt, 1989). The single-case study method was considered as being a potentially rich and valuable source of data, suited to exploring relationships between variables in their given context (Benbasat, Goldstein, & Mead, 1987). Remenyi (1998) argues that a particularly strong tactic in ensuring the validity of research is to use multiple sources of evidence when conducting a single-case study as this helps ensure validity. Furthermore, the use of multiple data sources (interviews, documents, questionnaires) in our study provided more convincing and accurate evidence on findings from industrial research. In our data collection efforts, we used face-to-face interviews and documentary materials as the primary source of data. We conducted private interviews with each of the 10 focus group members from the paper-manufacturing company, customer printing houses, and wholesaler. The official documents we received (a standard flow diagram and summaries of complaints handled in the year 2003) gave us a picture of how the complaints had been handled by the paper-manufacturing company. As an independent external facilitator of the process, the author of this study conducted a series of expert surveys to enable customers, service technicians, and management at the mill to provide data for defining problems and opportunity situations regarding customer service. Responses were obtained from 16 experts. The data-collecting process, covering the stages of feasibility and development through to preliminary experiment, took 1.5 years. During the feasibility study, which included eliciting customer requirements, a strategic decision was made to focus on enhancing customer care solutions as handling complaints proved to be the most critical business process. The ultimate goal of the project was to design a mobile system prototype that would provide access to customer care services via mobile devices at the point of need and would also benefit B2B relationships in the value chain for fine paper. From the beginning of the project in March 2003 until the final prototype was delivered in June 2004, potential users from all three companies were constantly involved in the process of interface design and influenced its major features during monthly project meetings and workshops.

The findings reported here were obtained from an in-depth case study that was the by-product of a research and development process where a team of four people at a research organisation participated in the actions of the target organisation as consultant body. For this purpose we have focused on developing mobile solutions and used constructive research methods based on action research methodology. We have to admit that the action research method has not proved very popular among North American IS researchers famous for setting the rules of the game in scientific research (Baskerville & Wood-Harper, 1996). Critical reviewers have expressed serious doubts about the appropriateness of research into IS because of its poor instruments and measurement validity, unclear objectives, lack of control and statistical power, as well as its tendency to ignore methodological issues. In spite of this, action research is a growing research method in Scandinavian IS literature. In particular, we argue that, in light of the widespread adoption of mobile technology, action research is not only applicable to IS research but also deserves particular attention. One of the major reasons is that it aims to bridge the gap between academics, who generate predictions and intentions concerning the future of the mobile information society, and practitioners, who actually create the mobile information society. Needless to say, research in the mobile business domain sometimes seems to be losing its leading role in the field. It is currently striving to catch up with the rapid development of practical mobile solutions driven by global telecommunications vendors. In most cases, IT consulting companies and vendors tend to dominate the question of evaluating the business benefits accruing from investment in mobile technology, and it is not surprising that they tend to overhype its true essence. Some business managers have fallen for these highly exaggerated stories without considering the critical points of mobile information systems initiatives in the company's operating context. As a result, the majority of projects may fail to deliver the expected results, as has happened with previously overhyped information systems. In order to avoid this, the field of mobile commerce needs to be supported critically by research and needs an appropriate academic focus in parallel with its practical development. Action research methods are particularly suitable here, as they allow organisational changes to be studied close up (Benbasat et al., 1987) and place IS researchers in a "helping-role" within the organisations being studied. In the long run, well carried out action research leads to best business practices and success stories, which is one of the most effective ways of educating the business community. By producing research findings that are relevant to practice, we argue the action research is the most suitable method to mix and match findings between IS research and business.

Research Findings

The following section describes the industrial case in more detail with examples from three different levels of the organisation and shows how to make the best possible use of mobile technology in business customer care.

Relationship Freedoms: The Struggle for Survival

It was evident by 2002 that the fine-paper industry in Europe was not achieving its desired level of performance and was unlikely to achieve its stated aim of increasing profits by 2 more years. The companies recognized that the demand for fine-paper products by both resellers (wholesalers, retailers) and end users was very sluggish and forecast that growth in paper prices was unlikely to be realised because of the ongoing economic downturn. The fine-paper branch of the pulp and paper industry produces a multitude of paper-based products. It is a complex industry and depends on good customer relations with a great variety of players, ranging from paper mills and printers to retailers and consumers. A successful commercial publication is the sum of many different elements: content, layout, illustrations, font, and often very crucially, paper and print quality. Despite the complexity, there are certain key guides to how to become more successful on the market. In the supply chain you need your customers' business. There is much talk in the industry about focusing on customer service in manufacturing settings, but surprisingly few companies have visualised where to start and how to accomplish this. Our case is no exception, and critical incidents are often overlooked. Critical incidents related to fine-paper products consist of feedback and claims, which together are referred to as complaints. A call centre cannot provide the answer since resolving the conflict requires group decision making and follows complex checkup procedures on the part of the technical customer service working at the mill and of the personnel maintaining the printing machines at the customer site. Besides being a complex activity that requires group decision making, coordination, and information sharing, the handling of complaints related to fine-paper products is also a seasonal activity that puts additional pressure from time to time on both customers and technical service personnel. Another factor is related to the timeliness of the product to the end user, as commercial prints and reports with "old news" do not sell to the consumer.

Despite being considered a very conservative industry, paper companies have begun to look for state-of-the-art solutions in customer care and more attractive ways of resolving conflicts. For major processes in the supply chain, such as handling complaints, the cooperation takes place mainly through complex communication and understanding. This suggests that an interactive medium that enables instant information delivery, easy access to that information flow and sharing among participants should be considered an appropriate communication artefact for resolving industrial complaints. The latest "lean" media today are related to mobile communication technologies. Mobile technology indeed is the most efficient communication medium so far: it is real time, always and everywhere with the user, fast, and easy to use. In the era of the mobile Internet, solving a certain complaint or paying off claims becomes a matter of minutes. But is it enough to rely upon a stand-alone mobile communications medium when resolving customer complaints? It is not likely that the adoption of mobile communication devices will automatically induce order and discipline within groups. But both task demands and individual differences may affect the way groups use the media available to them. However, the picture becomes quite complicated in our case. Many actors from the different organisations that make up the fine-paper supply chain are involved in handling complaints. They have major differences in their working structures and therefore have individual preferences, as well as the freedom to serve their customers by the medium of their choice. Customer service can indeed be reinforced by the mobile use of Web technology if users organise their activities in a collaborative manner with dynamic membership. Collaborative customer service implies that communication of critical information and instant interactions are established at the point of the critical incident, where people from different organisations are working on critical incidents through dynamically established collaborative groups. Up to now, the fine-paper industry has been principally analysed using the value chain framework. This framework was well suited to analysing traditional manufacturing industries in the past. But most printers and end users today want more flexible service and more transparency in the supply chain. Companies should therefore be willing to work more closely in planning production, organising logistics, and handling complaints to this end. The industry needs a new collaborative form of customer service where activities related to handling complaints take place in a network rather than in a conventional linear way (see Figure 1).

Modern and complex organisations have a typical value network configuration (Stabell & Fjeldstad, 1998) in their main activities. Business relationships stem from the wisdom that in the competitive and turbulent business environment of the present day companies cannot expect to grow by acting alone. Both companies in a relationship actually bring their entire relationship network to focus on developing their core competences in order to be competitive in the market. Technology in use and technical know-how are important to business activities and, if mastered by cooperating companies, offer a strategic competitive edge over competitors. With increasing market pressure and higher customer requirements the fine-paper industry needs a completely different mind-set in the

Figure 1. Linear flow of complaint information in a supply chain (left) vs. information flow in a collaborative network (right)

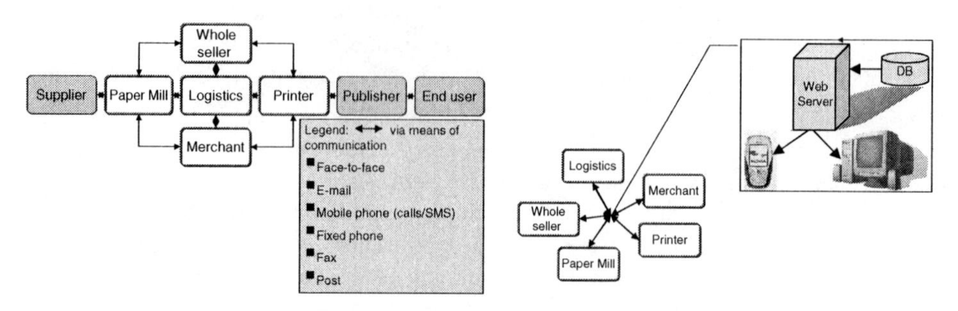

customer care domain from what it had before. Today paper producers need a network of partners and customers in customer service, so they can implement wide-scale mobility when handling complaints and consequently offer a higher degree of flexibility and a more immediate response to the end user. Paper producers need to network with printers, sales offices, and wholesalers who should be also involved in problem resolution in order to avoid communication inefficiencies and indirect interactions. Moreover, mobile services reside on a mobile network where the effectiveness of service will depend on the size of the network and need an industrial structure that supports networking possibilities. The emergence of mobile communications technologies provides the relevant support for flexible forms of organisation and temporal group work in producing a service. Systems for complaint-centred collaboration that might also incorporate mobile technologies would appear to correspond to and facilitate new forms of customer service, offering the same kind of support for all geographically dispersed group members. A Web-based complaints handling system can be accessed via a mobile device and the Internet accompanies its user everywhere, regardless of the type of device used. The system can work on any html/xhtml-supporting mobile device. In the final product the Web interface and mobility feature should be integrated into the existing corporate enterprise resource planning (ERP) system. Ubiquitous access to time-critical information by partners, customers, and employees offers capabilities that go beyond the traditional e-business service concepts. The principal value for the customer will be the transparency of the handling process, which is critical to them—the ability to track the complaint in real time and to have control over the process from the date the complaint was instigated.

Process Freedoms: Achieving Competitive Process

Every company strives to have the best business processes, be the "best in class" and follow the best practices. Business processes are the means by which a company conducts its business to produce value for its customers. Hammer and Champy (1993) define a business process as a collection of activities that takes one or more kinds of inputs and creates an output that is of value to the customer. In their definition the focus is on value for the customer. In our case e-business applications are currently being integrated across companies, and complaints handling is no exception. We believe that a competitive business process should aim at increasing the speed and quality of customer care. When the process is characterised by time-sensitive decision making and collaborative work, the value that comes from mobility is expected to have an especially high impact here. The challenge remains to understand that a critical business process should carry an appropriate customer–employee interaction where the value of mobile technology can be generated. To do so, we have to understand what constitutes the effective design of a business process. Business processes typically have a specified duration and place, a beginning and an end, as well as inputs and outputs. This orientation is all about identifying an excellent business process and concerns removing activities that do not add value and ensuring the efficiency and effectiveness of those activities that do add value. Meyer (2001) goes even further in the process performance; he suggests that the company should add value to the customer experience if it cannot reduce the value-added time anymore.

In our study we used standard modelling principles and were able to illustrate the complaints/claims process in a more standardised way, so that the working activities, actors, and input and output objects became more visible. By adding location, time, and information dimensions, it was possible to model the working activities, information objects, and actors and identify opportunity areas in the process by changing their mobility parameters.

Fragmentation of working time, information flow, and working space were used as measures to reveal shortcomings within the existing process. The existing level of

Figure 2. Preparing the complaint handling process for redesign: Workflow diagram before (left) and after modelling (right)

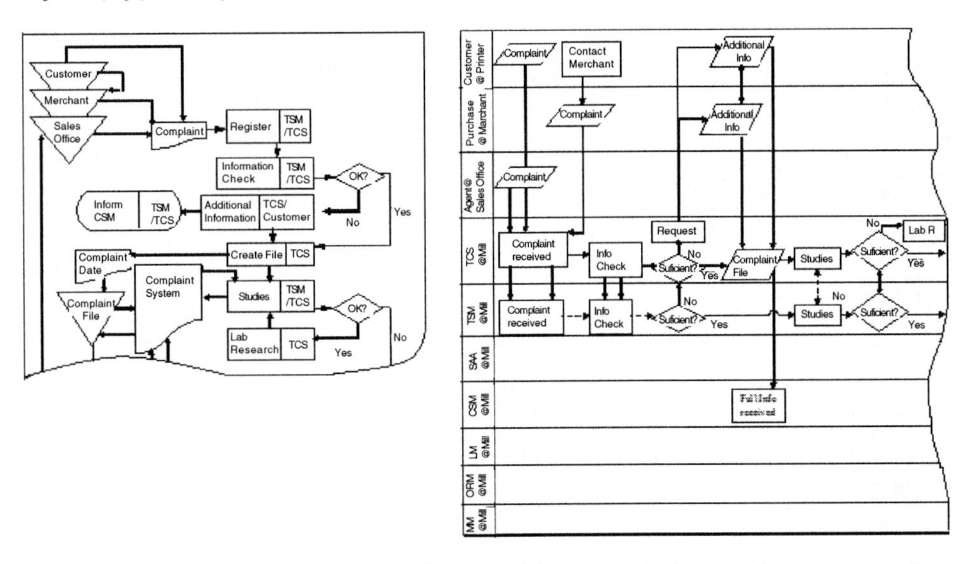

Note: the vertical axis indicates the static location of the actors, the horizontal—time; rectangular—complaint process activities; parallelogram—input and output information; two-sided arrows—activity is coordinated between actors; dashed line—partial involvement, not always required.

customer service called for improvement and process redesign. Each paper-manufacturing mill in the case company relied on a different database to store incoming complaints and most of the work in the customer service department was done manually. Customers as well as partners often became frustrated because they were unable to contact the company via a single access point, as information was often divided between different actors and communicated via multiple devices. The varying communication media used slowed the information flow process and made it inefficient (knowledge workers spent 15–30% of their day requesting and providing information manually where much copy/paste work was involved and lost about 6 working hours per complaint because of interruptions to their work). The existing complaints handling process suffers from coordination difficulties when the process actors are working remotely. These people are located in different offices across the globe and use various technological artefacts to communicate. Not all the actors, however, are equally mobile. Technical managers at the paper-producing company are key decision makers in solving complaint-related problems, but they are highly mobile and work at least twice a week away from desktop computers. Engineers, on the other hand, are involved in routine work on the mill premises, which is characterized as micro mobility. These people are most often overloaded with customers' questions, are heavily involved in the communication process and need a good visual interface when describing laboratory results to the customers. Effective customer response should be immediate. No matter how mobile and where the customers are, they expect to have solutions to their problems online (PC and/or mobile device) and in real time. Wireless has value for managers on the move, whereas high-resolution desktop computers are mandatory for engineers. The complaint handling process was redesigned by making fully mobile as many as possible of the activities, actors, and information items, where a combination of mobile and Web technology served as enablers.

As a rule of thumb, the redesigned process has been innovated from the customer perspective, as is suggested in BPR (business process reengineering) classical theories. The initiator of the process is a customer who encounters a problem that has already occurred in the past, so he/she knows what information the technical people at the paper mill will need. In an ideal case a customer will create a complaint file as soon as a problem was identified and provide as much structured information as possible about the problem. The person lodging the complaint will then have the option of choosing the type of complaint and filling in the required information fields according to the reason for the complaint. Mobile technology eliminates inefficiencies associated with double data handling. By taking digital pictures at the printing house, customers are able to capture critical information in real time and deliver them via a Web-based system to the customer service technicians. Greater mobility of time-critical information significantly reduced travelling costs. Customer transmission of documents at the point of the problem and the ability of technical people to apply their knowledge to solving them in respect of place reduced overall response times. A password is be required to log into the Web-based system, so paper-mill employees working with complaints will know who the customer is. The new way of handling customer complaints has resulted in lower costs by solving problems related to place and time of work and fragmentation of information, and increased flexibility of work. The system incorporates instantaneity and supports automatic notification; thus participants perceive each other's actions with no perceptible delay. Depending on work mobility, a user may select whether he/she wants to receive e-mail notification or an SMS. Such an instant and transparent way of handling critical incidents provides the key to a long-term customer relationship based on trust, customer empowerment, and smooth service. Combining the distinctive features of the mobile channel with the advantages of e-commerce services would lead to substantially more value delivered to the customer. Better collaboration means more mobile technology value, leads to more up-to-date information, makes resources continuously available, cuts idle time and shortens response time in critical business incidents, and facilitates producer–purchaser interaction.

Knowledge Freedoms: Improved Individual Performance

With automated proactive service mechanisms carrying more of the burden and being always accessible, the system is a promising tool that allows customer service technicians at the mill to support more customers and resolve more complaints with existing staffing levels. Technical knowledge is mobilised by equipping customer service technicians with smart phones including browser with PDF support, an integrated mega-pixel camera, and e-mail attachments (xls, ppt, doc) with document viewer applications that work anywhere in the world. All these are necessary to provide decisions in a simplified version of office operations similar to PC alternatives. On the other hand, customers and partners also contribute essentially to knowledge freedoms by bringing time-critical and rich-content information via compatible smart phones. A well-structured and transparent way for handling routine issues also frees technical service experts to focus on more complex and/or urgent problems that truly warrant their personal attention and knowledge workers' skills.

Despite the obvious positive effect of using a mobile Web-based system for handling complaints, various barriers do exist to its widespread adoption and use. The client of a new system is the worker in the customer service department, only indirectly the organisation. The major barrier to introducing a state-of-the-art technology in an organisation is the learning effort required: this is why we need "usable" systems and devices. People are likely to learn how to use new technology if they find it useful. On the other hand, a successful design is one that increases the group members' satisfaction and improves their attitudes towards work. A mobile device is also a tool providing access to corporate networks and complex ERP systems; therefore optimal interface design and device capabilities are needed. The value proposition for individual users is to provide the physical mobile devices to end users that enable them to access a mobile network and to run mobile applications in an easy and user-friendly way. The main question here is not only related to the speed of the mobile network, and quality and usability of the mobile service interface, but also what mobile tool could be a device of their choice. Will they be successfully adopted by distance-working employees as a primary service tool or fail to be adopted by employees who also resisted using laptop computers? With existing offerings on the market, it is obvious that business users will favour using smart mobile phones rather than simpler mobile devices, and would rather go for the former than a combination of PDAs with wireless cards. However, even today we have a wide range of smart phones available on the market; these have similar functions but differ in interaction, the navigation effort required and price.

Our recent studies have focused on evaluating the usability of the working prototype in laboratory settings with technical customer service technicians and business customers. In particular, we have directed attention to trying to find out what type of device is most likely to be adopted by different team members. This form of evaluation is needed to implement the system and to ensure it is usable and easy to use in order to create knowledge freedoms. Usability and user experience are factors that must complement each other to ensure successful acceptance of mobile applications. However, when it comes to mobile business applications, the dividing line between these two categories often becomes blurred because corporate users are even more critical in their acceptance decisions, as they want to get the most out of the product they use in their daily work as well as for personal activities. Therefore, the goal of designing mobile business applications is to ensure that the end product will improve the user's working experience and performance. The latter is known as product usability: this can be assessed in terms of efficiency during test sessions. However, the assessment of user experience is less formalised and straightforward since it is more difficult to define. The average age of the selected test group falls within the interval of 36–62 years. Thus it clearly represents an older segment of the "mobile" population that has major expectations when it comes to the ease of use and functional capabilities of the devices. It is important to note that the paper-manufacturing industry is considered to be rather conservative. Its corporate policies are unlikely to embrace the latest state-of-the-art technology for communication and administrative purposes unlike a service or ICT manufacturing industry. In this context, decisions by senior managers on investment in mobile devices might seem insignificant or accompanied by a high degree of uncertainty. We also accepted the fact that none of the test participants had any previous experience of smart phones or PDAs that operate in similar fashion. We therefore wanted to organise good training sessions especially for those felt most likely to be sceptical of mobile technology. Although in the early stages of the tests, most of the users were quite worried about their ability to use unfamiliar yet cutting-edge technology, their confidence grew along with their experience. By the end of the tests, most users felt they had control over the phenomenon. Therefore industrial operating companies should take into account that their employees, even those who are most technology-phobic, naturally like to use new tools and modern technology provided that adequate training is provided. Two participants, who were much less familiar with mobile technology than others and were mostly worried about making fools of themselves during the test, admitted to being real sceptics of mobile phones mainly for self-protection. By the end, however, they had become enthusiasts of mobile technology and were talking about a "mobile future" and clearly expressing their emotions and desires to own a preferred type of device.

The results of the test are interesting in the sense that for the test participants, as business users, the design and aesthetics of the smart phones were equally or sometimes even more important than their functional properties. For example, the appearance of the smart phone has a strong influence on employees' preferences when choosing a device for daily work activities. This once again confirms the fact that mobile technologies are increasingly personal social technologies supporting work interaction and collaboration. Järvenpää (2000) points to how mobile commerce services and devices have already surpassed the boundaries between industrial buyers and individual consumers. In Europe, many employers pay for their employees' mobile phone use, including personal use. At least in our case this is true, as three Finnish-based companies liberally support their employees' personal use, which generally comprises about one third of total mobile phone use. Because of the relatively small costs, it is not worthwhile for large organisations to devote their time and resources to identifying and recalculating nonwork-related mobile use from the total mobile phone bill per user. Furthermore, this overlooks the fact that multiple device ownership—an idea much embraced by mobile phone vendors and which means that one user has several mobile devices for different social and work activities—may become popular among the working population in the near future. The reality in our example is that single SIM card ownership is very popular among the working population, partly because it is sponsored by employers and partly because it obviates the inconvenience and time spent in switching the card to another device. Moreover, cost savings and device compatibility issues also make companies favour a single, corporate-wide mobile device, without their considering individual employees' choices.

As our results showed, employees in two of our case organisations felt not only more comfortable with the devices of their choice, but also more empowered at work and confident when meeting customers. Needless to say, these mobile smart phones also serve as personal communication devices; it is therefore important to take into account employees' personal choices regarding a device and its job-related fit before deciding to purchase a single corporate-wide mobile phone model. When given a new device that not only fits with job-related tasks but also suits their personal choice, employees feel instant gratification. This is one of the ways that senior management can show that they care about their employees and make their industry more attractive to talented recruits. In our case, the test users were concerned about the phone's outer appearance and its navigation methods during meetings with customers. They wanted to be sure that the impression given to outsiders would be suitably impressive and positively contribute to the corporate image. This indicates that equipping customer service people with brand-new sophisticated mobile devices is also a question of corporate image and may have unique marketing power for end users. It assures customers that they are dealing with a company that is modern in terms of technology and the people who use it.

Mobile technology adoption processes are complex processes that rely on factors other than utility and performance measures. In particular, we argue that individual user experience has an important influence on further adoption of the device in daily work context. Unfortunately, where device-related costs and quantitative measures of usability are easier to incorporate into a tangible cost–benefit analysis related to mobile technology investment, this is not easily the case with user experience. Consequently, there is a risk that the impact of user experience evaluation might be ignored.

Concluding Remarks

This chapter is based on realisation of main conceptual ideas of the Freedom Economy when designing mobile B2B and B2E applications for customer care in the fine-paper industry. In order to fully realise the added value of mobile technology, these rules are seen as separate initiatives to be taken at three different organisational levels: individual (e.g., technical customer service employees, purchasers, technicians at the printers), organizational (complaint handling process scaled across three organisations), and supply chain (paper-producing mill, merchant organisation, and printing house at the customer site). Even though the process and technology fit is a necessary first step for transforming organisations to the next generation stage, we estimate that two other levels of alignment remain essential in achieving best business practice in the dynamic, complex, and uncertain business environment of fine-paper industry. In addition to this, different concepts and theories in the frame of reference can be suggested for each level, where technology-acceptance model and concept of user experience can be considered in more detail at the individual level, business process reengineering and process transparency issues at the organisational, and transaction cost and network theories at the value-chain level. In the future, customer care applications in the corporate intranet and extranet of the fine-paper supply chain will have mobile extensions, enabling employees, partners, and customers to access and send time-critical information regardless their location. Mobile technologies will also transform organisations within the fine-paper industry causing impulsive communications, dynamic interactions, and faster execution of other critical business processes as knowledge and expertise imbedded in people can be reached via smart phones regardless time and place constrains. We believe the framework that emerged in this paper will provide guidelines for companies to successfully design mobile business services in other industries and enable them to achieve best business practice in the next generation economy.

References

Baskerville, R.L., & Wood-Harper, A.T. (1996). A critical perspective on action research as a method for information systems research. *Journal of Information Technology*, (11), 235–246.

Benbasat, I., Goldstein, D.K., & Mead, M. (1987). The case research strategy in studies of information systems. *MIS Quarterly, 11*(3), 369–386.

Bendeck, F., Kötting, B., Schaaf, M., Maurer, F., Valenti, M., & Robert, M. (2001). *Engineering of e-business applications & infrastructure and applications for the mobile Internet*. Proceedings of the 10th International Workshops on Enabling Technologies.

Carlsson, C., Walden, P., & Veijalainen, J. (2004). Mobile commerce: Core business technology and intelligent support: Minitrack introduction. *Proceedings of the 37th Annual Hawaii International Conference on System Sciences.*

Eisenhardt, K.M. (1989). Building theories from case study research. *Academy of Management Review, 14*(4), 532–550.

Hammer, M., & Champy, J. (1993). *Reengineering the corporation: A manifesto for business revolution*. New York: Harper Business.

Järvenpää, S. (2000). *Internet goes mobile: How will wireless computing affect your firm's Internet strategy?* (Working Paper).

Keen, P., & Mackintosh, R. (2001). *The Freedom Economy: Gaining the m-commerce edge in the era of the wireless Internet.* Berkeley, CA: Osborne/McGraw-Hill.

Kuuluvainen, J. (2002). *Business cycles, information technology and globalisation in the forest sector.* Final report on Finnish Forest Cluster Research Programme: Wood Wisdom.

Lehmann, H., Kuhn, J., & Lehner, F. (2004,). The future of mobile technology: Findings from a European delphi study. *Proceedings of the 37th Annual Hawaii International Conference on System Sciences.*

Meyer, C. (2001). While customers wait, add value. *Harvard Business Review, 79*(7), 24.

Müller-Veerse, F. (1999). *Mobile commerce report.* London: Durlacher Research.

Paavilainen, J. (2002). *Mobile business strategies: Understanding the technologies and opportunities.* UK: Addison-Wesley, IT Press.

Remenyi, D. (1998). *Doing research in business and management: An introduction to process and method.* London: Sage.

Ritz, T., & Stender, M. (2003). B2B mobile business processes: Scenarios and technologies. *Proceedings of the 14th International Workshop on Database and Expert Systems Applications,* (pp. 885–889).

Stabell, C., & Fjeldstad, D. (1998). Configuring value for competitive advantage: On chains, shops, and networks. *Strategic Management Journal,* (19), 413–437.

Stafford, T.F., & Gillenson, M.L. (2003). Mobile commerce: What it is and what it could be. *Communications of the ACM, 46*(12), 33–34.

Chapter X

Bringing the Enterprise System to the Front Line: Intertwining Computerised and Conventional Communication at BT Europe

Alf Westelius,
Linköping University and Stockholm School of Economics, Sweden

Pablo Valiente,
Stockholm School of Economics, Sweden

Abstract

This paper draws on the need to understand how mobile technology is implemented and used at the organisational level. IT is a general-purpose technology and its use involves a high degree of uncertainty. Therefore, managers have trouble in identifying the real scope, the functionality, and the impact of new mobile applications. However, these three types of uncertainties need to be handled in change management projects where new information technology is involved. Gradual uncertainty reduction at these three different levels—that is, what technology can do, will technology work, and will users adopt it—is studied in this chapter. This is achieved through an analysis of the implementation process of an information system at BT Europe, a leading supplier of forklift trucks. The analysis shows how the computerised parts of the information system are complemented by mindful intertwining with the noncomputerised communication and manual data processing in order for the information system to work.

Introduction

The possibilities of mobile technology continue to broaden and expand. Many organisations have invested, or are considering investing, in this technology. The present slowdown in investments is expected to be temporary, and hopes for the future are high. One area that is attracting attention is the use of mobile terminals that can give mobile employees access to central information systems, such as enterprise resource planning (ERP) systems.

At the level of the business enterprise, investment decisions related to new technology in general and mobile technology in particular are usually fairly challenging. Some industry analysts and telecommunications providers claim dramatic business improvements, but many IT and business managers have expressed concern that the business value of mobile technology may not be quite as substantial as suppliers would like them to believe. One reason for this may be IT's "ambiguous" (Earl, 2003) and open-ended character (cf Asaro 2000; Orlikowski & Hofman, 1997). Earl (2003) proposes that three important sources of ambiguity are uncertainty regarding what technology can do, whether the technology will work, or if it will even be adopted. The aim of this chapter is to describe and analyse the uncertainty resolution process in a large-scale implementation of mobile technology, and the interplay of new and old technology and organisational solutions. This chapter analyses the implementation process of an information system (EASY—Engineer Administration SYstem) at a leading supplier of forklift trucks where mobile terminals are used to give service technicians access to the ERP system. The mobile terminals provide service technicians with an interface to structured, written data in order to rationalise the entire service order process. The project was carried out as a follow-up to a business process reengineering (BPR) project that resulted in the implementation of a common ERP system across the European division of the company.

The chapter is organised as follows. First, we present models central to the article, followed by a brief account of the research method. Then the implementation of the mobile application is described. The description includes the project background, a number of identified benefits, the project itself, the organisational impact and the management of change within the project, and finally some technical considerations. We conclude the paper with a discussion of three different kinds of ambiguity of IT implementations that are based on open-ended technologies, how the uncertainties interact, and how computerised and manual parts of the information system intertwine to create a functioning system.

Utilising Ambiguous Technology

In the research literature, it has long been acknowledged that technology interacts with other aspects of organised work, and that change in one aspect affects the others. Leavitt's "diamond" (1965) suggests that interaction with and among people, organisation, and task is important when dealing with technological change. Lundeberg's "levels of abstraction" (1993), dealing specifically with IS-related change, further divides technology into hardware and software, and distinguishes between information and activities. Talking of "results," rather than "tasks," perhaps reflects the rhetoric of the 1990s, while "people" and "behaviour" are seen as two important aspects pertaining to the individuals who are involved or affected. Both these models, and others like them, specify important subsystems that need to function in ongoing operation, and that require careful planning and consideration.

Orlikowski and Hofman (1997) developed the idea that IT-related change, especially regarding partially novel and "open-ended" technology, cannot be fully planned. Important opportunities—and obstacles—will arise, and be noted, over time, as people learn about the technology being applied, and reflect on the interaction between the computerised and noncomputerised aspects of the work being performed. An opportunity-based strategy for change acknowledges, and tries to benefit from, the emergent and changing understanding of IT-related change over time.

Earl (2003) further develops the notion of uncertainties connected with the use of ambiguous technology, suggesting that the actual benefits derived from the use of IT in an organisation will always be uncertain when envisaged in advance. He proposes three important types of uncertainties: enabling uncertainty, what technology can do; commissioning uncertainty, whether the envisaged application can be built; impact uncertainty, whether the application will be adopted and gainfully used. Robey, Schwaig, and Jin (2003) finally turn the focus back to the interaction between the virtual and the material. They suggest that full digitalisation is probably neither desirable nor feasible. The objective should instead be to develop a mindful intertwining of virtual and material communication. We will draw on these ideas of interrelations and interactions between aspects of technology and the use of it, of different types of uncertainties, and on the intertwining of computerised and noncomputerised aspects when analysing how a vision of computer-supported work is turned into practice.

Method

The BPR project and the hardware and ERP projects that formed the background for the EASY project were studied through interviews and project document studies. Close to 70 interviews with people in these projects, at headquarters and in three of the market companies, were performed, and extensive project documentation was made available to that research team, which included one of the present authors and three additional researchers: Linda Askenäs, Klas Gäre, and Cecilia Gillgren. The other present author has met with the supplier of the EASY application software. The present authors have in addition interviewed representatives of the EASY project team: one from the service process side at headquarters, one from the division's IT department, and a service market manager from a market company. The authors have also been able to study the application itself, and have been given access to training material and project evaluations from a market company. The interviews have been semistructured (Patton, 1990) and lasted 2–3 hours each. They have been tape recorded and transcribed in their entirety. The material has been analysed in a qualitative tradition. The interviews and documents relating to EASY have been searched for indications of uncertainties and trade-offs in the development and implementation process. These indications have then been used to explore connections between levels of uncertainty and concerning the intertwining of computerised and noncomputerised aspects of the administration process, when viewed as an information system.

Implementing Easy: Computerising Service Technicians Across Europe

Project Background

The organisation in focus for this study is BT Industries, part of the Toyota group. BT Industries is a leading supplier of forklift trucks, with a world market share of more than 20%, annual sales of euro 1.2 bn, and 8,000 employees. The company offers a wide range of forklift trucks plus servicing facilities and has manufacturing locations in Sweden, Italy, Belgium, United States, and Canada. From the middle of the 1990s onwards, the European division, BT Europe, embarked on an ambitious computerisation venture. A BPR planning project to explore IT-enabled business change was carried out. A shared hardware and communication platform was designed and implemented, and after thorough evaluation, an ERP system (Movex) was chosen and rolled out across Europe in a strategic partnership with an ERP supplier.

Having installed a shared information platform, new ideas related to technology-enabled projects started to appear in the organisation. In 2001, BT Europe decided to further improve quality in its customer offering through rationalisation of its service order process. Moreover, the common ERP platform could enable a pan-European project making the exchange of ideas between different local market units possible.

A project group with approximately a dozen participants was formed. The members had different background, coming from local operations, central staff personnel, and technology consultants.

Identified Possible Benefits

The group identified a number of possible administration-related cost reductions. BT Europe's field service engineers were processing 5,000 assignments daily. Annually, 1.2 million handwritten work orders were delivered on paper to the administrative offices, and fed into BT Europe's back-office ERP system. This was a costly and time-consuming process. BT Europe's service process management decided to implement an automated solution, which extended the back-office system to the field service force by providing them with handheld mobile devices for access to the job, contract, or product information required.

The original idea was to use information technology and to modify the business processes in order to benefit BT's customers and the company itself. Faster service routines could lead to more efficient service operations. Customers would benefit from less risk of human error, faster and more accurate communication and repairs, and more motivated and informed technicians. The technicians could reduce time-consuming processes such as work-order and service-contract processing, and they could access data about products and customers resulting in more informed employees. This could eventually change the role of the technicians to becoming more business oriented. Moreover, increased flexibility to plan their work could lead to more motivated staff. The mobile access to the ERP application would provide technicians with the possibility to plan their work based on the pending work orders, report service orders and assignments, order spare parts, review service contracts, and access extensive data about products and customers. Online reporting would also speed up the invoicing process, thus reducing the amount of capital bound up in the service process.

Service technicians would not only change role but also work practices. The physical work reports manually delivered on paper to the administrative office would be substituted with virtual work reports directly fed into the ERP system resulting in automatically completed invoices. A shift from physical work practices to virtual ones would be required when interacting with the mobile terminals. Automated service order process solutions could moreover reduce the administration workload by elimination of manual routines for invoicing, processing work reports, and service assignment scheduling. Instead of dedicating time to copy invoices into the system, administrative personnel could focus on giving feedback to service technicians when necessary. A consequence of these new work practices would be to rationalise the back-office function, enabling a reduction of the number of employees.

IT projects can provide organisations with an opportunity to revise work practices and business processes (Davenport, 1997), although the opportunity is not always utilised (Asaro, 2000). As indicated above, revising work processes in BT Europe was an important aspect of the EASY project. The pan-European character of the project could moreover facilitate benchmarking between local markets and improvement of the service order processes to reflect company best practice. There were, thus, a large number of potential benefits to be derived from the project. However, the process of getting there was not streamlined. A number of issues had to be dealt with during the project.

The EASY Project

Although the idea to give service technicians access to the ERP system had appeared already during the BPR study in the middle of the 1990s, the time had not then been considered ripe. Partly, the reason was that reliable and cost-efficient mobile platforms were not yet available on the market. One of BT's main competitors had equipped its service technicians with laptops and printers accessible at the service vans. However, that project was believed to be expensive and did not show the benefits the company had hoped for. Managers in BT were pleased that they could learn from the "bleeding edge" experience of others. Part of the lessons learned from that project was that service technician computerisation should be based on easy-to-use, robust, handheld devices.

Another reason why the project had not begun earlier was the lack of a common hardware and software platform in the company. Then, by the year 2000, mobile terminal development and the development of administrative software for such terminals had progressed to the point where computerisation of the service technicians' administrative tasks seemed feasible. Local tinkering had begun (cf Ciborra, 1994), but centrally placed imaginators (cf Hedberg, Dahlgren, Hanson, & Olve, 1994) envisaged a more thoroughly transformed organisation than they believed the local imaginators did. Although the centrally placed imaginators did not envisage a business-logic transformation that would revolutionise the industry (cf Cross, Earl, & Sampler, 1997; Hopper, 1990), they believed substantial benefits would be achieved through coordinated action, and state-of-the-art use of mobile technology. To achieve potential benefits of streamlining the service process across the entire division, and to be able to split the development cost, a joint project was initiated by divisional headquarters. Not only had mobile technology developed far enough, but the division also had a sufficiently common administrative platform in place to make a joint service administration possible. Unlike in other cases where ERP implementation has been found to *prevent* business changes (e.g., Hanseth & Braa, 1998), in BT Europe the shared Movex platform was *a prerequisite* for further change.

The idea to make BT Europe's 1,150 mobile service technicians more effective was therefore discussed at an annual brainstorming meeting of market and service development held centrally in 2000. Mobile technology was by that time often debated in media and mobile technology opportunities created visions about cost reductions and improved efficiency. Many people at BT regarded the project as a prestigious one. EASY could profile the organisation as a company at the cutting edge of new technology.

From the beginning, BT managers hoped to be able to learn from other companies where similar PDA-based systems for the service function had been implemented. However, such implementations on a multicountry scale could not be found, and the company was obliged to become a first mover to some extent. BT Europe managers did not want to undertake such a large-scale and innovative project purely on faith, and therefore asked for business case calculations from the market companies. As they had hoped, it turned out that there was a sound business case for a joint project, but that no market company could muster one for developing such a solution on their own. Based on the business case, the EASY project received a go ahead, but the consequences and the details were not yet fully known. Among other things, a degree of uncertainty regarding what technology could do was certainly present. Moreover, functionality to be implemented also evolved as the project matured. As an example, it was not until after the system launch that service managers realised the potential to automatically attach marketing flyers to work order notifications e-mailed to customers directly after performed repairs. This unforeseen, new communication channel also gave EASY marketing potential.

Organisational Impact

The intended change was substantial, and for the back-office function it was truly of a reengineering scope (cf Cross et al., 1997; Davenport & Short, 1990; Hammer, 1990; Hammer & Champy, 1993). Therefore, the development team had strong participation from back-office functions to access relevant operational knowledge and to achieve organisational credibility (cf Asaro, 2000; Westelius, 1996). Those responsible for the project certainly did not want to provoke sentiments of headquarters manipulating the local companies (cf Markus & Pfeffer, 1983), and therefore saw to it that the designing members of the team came from local operations, and that consultants and central staff personnel were only in the team as support to the ones with hands-on knowledge of the business process. In this way, the developers were to a large extent developing their own future, rather than being a specialist team stuck in the middle between demanding management and an oppositional workforce (cf Howcroft & Wilson, 2003).

A lesson from the ERP project was that without a strong enough focus on designing common business processes, a truly shared computer application would be difficult to achieve (cf Gäre, 2003). The project team thus spent close to a year on mapping the present processes in three countries, devising a redesigned process, and then requiring the service managers in the other countries to perform a gap analysis between the redesigned process and their existing ones. They were required to state what needed to be changed in the current process to implement the new one, and if there were aspects of the existing process that were essential to keep, and would require modification of the new process. In the end, just a couple of almost 70 *use cases* in the new application supported alternative subprocesses. All the others presupposed one common way of carrying out the administrative process.

It was obvious that the EASY project would require a change in the role(s) of service technicians. To some extent, their present worldview and that of back-office personnel were different (cf Checkland & Scholes, 1990). The typical service technician was viewed as having a strong customer focus—the goal was to keep the customer happy by keeping the customer's forklifts operating as reliably as possible and by getting them back into operation as quickly as possible in case of breakdown. However, technicians typically had little idea of the economic aspects of a customer's service contract. Nor were they expected to sell trucks, service contracts, or consider when it was time from a BT perspective to replace rental trucks. The role the service technician was moving into, with the introduction of EASY, would be more businessperson-like. However, rather than replace the customer-oriented perspective, the ideal would be to let it coexist with a businessperson perspective and an administrator's perspective. The new service technician role would be a multiple-identity one (cf Foreman & Whetten, 2002; Pratt & Foreman, 2000). In addition, administrative accuracy and correct filling-out of forms would now be expected from each service technician. They would in the future take full responsibility for quickly and reliably feeding the ERP system with data on service work (cf Petri, 2001), without back office serving as support and filter, cleaning data and translating between the service technician's world and the computer system's.

Concerning information provision to the service technician, the objective in EASY was certainly not to create ambiguity to make them think creatively (cf Hedberg & Jönsson, 1978), but rather to supply them with as well-tailored data as possible, to reduce the interpretive space needed to process the data (cf Thompson, 2002). The service technician's focus should initially still be on servicing trucks efficiently, not on making creative interpretations of the data they entered and retrieved. However, over time, the development of a businessperson-like role would require more interpretation of data, such as figuring out when it is time to sell the customer a new truck, or realise that the repair history of this rental truck suggests that we (BT) replace it to lower our total cost, and so forth.

Change Management

Change initiatives expecting a change in perceptions of identities or regarding tasks or even the existence of work, can be expected to evoke strong reactions (Checkland & Scholes, 1990; Fiol & O'Connor, 2002; Huy, 1999). The need to feel support in taking the step into the unknown is then highly important (Huy, 1999; Schein, 1993). In BT Europe, they have tried to achieve this through ambitious training programs aimed at back-office personnel and service technicians, and have provided a filtering function that buffers for errors the service technicians make in handling the administrative software, until the service technician masters the application at a virtually error-free level. Regarding the back-office personnel, the strategy has been to cut the number of staff at the beginning of the implementation, partly in order to get the remaining staff highly motivated to give feedback to the service technicians so they handle the application with fewer and fewer errors. A decreasing amount of errors from service technicians will reduce the workload for back-office personnel. The drawback of this strategy is that the workload on the (reduced) back-office staff is high and it is difficult for them to give adequate support to the service technicians. The learning period for the technicians is thus considerably longer than it would have been, given ample feedback. On the other hand, having kept people in back office who would ultimately be laid off, would give them an incentive to prove that they are needed, and thus maybe keep their job, by showing that the service technicians cannot handle the application well enough on their own. The equation is thus not an easy one, and it is not obvious that it will ever be possible to determine if the path chosen was better or worse than an alternative one.

Further development of the application and its use would be based on a mix of planned changes and opportunity-based change (cf Mintzberg, 1989; Orlikowski & Hofman, 1997). Some further development could already be foreseen, based on ideas that had been deferred from earlier stages for budget reasons or in order not to risk complicating the application design that had been planned for that developing stage. In an organisation, change is taking place continuously, at a micro level, because people are not machines, and do not faithfully repeat a process in an unchanging manner forever (cf Tsoukas & Chia, 2002). The challenge in the EASY project (and for future operation) was on the one hand to stop change, getting people to adhere faithfully to the carefully designed service administration process, while not stifling initiative (cf Galgano, 2002) and instead channelling it into a "versioning" system. The process should be carried out in the agreed manner. Change ideas should be submitted to the process owner, evaluated, prioritised, and if accepted by the process owner, sorted into the next or a future version of the service process. Preserving change initiative while not being in charge of deciding on implementing the changes is challenging, and can lead to reenactment rather than change becoming the ideal people follow (Westelius & Askenäs, 2004), or require strong and visible feedback on the handling of the proposed changes (Borovits & Neumann, 1988; Petri, 2001).

Technological Considerations

At the beginning of the project, few decisions regarding the technological platform were made. However, two agreements were made relatively early. One was that the operating system had to be future proofed and relatively well established so that competition could be guaranteed. To provide 1,150 technicians with portable devices could otherwise become expensive. At the decision point, Palm OS enjoyed a larger installed base. However, terminal suppliers consulted by BT managers foresaw Pocket PC to become more available in the future. This led to the choice of Pocket PC-based solutions, because that platform seemed to offer a lower risk of terminal supplier lock-in than other competing solutions. The other was to build the application based on offline synchronisation update methods and GSM networks. Tests had shown that connectivity was often a problem and that GSM was the only type of mobile network that offered a relatively extensive coverage in the dozen European countries where the application would be implemented. Technicians often work in areas with bad network connectivity, such as in rural areas, and inside industrial buildings with massive cement walls.

Using GSM networks, users at BT Europe can synchronise data, work offline, and then synchronise data again afterwards. The solution provides access to all necessary information through the PDA, so service technicians can carry out assignments independently of external factors, such as the lack of a network connection. However, service technicians are expected to synchronise several times a day to update data both in the handheld devices and in the ERP system. To facilitate this, a one-button synchronization feature was built. Because of the selected networking solution, connection between PDA and the ERP system via middleware has to be initiated from the PDA. However, the mobile telephones the technicians carry provide back office with a way to reach a specific technician. Back office can send an SMS to a technician's cell phone to indicate when a new urgent work order needs to be downloaded and taken care of. If the technician has not synchronised the PDA within a certain period of time, a follow-up mail is sent to back office, where further action to allocate the service task can then be undertaken.

When the system had been implemented, service technicians began to experience the terminals as slow. The application is uploaded from resident to primary memory and runs in primary memory. It can take several seconds to change screen image. At the beginning, this was not an issue. However, as technicians became more acquainted with the tool, waiting for the next screen to load was experienced as highly annoying, and gave the impression that filling out electronic forms was more cumbersome and took more time than filling out the paper-based version. However, recent measurements have shown that the computerised process seems to be no worse, and perhaps even faster overall than the paper-based process, but the stress of not being able to control the progress yourself leads to a subjective evaluation that differs from the measurements.

In general, during the pilot installations, the PDAs have proven reliable, the offline work mode and synchronisation processes have worked, and the training of the service technicians and back-office personnel has gradually led to the establishment of a new work process that is close to the intended one. Believing in the validity of the business case, the service process management of BT Europe then decided on a complete rollout of the application to all mobile service technicians in all the market companies. The rollout progressed according to plan, and now, March 2004, the 1,150 mobile service technicians use EASY. At the same time, work on enhanced versions of the application and the administration process is going on. Over time, the initial uncertainty will resolve, while new opportunities—and problems—will appear and be addressed.

Discussion

The story of EASY could be viewed as an account of deployment of mobile technology. However, as the case illustrates, that would be a too limited scope. It is rather a story of change management, where mobile technology is an important aspect, but one that has to intertwine with work processes and other technologies.

Moreover, a challenge that managers face when deciding on new technology investments is that change management involving open-ended technologies is an uncertain and ambiguous process. This is so because change related with open-ended technology is an ongoing process rather than an event with an endpoint after which the organisation can expect to return to a reasonably steady state (Orlikowski & Hofman, 1997). In the EASY case, this shows up as a number of projects that follow on each other, and as a learning process regarding the use of the EASY application. This learning process is far from finished today, and will also surely come to incorporate new steps from as yet unforeseen follow-on projects. Some uncertainty can be resolved through planning at an early stage, but some remain and will only resolve gradually as time passes.

In addition, new challenges will arise as the use of the open-ended technology develops. Orlikowski and Hofman (1997) distinguish between emergent and opportunity-based change. Emergent changes are changes that were not intended, but developed "spontaneously from local innovation." Opportunity-based changes are not anticipated either, but are purposefully implemented in the change process in response to an unexpected opportunity, breakdown, or event. A pan-European project, such as EASY, spanning different cultures, different organisational units, and a large number of users with little or no contact with most other users, is subjected to many forces pulling in different directions. It is then important to meet the unexpected with opportunity-based change rather than by relying on spontaneous, emergent processes. Otherwise the envisaged, shared service administration process will not be long lived, but soon dissolve into local variants, and then the shared mobile application will probably become a hindrance to development to local companies that cannot afford developing the application on their own.

To address this aspect, EASY has been set up with a process for handling new ideas, demands, and opportunities. Suggestions for changes and further development of the process or the computer application should be substantiated with a business case, and evaluated centrally in BT Europe. To allow for orderly development of the application and uniform implementation, a versioning strategy is followed, where suggested changes that would disrupt the present version are likely to be deferred to a later version. The submission of suggestions is not expected to develop unaided; an IS coordinator has been appointed and given the task of encouraging exchange of good ideas for use of the application. Earlier experience from impromptu modifications of software have made both central and local managers wary of unexpected complications, and the service process is believed to be mature enough to allow for a somewhat slower process for implementing good ideas.

As Earl (2003) has suggested, the ambiguities of an IT-related venture can be analysed in terms of three essential uncertainties: enabling, commissioning, and impact uncertainties. Resolving the *enabling uncertainty*, determining what could and could not be achieved with the help of mobile technology is a first step. The original vision behind the EASY project was to provide the service technicians with a direct, computerised link to the shared ERP platform, rather than having them rely on paper- and telephony-based communication with back-office personnel. The early vision of online access proved infeasible. GPRS coverage was not sufficient on a geographical basis, and local connectivity problems would add to make off-line solutions necessary. Thus, that part of uncertainty concerning what technology could do was resolved. The uncertainty concerning convenient portable terminals resolved itself in a more positive manner. Expensive and cumbersome PC terminals developed into more robust, less expensive, and more reliable PDAs. However, PDAs are developed for office workers, consultants, and so on, and are equipped with a number of applications that service technicians are unfamiliar with and that could cause confusion and prolong the training period needed. Part of making the PDAs suitable as terminals for EASY was then to block out standard functionality that was not required in the EASY application. Thus, the idea of portable terminals that all 1,150 technicians could use finally seemed to be feasible. The business case they had developed indicated that the technical capabilities expected from EASY had sufficient economic potential. This was then supplemented, for example, by the rather late realisation that the application delivering electronic work order notifications to customers could also be used to deliver campaign leaflets and offers. Further examples of resolving of the enabling uncertainty are certain to evolve over time. The challenge in the BT case will be to allow experimenting within the fairly strict, newly designed administrative process, and to identify and grow the good ideas into widely implemented practice.

The next level, *commissioning uncertainty*, was an issue right from the start. IT projects can easily escalate or fail to deliver altogether, and people within BT Europe wanted to avoid this as far as possible. One attempt was appointing an experienced project manager.

Starting with developing and agreeing on the new administrative process before finally choosing the actual handheld device and programming the application, and then keeping a strict regime concerning versions and changes to the specification was another. Yet another attempt was choosing a widely licensed technology from a powerful company like Microsoft, rather than proprietary technology from a smaller, specialised player. A final example was letting suppliers "go public" with the case, and thus making their public image dependent on the success of the development. So far, the strategy seems to have worked. The project has kept to the schedule, and it has also been possible to take advantage of an improved version of PDA that appeared on the market after the first pilot tests of the application.

There have also been parts of the commissioning uncertainty that have been solved through different kinds of intertwining (Robey et al., 2003) between the new and existing modes of communication. One is that error-free input by service technicians cannot be achieved through programmed controls alone. To deal with this, a filtering function was built in the middleware, giving back office an opportunity to set filtering conditions on an individual level for manually scanning transactions before releasing them to the ERP system proper. Another example is that due to the way the synchronisation is initiated from the PDA, not from the ERP system or from back office, efficient handling of rush orders includes the use of mobile telephones. The service technicians were already equipped with telephones, used them often, and will continue to use them. Calling the telephone or sending an SMS to it is thus a way to reach a specific technician. But informing of the details of a new job by telephone, rather than via the PDA would be an awkward duplication of the administrative process. Thus, the telephone is used to signal that it is time to synchronise the PDA. Similarly, since the EASY application is not built to give online access to stock levels of spare parts, that feature is solved by telephone when deemed important by the service technician. The order for the spare part is placed via the PDA, but when it is important to know if an unusual spare part is actually in stock, the service technician calls someone with direct access to the ERP system.

Enabling uncertainty and commissioning uncertainty can be viewed as mere hurdles. You have to get over them, but they do not guarantee success. Unless *impact uncertainty* is resolved in a satisfactory manner, the business case projections will not be met. In the EASY project, a start was to have frontline managers and users develop the specification with support from central managers, IT specialists, and consultants, rather than the other way round. The gap analysis in the companies that did not take part in the development project was another element in reducing the impact uncertainty. Here, a problem is that the enabling uncertainty is larger for someone who has not been part of the development project, and who has a less intimate understanding of what the actual technical flexibility will be, than the understanding possessed by those who have spent months or even years exploring the issue. There is thus a risk of not truly realising what gaps there will be between the actual, computer-supported process and the present, manual one. Since the implementation is well under way, there do not seem to have been any major surprises, but the late addition of at least one important use-case alternative indicates the presence of some impact uncertainty after it was believed that it had already been dealt with.

Strong focus on user training has been another way to reduce impact uncertainty, and so far it appears that is has been successful. Reportedly, customers have also been positive to the introduction of EASY, and want to see it developed even further. Perhaps the most challenging part regarding impact so far has been to get the feedback loop from back-office personnel to technicians to work. With the manual process, clerks at back office could correct much of the inaccuracies that existed in the service job reports that arrived on paper. For the new process to work as intended, technicians need to achieve error-free reporting. The filtering function described above gives back office the possibility to filter out transactions for control based on the data-entry proficiency of the individual service technician. But to learn from their mistakes the technicians need feedback concerning inaccuracies. Some of it can be handled by controls in the application itself, but some are less obvious. Should this job be charged to the customer or is it covered by a guarantee, or by a service contract? Is this replacement of this wheel noted on the right truck or on another of the same model? Is this spare part in my van really registered in the ERP system? These and a host of other questions need to be handled, and typically require communication between the technician and another human being, and maybe even repeatedly, before they become part of the active knowledge of the service technician. This again is an example of the need for intertwining between the new, mobile application and previously existing modes of communication. Given efficient feedback, this learning process will require many months. Given less efficient feedback, it will take longer or perhaps even result in a negative answer to the question: "Will EASY meet the high expectations?"

Conclusions

If we look at the conclusions the actors in the project reached and the decisions and actions they took at the different levels of uncertainty, interrelated chains become evident. At the enabling level, the wish for a light, robust terminal that would be likely to be useful in the service technician job and at a sufficiently low price finally found a match when PDAs where believed to meet the requirements. Nevertheless, actually having service technicians interact with the ERP system would require correct input of data from the service technicians. In addition, adoption would be unlikely unless the service technicians felt that they could master the application. The design of the PDA with a pen-like pointing device and touchscreen instead of keyboard seemed to match service technicians' present skills, but would require the construction of a menu-based interface. Such an interface has since been built, but by itself it does not guarantee complete and error-free input, and has also started to appear annoyingly slow to the more experienced users. *Thus, what at one point seems to lessen impact uncertainty can at another point in time increase it.*

To facilitate the role transition for the service technicians, the error-reducing capabilities of the software was complemented with a filtering function, prompting back-office personnel to manually check input from service technicians before passing the transactions on to the ERP system. If back office would also give feedback to service technicians and to trainers (outside the EASY application), the service technicians could with time improve their handling of the application. *Thus, attempts to improve the EASY application at the commissioning and impact levels had to be complemented by manual routines and conventional communication.*

Similarly, the synchronisation of the PDA and the ERP system relied on service technician initiative. At the commissioning level, it was possible to build an easy-to-use, one-button synchronisation function, but at the impact level, it rested on the routine that service technicians actually synchronise a number of times a day. In addition, to solve push of urgent service jobs from back office to service technicians, the PDA application had to be supplemented with SMSs to the service technicians' cell phones, prompting them to synchronise the PDA. Also, the replication of spare part availability data from the ERP system to the middleware being less than real-time, important spare part availability had to be checked via telephone, outside the EASY system. *Thus, consequences of design choices affected the impact level, and required intertwining of the EASY application with noncomputerised communication support to achieve certain functionality.* The analysis of chains linking different levels of uncertainty, intertwining computerised and noncomputerised communication between actors, modifying roles, and intertwining manual routines and IT processing of data could be carried to a greater depth, but the examples above illustrate our basic idea.

In technology-related change process literature, it has long been noted that successful change demands attention to the interplay between technology, tasks, and organisation, and with attention to the people who are involved in the change (e.g., Checkland & Scholes, 1990; Leavitt, 1965; Lundeberg, 1993). Often such literature has had a strong focus on planning, thereby trying to overcome obstacles and reduce uncertainty, while other authors have focused the emergent nature of change (e.g., Mintzberg, 1989; Orlikowski & Hofman, 1997; Tsoukas & Chia, 2002). In our analysis, we have attempted to combine these ideas, illustrating how the uncertainties of providing mobile service technicians access to a central ERP system have been resolved over time, while new problems and opportunities have arisen. The analysis has focused on three levels: uncertainty concerning what technology can do (enabling uncertainty), concerning if the envisaged application can be built (commissioning uncertainty), and concerning if the application will be gainfully used (impact uncertainty). We have also shown how these levels interact, and how the computerised parts of the information system are complemented by *mindful intertwining* of the computerised application and noncomputerised communication and manual data processing, in order for the information system to work as intended.

References

Asaro, P.M. (2000). Transforming society by transforming technology: The science and politics of participatory design. *Accounting, Management and Information Technologies*, *10*(4), 257–290.

Borovits, I., & Neumann, S. (1988). Airline management information system at Arkia Israeli Airlines. *MIS Quarterly*, *12*(1), 127–137.

Checkland, P., & Scholes, J. (1990). *Soft systems methodology in practice*. Chichester, UK: John Wiley.

Ciborra, C. (1994). From thinking to tinkering: The grassroots of IT and strategy. In C. Ciborra & T. Jellasi, (Eds.), *Strategic information systems: A European perspective* (pp. 3–24). Chichester, UK: John Wiley.

Cross, J., Earl, M.J., & Sampler, J.L. (1997). Transformation of the IT function at British Petroleum. *MIS Quarterly*, *21*(4), 401–423.

Davenport, T.H. (1997). *Information ecology: Mastering the information and knowledge environment*. Oxford: Oxford University Press.

Davenport, T.H., & Short, J.E. (1990). The new industrial engineering: Information technology and business process redesign. *Sloan Management Review*, *31*(4), 11–27.

Earl, M. (2003). IT: An ambiguous technology? In B. Sundgren, P. Mårtensson, M. Mähring, & K. Nilsson (Eds.), *Patterns in information management* (pp. 39–48). Stockholm: EFI.

Fiol, M., & O'Connor, E.J. (2002). When hot and cold collide in radical change processes: Lessons from community development. *Organization Science*, *13*(5), 532–546.

Foreman, P., & Whetten, D.A. (2002). Members' identification with multiple-identity organizations. *Organization Science*, *13*(6), 618–635.

Galgano, A. (2002). Quality: Mind and heart in the organization. *Total Quality Management*, *13*(8), 1107–1113.

Gäre, K. (2003). *Tre perspektiv på förväntningar och förändringar i samband med införande av informationssystem* [Three perspectives on expectations and changes when implementing information systems]. Linköping studies in science and technology, Dissertation No. 808, Linköping, Sweden: Linköping University.

Hammer, M. (1990). Reengineering work: Don't automate, obliterate. *Harvard Business Review*, *68*(4), 104–112.

Hammer, M., & Champy, J. (1993). *Reengineering the corporation: A manifesto for business revolution*. New York: HarperCollins.

Hanseth, O., & Braa, K. (1998, December 13–16). Technology as traitor. SAP infrastructures in global organizations. *Proceedings from the 19th Annual International Conference on Information Systems (ICIS)*, Helsinki, Finland.

Hedberg, B., Dahlgren, G., Hansson, J., & Olve, N.-G. (1994). *Imaginära organisationer* [Imaginary organizations]. Lund, Sweden: Liber-Hermods.

Hedberg, B., & Jönsson, S. (1978). Designing semi-confusing information systems for organizations in changing environments. *Accounting, Organizations and Society*, *3*(1), 47–64.

Hopper, M.D. (1990). Rattling SABRE—New ways to compete on information. *Harvard Business Review*, *68*(3), 118–125.

Howcroft, D., & Wilson, M. (2003). Paradoxes of participatory practices: The Janus role of the systems developer. *Information and Organization*, *13*(1), 1–24.

Huy, Q.N. (1999). Emotional capability, emotional intelligence, and radical change. *Academy of Management Review*, *24*(2), 325–345.

Leavitt, H.J. (1965). Applied organizational change in industry: Structural, technological and humanistic approaches. In J.G. March (Ed.), *Handbook of organizations* (pp. 1144–1170). Chicago: Rand McNally.

Lundeberg, M. (1993). *Handling change processes—A systems approach.* Lund, Sweden: Studentlitteratur.

Marcus, L., & Pfeffer, J. (1983). Power and the design and implementation of accounting and control systems. *Accounting, Organizations and Society*, *8*(2/3), 205–218.

Mintzberg, H. (1989). *Mintzberg on management.* New York: Free Press.

Orlikowski, W.J., & Hofman, J.D. (1997). An improvisational model of change management: The case of groupware technologies. *Sloan Management Review*, *38*(2), 11–21.

Patton, M.Q. (1990). *Qualitative evaluation and research methods.* New York: Sage.

Petri, C.-J. (2001). *Organizational information provision: Managing mandatory and discretionary utilization of information technology.* Linköping Studies in Science and Technology, Dissertation No. 720, Linköping, Sweden: Linköping University.

Pratt, M.G., & Foreman, P.O. (2000). Classifying managerial responses to multiple organizational identities. *Academy of Management Review*, *25*(1), 18–42.

Robey, D., Schwaig, K.S., & Jin, L. (2003). Intertwining material and virtual work. *Information and Organization*, *13*(2), 111–129.

Schein, E. (1993). How can organizations learn faster? The challenge of entering the green room. *Sloan Management Review*, *34*(2), 85–92.

Thompson, M.P.A. (2002). Cultivating meaning: Interpretive fine-tuning of a South African health information system. *Information and Organization*, *12*(3), 183–211.

Tsoukas, H., & Chia, R. (2002). On organizational becoming: Rethinking organizational change. *Organization Science*, *13*(5), 567–582.

Westelius, A. (1996). *A study of patterns of communication in management accounting and control projects.* Stockholm: EFI.

Westelius, A., & Askenäs, L. (2004). Getting to know your job—A social actor perspective on using information technology. *Information Technology Journal*, *3*(3), 227–239.

Chapter XI

Wireless Sales Force Automation in New Zealand

Brett Walker,
Centre for Interuniversity Research & Analysis on Organizations, Canada

Stuart J. Barnes, University of East Anglia, UK

Eusebio Scornavacca, Victoria University of Wellington, New Zealand

Abstract

Mobile and wireless technologies are deeply affecting the way many organizations do business. Among the several types of wireless applications, business-to-employee (B2E) applications have a strong potential to generate considerable value for organizations. Wireless sales force automation has been one of the most common applications found among companies that have adopted wireless B2E solutions. This chapter examines the impacts of wireless sales force automation on three organisations operating in food-related industries in New Zealand. The findings demonstrate that wireless technologies can enhance the benefits of traditional sales force automation, but that fundamental transformation of processes and value proposition is not yet apparent. The chapter concludes with recommendations for future practice and research.

Introduction

A trend that has been changing the landscape of information technology in recent times is the convergence of the Internet and wireless technologies. The developments of the Internet and mobile phones have followed two separate paths. Only in the past five years these technologies have converged, making possible a vast range of wireless data communication technologies such as the wireless Internet (Barnes, 2003). Much of the literature on mobile business has focused on consumer applications. However, according to studies published by the Boston Consulting Group (Manget, 2002) and AT Kearney (2003), the international market for business-to-employee (B2E) is expected to grow twice as rapidly as the market for wireless business-to-consumer (B2C) applications.

One wireless B2E application with the potential to generate considerable value for organisations is sales force automation (SFA). Donaldson and Wright (2002) point out that several authors have also noted this lack of a clear and convergent definition of SFA. Although SFA has been available for over two decades, the number of empirical studies about them is limited (Rivers & Dart, 1999). According to Morgan and Inks (2001, p. 463) these technologies involve "the use of computer hardware, software, and telecommunications devices by salespeople in their selling and/or administration activities." The introduction of information technology to the sales process has created significant benefits for salespeople and the organisations that employ them. The introduction of wireless technologies has the potential to enhance these benefits.

Organisations that have adopted relatively simple wireless applications for sales force automation have realised significant productivity gains. Developments in wireless technologies offer the potential to enhance the ability of salespeople to work more effectively away from the physical premises of their organisation. Increased mobility can increase the productivity and effectiveness of the sales force, as salespeople can remain in the field for longer periods of time. Other benefits include the ability for salespeople to remotely access back-office systems, the ability for organisations to deliver accurate information to employees in the field, improve communications with salespeople, reduce error rates, and improve order turnaround times. Wireless SFA can potentially impact an organisation's value chain and generate competitive advantage to organisations.

A better understanding of the strategic impacts of wireless SFA will aid organisations in realistically assessing expectations, investment decisions and implementation planning. There is a growing amount of literature on the actual and potential impact of mobile technologies on organisations. However, there is a gap in the literature examining the role of wireless technologies in sales force automation. The purpose of this study is to investigate the development and impact of wireless SFA solutions on organisations. In particular, we focus on three case studies within the food industry in New Zealand.

The remainder of the chapter will be structured as follows. The following section will present a review of relevant literature covering theory on traditional sales activities, sales force automation, and mobile business. This will be followed by an explanation of the research methodology that guided the current study. The results of the research are then provided for the cases, along with a cross-case analysis. The chapter will conclude with a discussion of the key research findings, limitations, and suggestions for further research and practice.

Background on Wireless Sales Force Automation

This section introduces the background literature on two key areas related to this research: sales activities and sales force automation in organisations, and mobile business and its application to sales force automation. Let us consider each of these in turn.

Sales Activities and Sales Force Automation

Given the aim of the research study is to determine the impact of wireless SFA, it is important to understand the processes and activities involved in the traditional sales function of an organisation. Moncrief (1986) compiled a taxonomy of sales activities by examining the literature and interviewing salespeople in 51 organisations. The study identified 121 activities that were grouped into 10 activity groups. The activity groups were selling function, conferences and meetings, working with orders, training and recruiting, servicing the product, entertaining, information management, travel, servicing the account, and distribution. Since Moncrief's study, sales activities and the selling environment have continued to evolve. These changes are due to the changing focus of sales organisations, greater customer sophistication, new selling methods, the changing demographics of salespeople, and the availability and use of new technologies (Marshall, Moncrief, & Lassk, 1999). Indeed, Marshall et al. (1999) discovered 49 new sales activities that had not been identified by Moncrief (1986). The 49 new sales activities are grouped into five categories: communication, sales, relationship, team, and database.

A common theme that emerged from the research of Marshall et al. (1999) was that developments in communications-related technology are a primary driver of changes in sales activities, and many of the new activities were being created because of new technologies, particularly sales force automation: the use of information and communications technology and devices by salespeople in their selling and/or administration activities (Morgan & Inks, 2001). This allows salespeople to do such things as access customer databases, track customer visits, record results of sales visits, send and receive sales data, and complete sales analyses (Engle & Barnes, 2000). Automating the sales force has the potential to produce several benefits for organisations, with improved effectiveness of the sales effort as the primary goal (Engle & Barnes, 2000). Benefits include the following:

- *Improved accuracy of information.* SFA solutions have the potential to improve the accuracy of information being gathered and communicated (Erffmeyer & Johnson, 2001). This is due to information being gathered directly from the customer, resulting in less rekeying of information.

- *Improved communication of information.* Salespeople are often the primary source of information exchange within a customer–seller relationship (Speier & Venkatesh, 2002). SFA can facilitate more rapid communication of information between salespeople and organisational systems, resulting in increased efficiency and faster turnaround in the sales process (Engle & Barnes, 2000; Erffmeyer & Johnson, 2001). The information gathered by salespeople can be utilised by other functions in the value chain without delay (Pullig, Maxham, & Hair, 2002).
- *More effective management of customers.* SFA solutions often allow salespeople to more effectively manage customers and potential contacts (Morgan & Inks, 2001). This is achieved through improved customer targeting, time management, and call planning (Engle & Barnes, 2000). SFA also offers the opportunity to increase customer contact and productive selling time.
- *Improved quality of communication with customers.* SFA can improve the quality of communication with customers during contacts as a result of communicating data in real time.
- *Increased closure rates of sales.* A result of enhanced interactions with customers is the potential to increase closure rates of sales and customer retention (Speier & Venkatesh, 2002). This comes partly as a result of the above factors.

Various ICT's are currently being applied to automate the sales process. Effymeyer and Johnson (2001) ascertained the extent to which various technologies are being used for SFA. The most popular investments were in fax machines (93%) and mobile phones (80%), followed by laptops and Internet usage (68% each). More recently, there is a growing proliferation of SFA solutions based on wireless Internet technologies, driving a new wave of mobile business.

Mobile Business and Wireless Sales Force Automation

Mobile (m-) business can be understood as "the use of the wireless Internet and other mobile information technologies for organisational communication and coordination, and the management of the firm" (Barnes, 2002). Siau and Shen (2003) identify three key drivers for mobile services: mobility, reachability, and personalisation. Mobility, what they consider to be the primary driver of mobile services, is the ability for users to access what real-time information they want, regardless of their location through Internet-enabled mobile devices. These devices include digital phones, wirelessly enabled personal digital assistants (PDAs), tablet PCs, and laptops computers. Reachability is the ability of a user to be in contact with and accessible to other people anywhere and at any time. Personalisation involves the filtering of relevant information and tailoring of services to the needs of particular users.

The implementation of mobile solutions within the business can lead to numerous benefits. Mobile solutions provide functionality that enables mobile workers to access information at the point of need (Siau & Shen, 2003). Effective mobile solutions enable the right information to be delivered to the right person at the right time (Yuan & Zhang, 2003), support work configurations and automate work processes, support the transient employee, and integrate remote, disparate or roaming employees into the corporate infrastructure (Barnes, 2003). Mobile services also present opportunities to better serve their customers and supply them with value-added services (Siau & Shen, 2003; Yuan & Zhang, 2003).

Several challenges must be overcome before the benefits of mobile business can be realised. These include integration of the solution with business processes and back-end systems (Johnson & Deighton, 2003), investment risk (Siau & Shen, 2003), the presence of appropriate mobile infrastructures within firms (IDC, 2000; Johnson & Deighton, 2003), the physical limitations of wireless devices (Siau & Shen, 2003; Johnson & Deighton, 2003), employees accessing solutions with incompatible devices (Mobilocity, 2001), and the limited performance and cost of using of wireless networks (Newman, 2003; Siau & Shen, 2003; Synchrologic, 2003).

Many of the benefits of SFA, mentioned above, are accentuated by wireless SFA. A primary benefit of mobile communication is that information can be delivered to salespeople in real time (Wireless Ready Alliance, 2003). Applications that require tight back-office application integration and real-time access to data are particularly suitable for delivery over wireless networks. These may include functions such as inventory availability information, price and delivery scheduling, credit information, messaging between salespeople and the organisation, appointment scheduling, and customer contact information (Close & Eisenfield, 2002). Mobile solutions can be used to facilitate such things as instant notification of sales leads, and downloading sales reports based on customer or geographical area. Real-time access to accurate information may be critical to building relationships with existing and prospective customers, answering queries, or closing a sale (Close & Eisenfield, 2002; Kalakota & Robinson, 2002).

Research Methodology

The purpose of this study is to investigate the strategic impact of wireless sales force automation solutions on organisations within the food industry. To address this objective the study attempts to understand how this technology is affecting core business processes, value propositions, and the mobility of salespeople in the organisations examined. These are important aspects in the development and impact of mobile B2E solutions.

This study is based on an exploratory multiple-case study methodology to discover how the introduction of wireless sales force automation technologies, a specific programme, impacts the organisation. Case research is appropriate for researching an area where theories are at formative stages, for emerging technologies, and where little research has been completed (Benbasat, Goldstein, & Mead, 1987). Case research is also appropriate for examining practice-based problems, since it allows a researcher to capture the knowledge of practitioners and use it to generate theory (Benbasat, Goldstein, & Mead, 1987).

Case organisations were identified opportunistically, due to the limited number of organisations that have implemented wireless sales force automation solutions. Where possible, cases with different technology platforms were selected. Participating organisations had to be (a) using a fully operational wireless SFA solution that was capable of wirelessly receiving and transmitting data to employees in the field, and (b) operating within the food industry in New Zealand. The following cases were studied (names changed for confidentiality):

- *Alpha*. Alpha is a snack food manufacturer. As of 2002 Alpha had sales of NZ$217 million (US$145 million) and employed approximately 800 people. Alpha's wireless SFA solution operates on a tablet PC. Salespeople have the ability to receive and transmit information wirelessly using wireless data cards. Approximately 45 salespeople are using Alpha's wireless SFA solution.
- *Beta*. Beta is the wholly-owned and independent subsidiary of a grocery distribution cooperative. It is the largest single source supply food service and route trade grocery wholesaler in the Lower North Island of New Zealand. Beta is a business-to-business operation. As of 2003, the grocery distribution cooperative that owns Beta had revenues of NZ$1.56 billion (US$1 billion). Beta's wireless SFA solution operates on laptop computers. The software is a replication of the company's internal order capture system. Salespeople have the ability to receive and transmit information wirelessly via a hardwire connection to a mobile phone. Approximately 15 salespeople are using Beta's wireless SFA solution.
- *Gamma*. Gamma is a fast-moving consumer goods importer and distributor with a focus on confectionary products. It is structured around three business units: grocery, confectionary, and wholesale confectionary. As of 2002 Gamma had sales of NZ$60 million (US$40 million) and employed over 160 people. Gamma's wireless SFA solution operates on iPaq PDAs. The software for the system is an in-house–developed Visual Basic system operating on Windows CE 2002. Salespeople have the ability to receive and transmit information wirelessly via a hardwire connection to a mobile phone. Approximately 35 salespeople are using Gamma's wireless SFA solution.

Semistructured face-to-face interviews were the primary method of data collection. Semistructured interviews are appropriate for the exploratory nature of the current research. Interviews were conducted for the three cases in July and August 2003. Interview participants had the roles of information technology managers and sales managers in their respective organisations.

Data analysis commenced with the coding of interview transcripts (Miles & Huberman, 1994). Initial codes were generated by identifying key themes that were present in the literature review. Pattern coding was then used to assimilate the initial list of codes into a smaller number of key themes. The pattern codes were applied to each of the three individual cases, as well as to locate themes that cut across cases (Miles & Huberman, 1998).

Results

In this section, we will examine the development and impacts of wireless SFA solutions on the three organisations studied via a cross-case analysis. To structure the analysis of the cases, the impacts of the mobile technologies on each organisation will be discussed according to the three constructs of the Mobile Enterprise Model (MEM) (Barnes, 2003): mobility, process, and market (shown in Table 1 and Figure 1). This model provides a basic framework for examining the impacts of mobile enterprise solutions.

Mobility

All of the organisations examined have achieved and surpassed the *transient* level of mobility in the Mobile Enterprise Model (MEM), as the wireless SFA solutions provide more than basic support to employees as they move from one location to another. All of the solutions provide functionality that allows salespeople to remain in the field for prolonged periods of time.

The solutions provide sufficient geographic independence to have achieved the *mobile* level of the Mobility axis on the MEM. The extent to which the organisations have reached the mobile level varies. The extent of prolonged geographic independence was dependent on how often salespeople were required to return to the company's physical office. The main reason for returning to company headquarters appeared to be the attendance of regular sales team meetings. Therefore, the regularity of these meetings significantly influences the level of salesperson mobility. Alpha requires salespeople to keep one day of the week free of scheduled customer visits to complete any unfinished administrative work. The regularity of their visits to the company office is higher, which indicates that their salespeople have a lower level of mobility than the other organisations.

Figure 1. Mobile Enterprise Model (MEM) (Barnes, 2003)

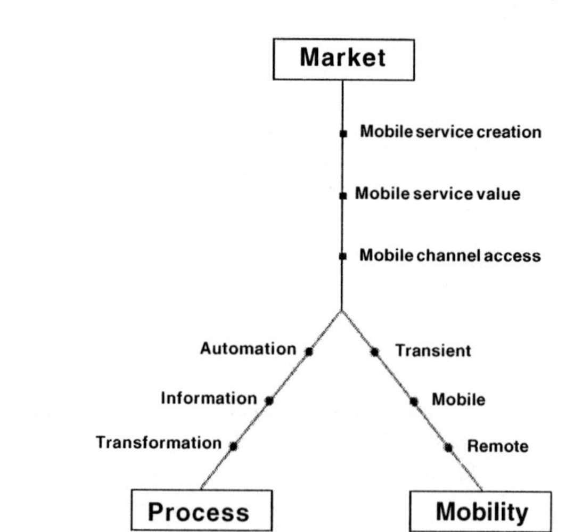

Table 1. Description of the Mobile Enterprise Model (Barnes, 2003)

Mobility	**The level of geographic independence of an organisation's employees, enabled by the wireless solution**
Transient	The wireless solution provides basic support to employees as they move from one location to another
Mobile	The wireless solution allows employees to have geographic independence for prolonged periods of time
Remote	The wireless solution allows employees to be almost completely removed from the company's corporate location
Process	**The change in work configuration and processes, enabled by the wireless solution**
Automation	The wireless solution results in efficiency gains in existing processes
Information	The wireless solution introduces a degree of effectiveness and knowledge work gains
Transformation	The wireless solution causes a fundamental degree of change in organisational processes
Market	**The value proposition in the marketplace**
Mobile Channel Access	The wireless solution is used primarily as a conduit of information for mobile employees
Mobile Service Value	The wireless solution is being used to add significant value to the market offering
Mobile Service Creation	The wireless solution is used to create entirely new service offerings or products

Meetings of the sales team at Gamma were less frequent than at the other two organisations. None of the wireless SFA solutions provided salespeople with enough geographic independence to be completely *remote* on the MEM.

The effectiveness of communication between headquarters and salespeople influenced both the reachability and mobility of salespeople. Incorporating voice and data communication functionalities into the solutions allows salespeople to remain geographically independent of their organisation for longer periods of time. The ability to spend more time in the field increases sales force productivity.

Alpha and Beta's wireless SFA solutions incorporated e-mail functionality, although neither gave salespeople the ability to receive or transmit attachments due to limited availability of bandwidth. Alpha and Gamma are utilising SMS as a means of communicating with salespeople while they are in the field. The ability for organisations to simultaneously send a consistent message to their whole sales team using e-mail and SMS was considered to be valuable.

Mobile phones were an important tool for communication between salespeople and company headquarters. Salespeople believed they could get a more immediate response by using their mobile phone. Mobile phones have greater reachability than the wireless SFA solutions at the present time.

It was noted that if mobility is influenced by the frequency of sales meetings, the mobility of salespeople could be improved by utilising wireless SFA solutions to distribute information that would ordinarily be communicated at sales meetings. This would also affect the productivity of salespeople. Salespeople generally did not use the wireless SFA solution to communicate with each other largely due to the fact that they are working in different territories.

Process

All of the wireless SFA solutions examined have resulted in efficiency gains in existing processes as a result of automation. This has occurred in a number of different areas:

- *Time to take orders.* The wireless SFA solutions examined produced savings in the time taken for salespeople to record orders. Gamma accentuated these time savings by incorporating bar code scanners in their devices. This removes the need for their salespeople to locate relevant product information before entering an order. Time savings in the recording of orders has resulted in improvements in employee productivity. The additional time has been used to visit more customers, or to carry out other activities within the store. The extent to which the organisations use the additional time to carry out activities in the store differed.
- *Time to complete orders.* The organisations examined have realised efficiency gains once the order has been transmitted to the organisation's back-office systems. More rapid communication of information between salespeople and organisational systems has increased efficiency resulting in faster turnaround of orders. The fulfilment of orders can proceed more rapidly, as the organisation's back-office systems receive and process order information immediately. All of the organisations regularly turn around orders within 24 hours of the transmission of an order.
- *Order accuracy.* The accuracy of information that is gathered and transmitted by salespeople has improved. Error rates have been reduced as a result of wireless transmission of information, eliminating the need to rekey order information. Information is entered by salespeople in the field and transmitted over wireless networks into back-office systems. This has removed opportunities for human error during the rekeying of information. Improved information accuracy and reduced error rates were viewed as an important benefit of wireless SFA by all of the organisations.
- *Other value chain impacts.* Improvements in order accuracy and the faster turnaround of orders have produced positive impacts for the service activity of the value chain. The wireless SFA solutions have had positive impacts on the inbound logistics activities in the value chains of the organisations' customers. Improvements in the effectiveness of sales activities are resulting in improvements in the effectiveness of outbound logistics through quicker turnaround of orders. There-

fore, the solutions are producing impacts in the organisations' wider value systems. The wireless SFA solutions can result in improved customer service and enhanced customer relationships.

Overall, wireless SFA technology has enabled improvements in the efficiency and effectiveness of sales activities. However, there is no evidence that the solutions have fundamentally transformed work configurations or processes at any of the organisations examined. Salespeople are still performing many of the traditional sales activities. More broadly, in terms of Marshall et al.'s (1999) expanded taxonomy of sales activities, the impacts of the wireless SFA solutions examined were most relevant to the Communications and Databases areas. Mobile devices are communications devices that have further enhanced the benefits included within the Communications area. Wireless SFA solutions have improved the reachability of salespeople, made it easy for salespeople to carry out a variety of tasks while away from the office, and diminished the downtime of employees. Impacts relevant to the Databases area include efficient access to information in company databases, and better ability to gather information and transfer it to company databases.

Market

All of the factors contributing to the value proposition of the solutions involve the delivery of information to employees in the field using the wireless SFA solution. All of the organisations examined have achieved the *mobile channel access* level of mobility in the MEM. The organisations have achieved important benefits of both mobile business and sales force automation. The solutions are providing access to real-time information at the point of need, and remote communication with back-office systems. The wireless SFA solutions enabled salespeople to efficiently access up-to-date information on customers, products, stock levels, pricing, and promotions.

Every one of the organisations has derived additional value from the ability to transfer information using the mobile medium. This indicates that they have attained the mobile service value level of the MEM's Market axis. Improvements in service value has been driven by, for example,

- *Order summary information.* Alpha's solution provides salespeople with order summaries. Order summaries can be used to verify an order with a customer prior to transmission. This helps to ensure that order information is accurate, and that customer satisfaction is maintained due to lower error rates. Gamma stated that it would like to incorporate this functionality into its solution in the near future.
- *Stock information.* The organisations are facilitating access to information related to the manufacturing activity of the value chain, in the form of information regarding the availability of stock. Both Beta and Gamma emphasised that the value of having access to this information in the field was the ability to offer customer's

alternative products immediately. Having access to this information in the field reduces the likelihood of inconveniencing customers. Therefore, the provision of out-of-stock information has a positive impact on customer service.

- *Pricing information.* All of the companies have utilised the mobile medium to ensure that salespeople have access to accurate pricing information. Beta and Gamma's solutions enable pricing information to be tailored to individual customers. Their solutions provide information on the trading terms and discount structures relevant to individual customers. This is an example of personalisation of mobile services. Beta emphasised the importance of timely and accurate pricing information to their business. It considered the ability to pass on pricing changes across the wider value system as quickly as possible to be a source of competitive advantage.
- *Promotional information.* The mobile medium is being used to make information regarding promotional opportunities available to salespeople. The solutions notify salespeople of a product promotion when a salesperson is entering an order. The wireless SFA solutions are helping to ensure that salespeople do not miss opportunities to up-sell. This improves the effectiveness of the sales and marketing activity.
- *Multimedia product information.* Alpha's wireless SFA solution has incorporated functionality for showing television commercials to its customers, loaded onto the devices using data cards. Alpha considers this to be an important tool when launching new products, as salespeople can demonstrate the media support the company will be giving these products. Gamma would like to see this functionality incorporated into its solution in the near future as they believe it could influence the likelihood of making a sale. The ability to demonstrate the media support of products in the field improves the effectiveness of the sales and marketing activity.
- *System updates.* Beta and Gamma are utilising the mobile medium to distribute files that update their wireless SFA solutions. This means problems with systems can be addressed quickly which minimises employee downtime, salespeople have a consistent version of the system, and salespeople are not required to visit the company's offices to update their systems. The delivery of system updates via the mobile medium minimises the amount of time salespeople are out of the field due to technological problems. This should in turn increase the productivity of the sales force.

None of the organisations examined were utilising the mobile medium to create entirely new products or services. Thus, they have not yet moved to mobile service creation in the MEM.

Discussion

The purpose of the research was to investigate the impact of wireless SFA solutions on organisations in the New Zealand food industry. The research shows that wireless SFA solutions have impacted several areas of the organisations studied. It has also been demonstrated that wireless SFA has augmented the benefits of traditional SFA solutions. Table 2 lists the benefits of SFA and mobile business that were outlined in the literature

Table 2. Comparison of results with factors contained in the literature

Sales Force Automation	**Source**	**Impact**
Improve the collection, processing, and distribution of information	Morgan & Inks (2001)	●
Improve productivity of the sales force	Morgan & Inks (2001)	●
Increased customer contact and productive selling time	Morgan & Inks (2001)	●
Enhance customer relationships	Morgan & Inks (2001)	◐
Ability to remotely communicate with a centralised system that is constantly being updated with current information that is of interest to the salesperson	Parthasarathy & Sohi (1997)	●
Improve the accuracy of information	Engle & Barnes (2000)	●
More effective management of customers	Engle & Barnes (2000)	◐
Improve quality of communication with customers	Engle & Barnes (2000)	◐
More rapid communication of information between salespeople and organisational systems	Engle & Barnes (2000); Erffmeyer & Johnson (2001)	●
Increased efficiency and faster turnaround in the sales process	Engle & Barnes (2000); Erffmeyer & Johnson (2001)	●
Improve the accuracy of information being gathered and communicated due to information being gathered directly from the customer, resulting in less rekeying of information	Erffmeyer & Johnson (2001)	●
Information gathered by salespeople can be utilised by other functions in the value chain without delay	Pullig et al. (2002)	●
Mobile Business	**Source**	**Impact**
Mobility – the ability for users to access what real-time information they want at the point of need, regardless of their location	Siau & Shen (2003)	●
Reachability – the ability of a user to be in contact with and accessible to other people anywhere and at any time	Siau & Shen (2003)	◐
Personalisation – the filtering of relevant information and tailoring of services to the needs of particular users	Siau & Shen (2003)	◐
Support the transient employee	Barnes (2003)	●
Integrate remote, disparate, or roaming employees into the corporate infrastructure	Barnes (2003)	●
The ability to provide better service to customers	Siau & Shen (2003); Yuan & Zhang (2003)	●

Key:

Impact occurred in some of the case study organisations

Impact occurred in all of the case study organisations

Figure 2. Map of causal benefits for wireless SFA

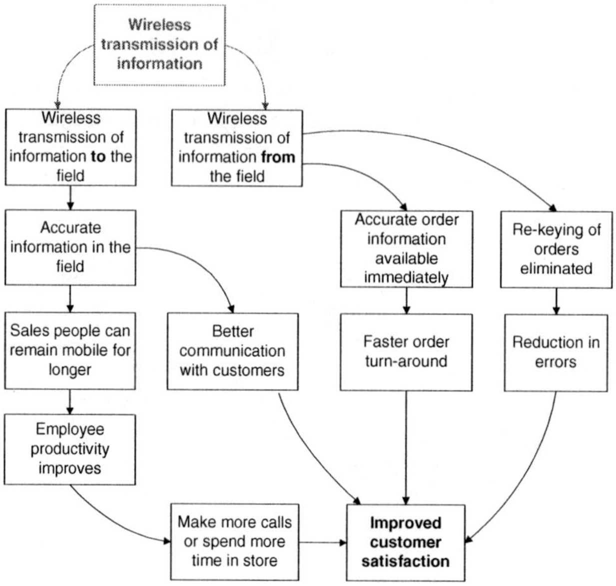

Wireless transmission of information
Wireless transmission of information **to** the field
Wireless transmission of information **from** the field
Accurate information in the field
Accurate order information available immediately
Re-keying of orders eliminated
Sales people can remain mobile for longer
Better communication with customers
Faster order turn-around
Reduction in errors
Employee productivity improves
Make more calls or spend more time in store
Improved customer satisfaction

review. The table illustrates that many of the benefits of SFA and mobile business are analogous. The case study organisations have attained the majority of benefits of SFA and mobile business contained in the literature.

Figure 2 shows the relationships between the various impacts and value drivers of wireless SFA. The impacts contained in the causal map, and the relationships between them, were identified from analysing the transcripts of interviewees. The majority of factors in the causal map are also contained in Table 2. Highlighting the relationships between the factors illustrates the influence of mobile business on enhancing traditional SFA. For example, the wireless transmission of orders from the field and immediate processing of order information results in faster turnaround in the sales process.

It is evident that the wireless transmission of information drives many of the impacts of wireless SFA solutions. The ability of the mobile medium to efficiently collect and distribute information is extremely important to deriving value. The causal map demonstrates that wireless collection and distribution of information ultimately leads to key benefits, including more effective communication with salespeople, time savings in the store and in the processing of orders, increased mobility and productivity of employees, and faster turnaround of orders.

All of the organisations examined are using their solutions as a conduit to deliver accurate information to salespeople at the point of need. The wireless SFA solutions have integrated remote employees into the organisation's infrastructure, enabling them to remotely access back-office systems. Communication functionality such as e-mail and SMS increase the mobility of salespeople, allowing them to remain in the field for longer periods of time. This, in turn, helps to improve the productivity of the sales force.

Overall, the use of wireless SFA does not appear to have fundamentally changed sales activities. None of the organisations experienced a fundamental transformation of processes. Several of the managers interviewed indicated that salespeople are performing the same activities that they were prior to the introduction of wireless SFA solutions. Thus, salespeople are still carrying out the many of the activities listed in Moncrief's (1986) taxonomy of sales activities. The primary difference is that the utilisation of wireless SFA technologies has improved the efficiency and effectiveness of such processes. As a knock-on effect, the implementation of wireless SFA has enabled a change of focus from activities related to recording order information, to value-added business development activities. One sales manager commented that "the role of my Territory Managers goes from being an order taker to an order maker."

In terms of specific efficiency gains, it is clear that the wireless SFA solutions have advantages in the recording of orders, and in the total time taken to fulfil an order. This is due to immediate order entry using devices while in the store, the wireless transmission of orders into back-office systems for more rapid fulfilment, and the elimination of order information rekeying. Error rates have also fallen as a result. Time savings within the store have resulted in significant improvements in salesperson productivity. For example, salespeople do not spend as long recording orders and can therefore visit more customers, or carry out other value-added activities in the store, such as arranging products or further building relationships with customers.

The gains in efficiency and the enhancements to value propositions experienced by the organisations are very dependent on the ability of the mobile medium to facilitate remote access to back office systems, thereby facilitating access to corporate systems regardless of location. Key enhancements to value propositions include increased information accuracy and a subsequent reduction in error rates, improved sales force productivity, and faster turnaround of orders. Value propositions have extended beyond sales and marketing activities to other areas of the traditional value chain (Porter & Millar, 1985), namely, outbound logistics and service.

Infrastructure also plays an important role in providing the platform for wireless SFA solutions. The effectiveness of the mobile medium as a conduit of information is tempered by the wireless bandwidth available to organisations. This is important as the ability to efficiently transmit information has a significant impact on the value that can be derived from a wireless SFA solution. Transmission of information over the mobile medium is also fundamental in generating the benefits that distinguish wireless SFA from traditional SFA.

Looking ahead, it is interesting to note the case organisations' planned improvements to the functionality of their wireless SFA solutions. Proposed improvements include wider use of the devices for showing television commercials, sending e-mail attachments, and shelf planning capabilities. These improvements in functionality are dependent on the expansion of wireless bandwidth (Kumar, 2004). The introduction of such new functionality will doubtless generate additional impacts above and beyond those mentioned above. Subsequently, further research should be conducted to determine the impact of more advanced wireless SFA solutions.

Conclusions

The convergence of the Internet and wireless technologies has led to the enhancement of traditional SFA solutions. This chapter has examined the impact of wireless sales force technologies on three case study organisations in the New Zealand food industry. The MEM was used to structure an analysis of the impact of wireless SFA technologies in the case study organisations, supplemented by other relevant literature to help explain the impacts experienced by the organisations. As a result of applying wireless technologies to their sales function, the three cases studied all experienced impacts that have led to improvements to sales force and overall organisational performance.

The research determined that the case studies had realised the majority of benefits and impacts outlined in the literature, but it also demonstrated how the impacts and benefits are related to each other. The identification of these relationships highlights which of the factors are particularly important; for example, several of the positive impacts are derived from mobile channel access, or the mobile medium's ability to act as an efficient conduit of information. Additional insight was achieved by examining enhancements to traditional SFA; for example, the solutions are providing better remote access to back-office systems, more efficient provision of up-to-date information, and improved ability to communicate with salespeople.

Interestingly, the development of wireless SFA solutions in the case studies has been limited to the improvement of existing processes. The assessment of the wireless SFA solutions using the MEM has shown that they could all be developed further to provide more value to their organisations. None of the companies reached the outmost levels on each of the three constructs of the MEM. One of the sales managers interviewed commented that the value of their wireless SFA solutions was limited if it was used solely as an order gathering device; to provide maximal benefits from this new medium organisations need to identify how their solutions can transform processes and create entirely new products and services for their customers.

Overall, this research has aimed to advance the understanding of wireless SFA, which is one of a number of B2E wireless application areas. This research has made a small contribution towards rectifying the gap that exists in the academic literature regarding the impact of wireless SFA, and demonstrated that wireless technologies have the ability to enhance the benefits of traditional SFA. This research provides further justification that mobile business is a worthwhile focus for academic research. The wireless SFA solutions being operated in the organisations examined have not achieved their full potential. Future research will need to focus on assessing the impacts of future developments in wireless SFA technology.

Wireless technologies are developing rapidly, and the future impacts of wireless SFA solutions will mirror the dynamism of wireless technology. For one, the development of wireless SFA solutions is quite dependent on the performance of mobile networks and available bandwidth. While there is sufficient bandwidth to facilitate the basic but fundamental functionalities of wireless SFA, many of the proposed improvements to current wireless SFA solutions are dependent on increased bandwidth. As a result of further infrastructure and device improvements, the fundamental benefits that are being achieved using current wireless technology will be surpassed by even more significant benefits in the future (Panagiotakis, Koutsopoulou, Alonistioti, Houssos, Gazis, & Merakos, 2003).

The research was based on three case studies of organisations within the food industry in New Zealand. The impacts of wireless SFA solutions described in this research, while generalisable to this context, must be closely scrutinised in their application to other contexts. Further qualitative research will be conducted to investigate the impact of wireless SFA solutions in other industries. In addition, the research was focused at an overall strategic level; data collection concentrated on management perceptions of impacts caused by wireless SFA solutions. Future research could also investigate the effect of wireless SFA on the role of the individual salesperson.

References

AT Kearney (2003). The new mobile mindset. Retrieved October 14, 2003, from *www.atkearney.com/shared_res/pdf/Mobinet_Monograph_S.pdf*

Barnes, S.J. (2002). Unwired business. *E-Business Strategy Management*, *4*(1), 27–37.

Barnes, S.J. (2003). Enterprise mobility: Concept and examples. *International Journal of Mobile Communications*, *1*(4), 341–359.

Benbasat, I., Goldstein, D., & Mead, M. (1987). The case research strategy in studies of information systems. *MIS Quarterly*, *11*(3), 369–386.

Close, W., & Eisenfield, B. (2002). *Mobility matters for sales*. Stamford, CT: Gartner Research.

Donaldson, B., & Wright, G. (2002). Sales information systems: Are they being used for more than simple mail shots? *Journal of Database Marketing*, *9*(3), 276–284.

Engle, R.L., & Barnes, M.L. (2000). Sales force automation usage, effectiveness, and cost-benefit in Germany, England and the United States. *The Journal of Business and Industrial Marketing*, *15*(4), 216–241.

Erffmeyer, R.C., & Johnson, D.A. (2001). An exploratory study of sales force automation practices: Expectations and realities. *The Journal of Personal Selling and Sales Management*, *21*(2), 167–175.

IDC. (2000). Growing mobile workforce demands application access. Retrieved March 28, 2003, from *www.axcentsolutions.com/learningcenter/whitepapers/citrix/growmw.pdf*

Johnson, G., & Deighton, N. (2003). *Key issues for mobile business in 2003*. Stamford, CT: Gartner Research.

Kalakota, R., & Robinson, M. (2002). *M-business: The race to mobility*. New York: McGraw-Hill.

Kumar, S. (2004). Mobile communications: Global trends for the 21st century. *International Journal of Mobile Communications*, *2*(1), 67–86.

Manget, J. (2002). Competitive advantage from mobile applications. Retrieved October 14, 2003, from *www.bcg.com/publications/files/Competitive_Adv_Mobile_Apps_OfA_Feb02.pdf*

Marshall, G.W., Moncrief, W.C., & Lassk, F.G. (1999). The current state of sales force activities. *Industrial Marketing Management, 28*(1), 87–98.

Miles, M.B., & Huberman, M. (1994). *Qualitative data analysis: An expanded sourcebook.* London: Sage.

Miles, M.B., & Huberman, M. (1998). Data management and analysis methods. In N.K. Denzin & Y.S. Lincoln (Eds.), *Methods of collecting and analysing empirical materials* (pp. 428–444). London: Sage.

Mobilocity, Inc. (2001). Fundamentals of m-business: An m-business 101. Retrieved May 17, 2003, from *www.mobilocity.net/mi/white.php*

Morgan, A.J., & Inks, S.A. (2001). Technology and the sales force. *Industrial Marketing Management, 30*(5), 463–472.

Moncrief, W.C. (1986). Selling activity and sales position taxonomies for industrial salesforces. *Journal of Marketing Research, 23*(3), 261–270.

Newman, K. (2003, January 2). Wireless deployment—your biggest challenge in 2003: Wild ride for wireless. *Managing Information Strategies*, New Zealand edition, pp. 48–51.

Panagiotakis, S., Koutsopoulou, M., Alonistioti, A., Houssos, N., Gazis, V., & Merakos, L. (2003). An advanced service provision framework for reconfigurable mobile networks. *International Journal of Mobile Communications, 1*(4), 425–438.

Parthasarathy, M., & Sohi, R.S. (1997). Salesforce automation and the adoption of technological innovations by salespeople: Theory and implications. *Journal of Business and Industrial Marketing, 12*(3), 196–208.

Porter, M.E., & Millar, V.E. (1985). How information gives you competitive advantage. *Harvard Business Review*, 63(4), 149–160.

Pullig, C., Maxham, J.G., & Hair, J.F. (2002). Salesforce automation systems: An exploratory examination of organisational factors associated with effective implementation and salesforce productivity. *Journal Business Research, 55*(5), 401–415.

Siau, K., & Shen, Z. (2003). Mobile communications and mobile services. *International Journal of Mobile Communications, 1*(1/2), 3–14.

Speier, C., & Venkatesh, V. (2002). The hidden minefields in the adoption of sales force automation technologies. *Journal of Marketing, 66*(3), 98–111.

Synchrologic. (2003). The future of enterprise mobile computing—an updated review from 2002 on the trends driving enterprise mobile strategy. Retrieved March 30, 2003, from *www.synchrologic.com/demo/whitepapers.html*

Wireless Ready Alliance. (2003). Why wireless for sales force automation/customer relationship management? Retrieved May 17, 2003, from *www.wirelessready.org/automation.asp*

Yuan, Y., & Zhang, J.J. (2003). Towards an appropriate business model for m-commerce. *International Journal of Mobile Communications, 1*(1/2), 35–56.

Chapter XII

A Mobile Portal Solution for Knowledge Management

Stefan Berger, Universität Passau, Germany

Ulrich Remus, University of Erlangen-Nuremberg, Germany

Abstract

This chapter discusses the use of mobile applications in knowledge management (mobile KM). Today more and more people leave (or have to leave) their fixed working environment in order to conduct their work at changing locations or while they are on the move. At the same time, mobile work is getting more and more knowledge intensive. However, the issue of mobile work and KM is an aspect that has largely been overlooked so far. Based on requirements for mobile applications in KM an example for the implementation of a mobile KM portal at a German university is described. The presented solution offers various services for university staff (information access, colleague finder, campus navigator, collaboration support). The chapter is concluded by outlining an important future issue in mobile KM: the consideration of location-based information in mobile KM portals.

Introduction

Today many working environments and industries are considered as knowledge intensive, that is, consulting, software, pharmaceutical, financial services, and so forth. Knowledge management (KM) has been introduced to overcome some of the problems knowledge workers are faced by handling knowledge, that is, the problems of storing, organizing, and distributing large amounts of knowledge and its corresponding problem of information overload, and so forth. Hence, KM and its strategies aim at improving an organization's way of handling internal and external knowledge in order to improve organizational performance (Maier, 2004).

At the same time more and more people leave (or have to leave) their fixed working environment in order to conduct their work at changing locations or while they are on the move. Mobile business tries to address these issues by providing (mobile) information and communication technologies (ICT) to support mobile business processes. However, compared to desktop PCs, typical mobile ICT, like mobile devices such as PDAs and mobile phones, have some disadvantages (Hansmann, Merk, Niklous, & Stober, 2001):

- Limited memory and CPU – Mobile devices are usually not equipped with the amount of memory and computational power in the CPU found in desktop computers.
- Small displays and limited input capabilities – for example, entering a URL on a Web-enabled mobile phone is cumbersome and slower than typing with a keyboard.
- Low bandwidth – in comparison to wired networks, wireless networks have a lower bandwidth. This restricts the transfer of large data volumes.
- Connection stability – due to fading, lost radio coverage, or deficient capacity, wireless networks are often inaccessible for periods of time.

Taking into account the aforementioned situation one must question whether current IT support is already sufficient in order to meet the requirement of current knowledge-intensive mobile work environments. So far, most of the off-the-box knowledge management systems are intended for use on stationary desktop PCs and provide just simple access from mobile devices. As KMS are generally handling a huge amount of information (e.g., documents in various formats, multimedia content, etc.) the management of the restrictions described above become even more crucial. In addition, neither an adaptation of existing knowledge services of stationary KMS nor the development of new knowledge services according to the needs of mobile knowledge workers is taking place.

The goals of this chapter are to identify the main issues when mobile work is meeting knowledge management. In particular the focus lies on mobile knowledge portals, which are considered to be the main ICT to support mobile KM. Further on the applicability of these suggestions is shown with the help of a mobile knowledge portal that was implemented at a German university.

The chapter is structured as follows: Section two will detail the understanding about mobile KM and derive important requirements to be fulfilled. In section three mobile knowledge portals are then described as main ICT to support tasks in mobile KM. As an example the mobile KM portal of the University of Regensburg is presented (section four) whereas section five shows location orientation as the next step in mobile KM. Finally, section six concludes this chapter and gives an outlook on future research issues within the field of mobile KM.

Knowledge Management Meets Mobile Work

A mobile working environment differs in many ways from desk work and presents the business traveler with a unique set of difficulties (Perry, O'Hara, Sellen, Brown, & Harper, 2001). In the last years several studies have shown that mobile knowledge workers are confronted with problems that complicate the fulfillment of their job (Figure 1).

Mobile workers working separated from their colleagues often have no access to the resources they would have in their offices. Instead, business travelers, for example, have to rely on faxes and messenger services to receive materials from their offices (Schulte, 1999). In case of time-critical data, this way of communication with the home base is insufficient. Bellotti and Bly (1996) show in their survey about knowledge exchange in a design consulting team that it is difficult for a mobile team to generally stay in touch. This is described as "Lack of Awareness." It means that a common background of common knowledge and shared understanding of current and past activities is missing. This constrains the exchange of knowledge in teams with mobile workers. In addition, mobile workers have to deal with different work settings, noise levels, and they have to coordinate their traveling. This "Logistics of Motion" lowers their ability to deal with knowledge-intensive tasks (Sherry & Salvador, 2001) while on the move. The danger of an information overflow increases.

Figure 1. Problems related to mobile work

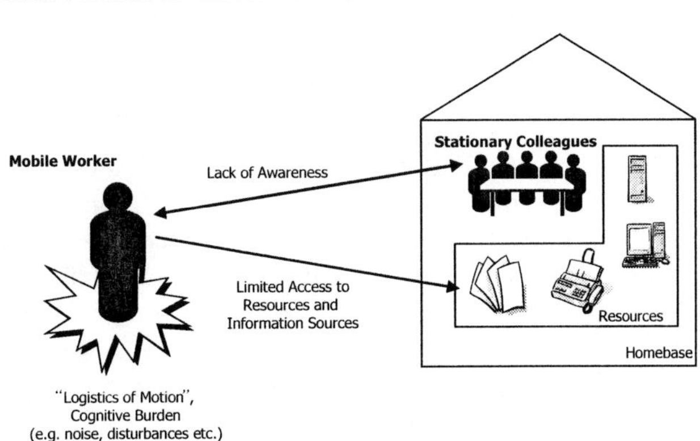

Mobile KM is an approach to overcome these problems. Rather than adding to the discussion of what actually is managed by KM—knowledge workers, knowledge, or just information embedded into context—in this chapter, mobile KM is seen as KM focusing on the usage of mobile ICT in order to:

- provide **mobile access** to KMS and other information resources;
- generate **awareness** between mobile and stationary workers by linking them to each other; and
- realize **mobile KM services** that support knowledge workers in dealing with their tasks (Berger, 2004, p. 64).

The next section reviews the state of the art of KMS and reviews if it meets these requirements.

Mobile KM Portals

Currently, many KMS are implemented as centralistic client/server solutions (Maier, 2004) using the portal metaphor. Such knowledge portals provide a single point of access to many different information and knowledge sources on the desktop together with a bundle of KM services. Typically, the architecture of knowledge portals can be described with the help of layers (Maier, 2004). The first layer includes data and knowledge sources of organizational internal and external sources. Examples are database systems, data warehouses, enterprise resource planning systems, and content and document management systems. The next layer provides intranet infrastructure and groupware services together with services to extract, transform, and load content from different sources. On the next layer, integration services are necessary to organize and structure knowledge elements according to a taxonomy or ontology.

The core of the KMS architecture consists of a set of knowledge services in order to support discovery, publication, collaboration, and learning. Personalization services are important to provide a more effective access to the large amounts of content, that is, to filter knowledge according to the knowledge needs in a specific situation and offer this content by a single point of entry (portal). In particular, personalization services together with mobile access services become crucial for the use of KMS in mobile environments.

Portals can be either developed individually or by using off-the-shelf portal packages, for example, Bea WebLogic, IBM Portal Server, Plumtree Corporate Portal, Hyperwave Information Portal, or SAP Enterprise Portal. These commercial packages can be flexibly customized in order to build up more domain-specific portals by integrating specific portal components (so called portlets) into a portal platform. Portlets are more or less standardized software components that provide access to a various amount of applications and (KM) services, for example, portlets to access ERP-systems, document management systems, personal information management.

Figure 2. Tasklist, Calendar, and Discussion Board of Open Text's Livelink Wireless (Open Text, 2003, p. 12)

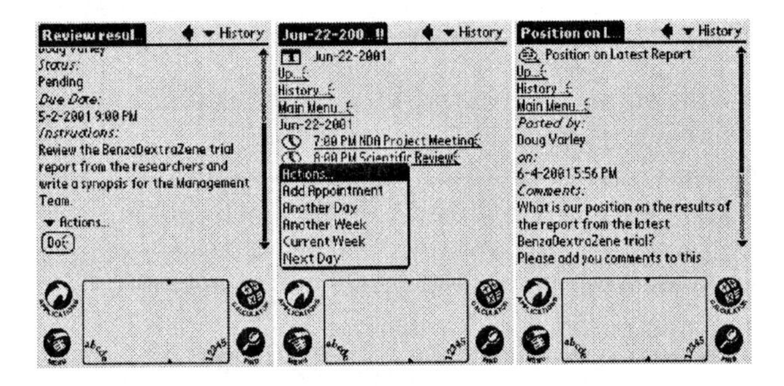

Figure 3. Automatic text summarization (Open Text, 2003, p. 11)

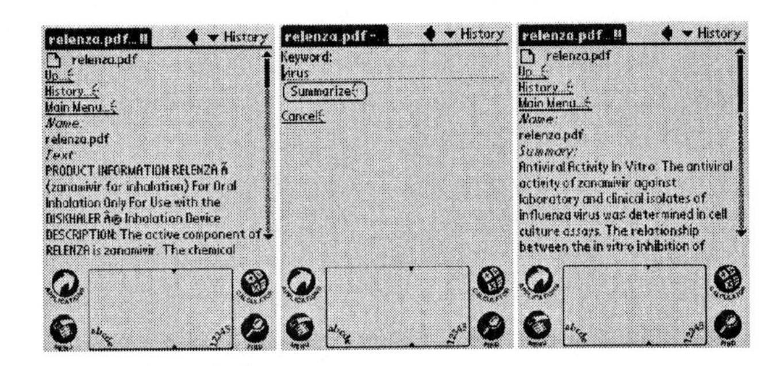

In order to realize mobile access to knowledge portals, portlets have to be implemented as mobile portlets. That means that they have to be adapted according to technical restrictions of mobile devices and the user's context. At the moment, commercial portal packages cannot fulfill sufficiently the needs of mobile KM. Most of the systems are enhanced by mobile components, which are rather providing mobile access to stationary KM services instead of implementing specific mobile KM services.

Hyperwave's WAP (Wireless Application Protocol) Framework, for example, enables mobile users to browse the Hyperwave Information Portal with WAP-enabled devices. The Wireless Suite of Autonomy is a WAP-based solution with the focus on awareness-generating features such as peoplefinder and community support.

At present, the most comprehensive support for mobile KM is provided by the Livelink portal from Opentext Corporation. With the help of the Wireless Server users can access discussion boards, task lists, user directories (MS Exchange, LDAP, Livelink User Directory), e-mail, calendar, and documents (Figure 2). In addition, it provides some KM services specially developed for mobile devices, for example, automatic summarization of text. Hence even longer texts can be displayed on smaller screens (Figure 3).

Example: A Mobile KM Portal for a German University

In recent years German universities, which are financed to a large extent by public authorities (federal states and federal government), have been severely affected by public saving measures. As a result lean, efficient administrative procedures are more important than ever. KM can help to achieve these objectives. One example is to provide easy accessible expert directories, where staff members with certain skills, expertise, and responsibilities can be located ("Person XY is responsible for third-party funding") in order to support communication and collaboration.

However, there are several reasons why the access to information of this type is limited at the University of Regensburg. First, there is the decentralized organizational structure. All together about 1,000 staff members are working in 12 different schools and about 15 research institutes at the university, serving about 16,000 students. Because most of the organization units are highly independent, they have their own administrations and the exchange of knowledge with the central administration is reduced to a minimum. Likewise there is hardly an exchange of knowledge between different schools and departments. As a result, knowledge that would be useful throughout the whole university is limited to some staff members ("unlinked knowledge," Figure 4).

A second problem is that many scientific staff members work on the basis of (short-term) time contracts. This leads to an increasing annual labor turnover, comparable to the situation that consulting companies are facing. Important knowledge about past projects, courses, and scientific results is lost very easily. Due to this fact, a high proportion of (new) staff members are relatively inexperienced to cope with administration processes, which can be described as highly bureaucratic and cumbersome.

To overcome these problems—the lack of communication between departments and the need to provide specific knowledge (i.e., administrative knowledge) for staff members—the University of Regensburg decided to build up a knowledge portal called U-Know (Ubiquitous Knowledge). U-Know is meant to be a single point of access for all relevant information according to the knowledge needs described above. When conducting a knowledge audit it became obvious that a large amount of knowledge is needed when knowledge workers are on the move, that is, working in a mobile work environment. Staff

Figure 4. Unlinked knowledge because of independent organization structures

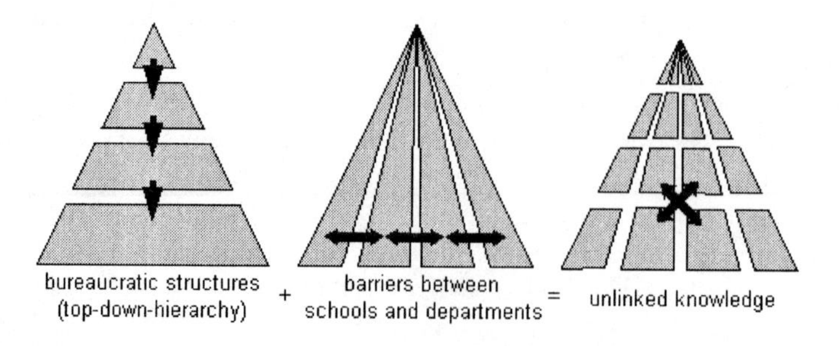

Figure 5. Knowledge demand in "mobile" situations

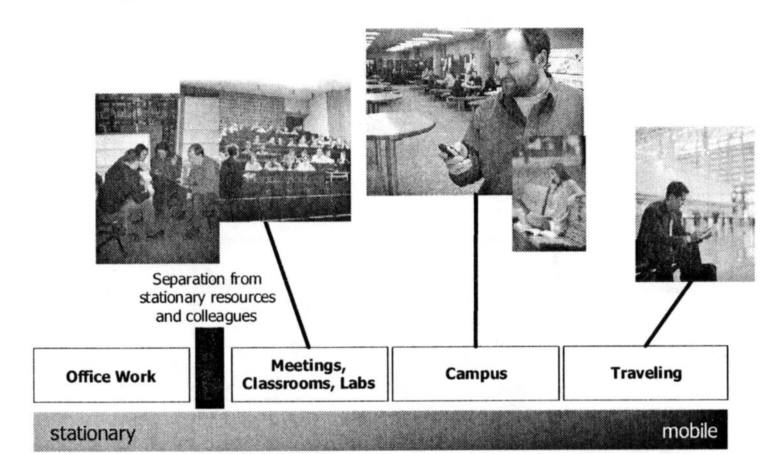

is frequently commuting between offices, meeting rooms, laboratories, home offices; they attend conferences; and sometimes they are doing field studies (e.g., biologists or geographers).

Hence the picture of one single resource-rich office has to be extended towards different working locations, where a large number of knowledge-intensive tasks are carried out as well (Figure 5). Consequently the considered solution should meet these "ubiquitous" knowledge needs of current work practices at a university.

The portal should support staff members by managing the following:

1. **Documented knowledge**: A knowledge audit was conducted in order to obtain a better picture of knowledge demand and supply. This was mainly done with the help of questionnaires and workshops where staff members were asked to assess what kind of (out-of-office) information is considered as useful.
2. **Tacit knowledge**: In order to support the exchange of tacit knowledge (which is difficult to codify due to the fact that this knowledge lies solely in the employees' heads, often embedded in work practices and processes), the considered KM solution should enable communication and cooperation between staff members.

In order to meet these requirements U-Know should offer the KM services in Figure 6.

The services can be categorized into information, communication, collaboration, and search. The first category comprises all services that are responsible to manage simple information in the knowledge base. By invoking these services staff members obtain the information they need to perform their daily tasks, for example, news, notifications about changes in rooms, or phone numbers. A very important part of this section is the yellow pages (Figure 7) where all staff members are listed. This list can be browsed by names, departments, fields of research, and responsibilities.

Figure 6. Features of U-Know

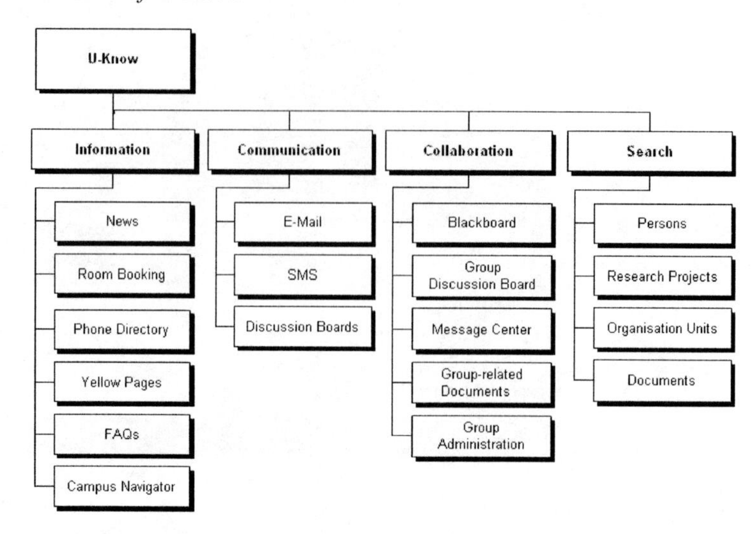

Figure 7. U-Know yellow pages

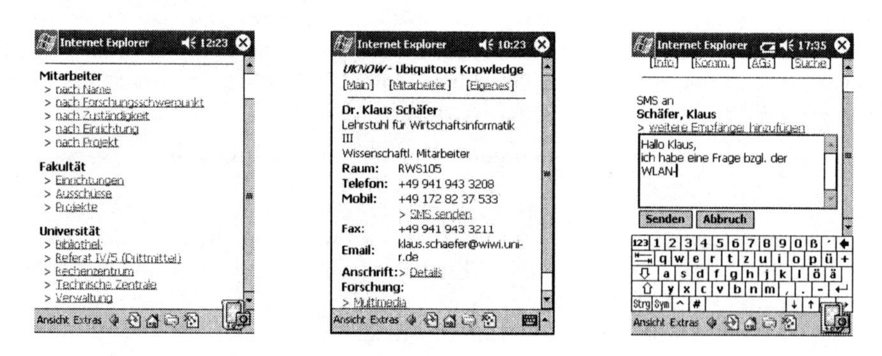

Frequently asked questions (FAQ) answer questions that are typically asked by new staff members. The Campus Navigator helps locate places and finding one's way around the campus. Each room at the university carries a doorplate with a unique identifier. After entering a starting point in form of the identifier and a destination in form of the name of a person, of an office (e.g., "Office for Third-Party Fundings," "Academic Exchange Service"), or just another room number, the shortest way to the destination is calculated and shown on maps of different sizes (Figure 8).

Communication-oriented features like e-mail, short message service (SMS), and discussion boards are intended to support the exchange of tacit knowledge between staff members.

To foster collaboration, for example, in temporary project groups, staff members can initiate workgroups by inviting colleagues via SMS or e-mail to join a virtual teamspace. After forming a workgroup the participants can use their teamspace for (electronic) group

Figure 8. U-Know Campus Navigator

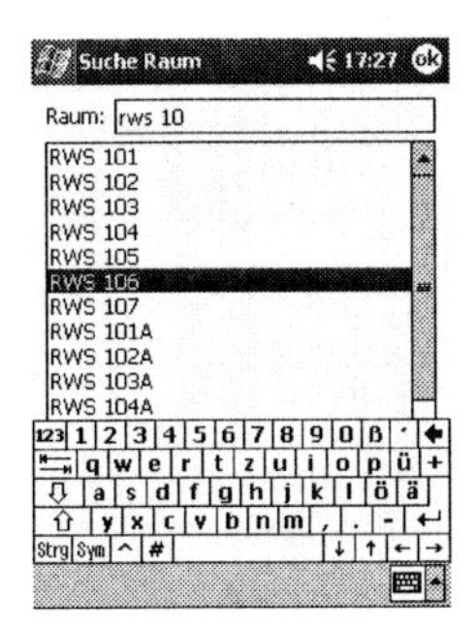

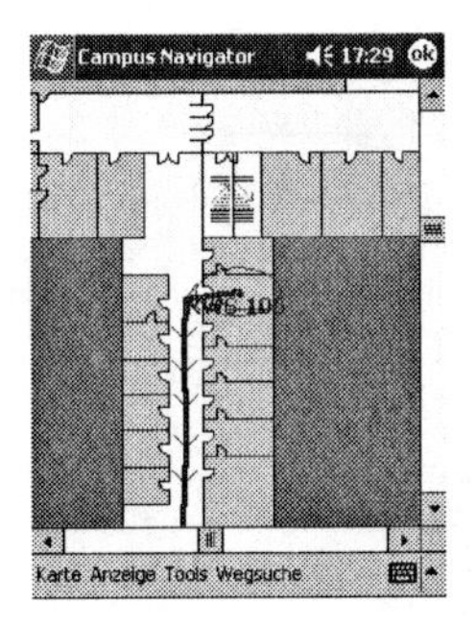

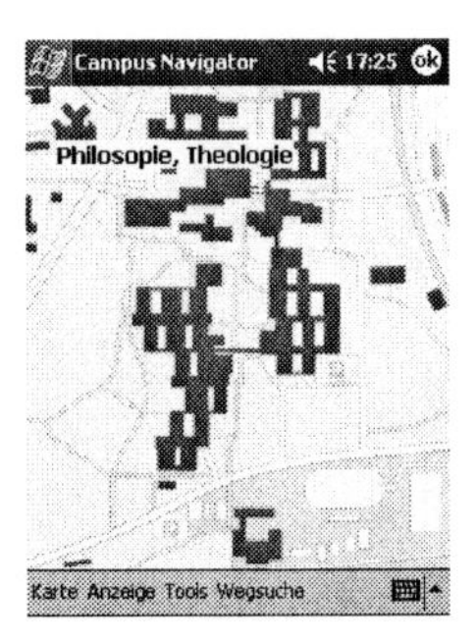

discussions and sharing documents. The blackboard displays all recent events, including new group members, new files, discussion entries, and administrative actions that are taken. In the search section queries can be limited to persons, research projects, organization units, or documents.

To support different networks there are several ways to access the portal. University staff can use the campuswide Wi-Fi network with Wi-Fi–capable devices. Users can also deploy a mobile phone and access the portal via a GSM-network and the Wireless Application Protocol (WAP). Hence it is possible to use the portal even when users are outside the university, for instance, at a conference. The phone directory or the yellow pages can be accessed via voice as the entry of longer words may be cumbersome in many situations. An integrated speech-recognition system "translates" the user's spoken words into database requests and the results back into speech.

Location Orientation as Next Step in Mobile KM

Generally, there is agreement about the distinction between human- and technology-oriented KM approaches which basically reflects the origin of the approaches. KM research should try to bridge the gap between human- and technology-oriented KM. Many authors have propagated a so-called "holistic" approach to KM. However, so far these authors leave it to the interpretation of the reader what such an approach might look like. The examples in the last column of Table 1 should be seen as a step towards detailing this approach which is called "bridging the gap" KM. In Table 1 this classification (Maier, 2004; Maier & Remus, 2003) is enhanced towards the consideration of mobile KM. As mobile KM is mainly focusing on instruments and systems, other dimensions like strategy, organization, and economics are not considered in this table.

In order to structure mobile KM, one can distinguish two dimensions: mobile access and location orientation. Mobile access is about accessing stationary KMS whereas location

Table 1. Mobile KM approaches (gray highlighted cell is covered by U-Know)

	Technology-oriented instruments and systems	**Human-oriented instruments and systems**	**Bridging-the-gap instruments and systems**
Mobile Access	Mobile access to content, for example, knowledge about organization (e.g., Campus Navigator), processes, products, internal studies, patents, online journals by using mobile devices focusing on services for presentation (e.g., summarization functions, navigation models) and visualization	Mobile access to employee yellow pages, skill directories, directories of communities, knowledge about business partners using mobile devices focusing on asynchronous e-mail, SMS, and synchronous communication (chat), collaboration and cooperation, community support	Mobile access to ideas, proposals, lessons learned, best practices, community home spaces (mobile virtual teamspaces), evaluations, comments, feedback to knowledge elements using mobile devices focusing on profiling, personalization, contextualization, recommendation, navigation from knowledge elements to people
Location Orientation	Adaptation of documented knowledge according to the user's current location	Locating people according to the user's location, for example, locating colleagues, knowledge experts	Personalization, profiling according to the user's location and situation, providing proactive mobile KM services

orientation explicitly considers the location of the mobile worker. The field of location-oriented KM draws attention from research in mobile KM, ubiquitous computing, location-based computing, and context-aware computing (Lueg & Lichtenstein, 2003).

So far, the implemented solution provides mobile access to a broad range of different knowledge sources in a mobile work environment. University staff can use the KM services provided by U-Know in order to access information, to find colleagues, to navigate the campus, to collaborate, and so forth. These KM services mainly support the human-oriented KM approach. In fact, typical knowledge services were adapted with regard to the characteristics of mobile devices, that is, small display, bandwidth, and so forth.

However, an adaptation of these services according to the user's location has not taken place yet, whereas a customization of services according to the location of the user would enable a mobile knowledge portal to supply mobile knowledge workers with appropriate knowledge in a much more targeted way. At the same time, information overload can be avoided, since only information relevant to the actual context and location is filtered and made available. Think of a researcher who is guided to books in a library according to his/her own references but also according to his/her actual location.

Currently, common "stationary" knowledge portals are ill-suited to support these new aspects of KM derived from a location-oriented perspective (Berger, 2004). One reason is that the context, which is defined by the corresponding situation (tasks, goals, time, identity of the user) is still not extended by location-oriented context information (Abecker, van Elst, & Maus, 2001).

Location-oriented knowledge services could contribute to

- **More efficient business processes**: Shortcomings arising from mobility can be compensated by considering location-oriented information. Times for searching can be reduced due to the fact that information about the location might restrict the space of searching (e.g., an engineer might get information about a system that he/she is currently operating). Possibly, redundant ways between mobile and stationary work place are omitted when the information is already provided on the move.
- **Personalization**: When considering the user's location information can be delivered to the user in a much more customized and targeted way (Rao & Minakakis, 2003). For example, an engineer in a production hall is seeking information about outstanding orders, whereas close to machines he might need information about technical issues or repair services. In addition, location-oriented information might be helpful to locate other "mobile" colleagues who are nearby.
- **New application areas**: The integration of common knowledge services together with location-oriented mobile services may also extend the scope for new applications in KM, for example, the use of contextual information for the continuous evolution of mobile services for mobile service providers (Amberg, Reus, & Wehrmann, 2003). One can also think of providing a more "intelligent" environment where information about the user's location combined with sophisticated knowledge services adds value to general information services (e.g., in museums, where customized information to exhibits can be provided according to the user's location).

To build up mobile knowledge portals that can support the scenario described above, mobile portlets are needed that can realize location-oriented KM services. In case of being implemented as proactive services (in the way that a system is going to be active by itself), these portlets might be implemented as push services. In addition, portlets have to be responsible for the import of location-oriented information, the integration with other contextual information (contextualization), and the management and exploitation of the location-oriented information. Of course, the underlying knowledge base should be refined in order to manage location-oriented information.

With respect to mobile devices, one has to deal with the problem of locating the user and sending this information back to the knowledge portal. Mobile devices might be enhanced with systems that can automatically identify the user's location. Depending on the current net infrastructure (personal, local, or wide area networks), there are many possibilities to locate the user, for example, Wi-Fi, GPS, or radio frequency tags (Rao & Minakakis, 2003).

Conclusions and Outlook

The example of U-Know shows some important steps towards a comprehensive mobile KM solution. With the help of this system it is possible to provide users with KM services while being on the move. With its services like yellow pages, messaging features, and so forth, it creates awareness among remote working colleagues and thus improves knowledge sharing within an organization.

With respect to the acceptance of U-Know, two user groups can be distinguished. The first group is characterized by users who already own a mobile device, especially a PDA, in order to organize their appointments and contacts (personal information management). They are the main users of the system because they perceive the additional KM-related services as an extension of the capabilities of their devices. In contrast, staff members who did not use mobile devices for their personal information management are more reluctant to adopt the new system.

The Wi-Fi access soon became the most popular way of accessing the system. This is because of several reasons. Most of the staff members are actually working on the campus and the Wi-Fi access is free of charge for university members. Another reason is probably the higher bandwidth (and therefore faster connections) of Wi-Fi in comparison to a GSM-based access via WAP. Nevertheless, it can be assumed that decreasing connection fees and higher bandwidths of 3G-Networks (UMTS) would encourage staff to use the system from outside the university.

However, in order to fully meet the requirements of mobile KM in the near future, mobile KM portals have to be enhanced with mobile knowledge services that consider location-oriented information. Current work needs once more to address the adaptation of mobile services, the consideration of the user and work context for KM, and the design of highly context-aware knowledge portals.

References

Abecker, A., van Elst, L., & Maus, H. (2001, July 13–16). *Exploiting user and process context for knowledge management systems.* Workshop on User Modeling for Context-Aware Applications at the 8th International Conference on User Modeling, Sonthofen, Germany.

Amberg, M., Remus, U., & Wehrmann, J. (2003, September 29–October 2). Nutzung von Kontextinformationen zur evolutionären Weiterentwicklung mobiler Dienste. *Proceedings of the 33rd Annual Conference "Informatics 2003," Workshop "Mobile User - Mobile Knowledge - Mobile Internet,"* Frankfurt, Germany.

Belotti, V., & Bly, S. (1996). Walking away from the desktop computer: Distributed collaboration and mobility in a product design team. *Proceedings of CSCW '96* (pp. 209–218). Boston: ACM Press.

Berger, S. (2004). *Mobiles Wissensmanagement. Wissensmanagement unter Berücksichtigung des Aspekts Mobilität*. Berlin: dissertation.de.

Grimm, M., Tazari, M.-R., & Balfanz, D. (2002). Towards a framework for mobile knowledge management. *Proceedings of the Fourth International Conference on Practical Aspects of Knowledge Management 2002* (PAKM 2002), Vienna, Austria.

Hansmann, U., Merk, L., Niklous, M.S., & Stober, T. (2001). *Pervasive computing handbook*. Berlin: Springer.

Lueg, C., & Lichtenstein, S. (2003, November 26–28). *Location-oriented knowledge management: A workshop at the Fourteenth Australasian Conference on Information Systems (ACIS 2003),* Perth, Australia.

Maier, R. (2004). *Knowledge management systems, information and communication technologies for knowledge management*. Berlin: Springer.

Maier, R., & Remus, U. (2003). Implementing process-oriented knowledge management strategies. *Journal of Knowledge Management, 7*(4), 62–74.

Open Text Corporation. (2003). *Livelink Wireless: Ubiquitous access to Livelink Information and Services* (White paper). Waterloo, Canada: Author.

Perry, M., O'Hara, K., Sellen, A., Brown, B., & Harper, R. (2001). Dealing with mobility: understanding access anytime, anywhere. *ACM Transactions on Human-Computer Interaction, 8*(4), 323–347.

Rao, B., & Minakakis, L. (2003). Evolution of mobile location-based services. *Communications of the ACM, 46*(12), 61–65.

Schulte, B.A. (1999). *Organisation mobiler Arbeit. Der Einfluss von IuK-Technologien.* Wiesbaden, Germany: DUV.

Sherry, J., & Salvador, T. (2001). Running and grimacing: The struggle for balance in mobile work. *Wireless world: Social and interactional aspects of the mobile age* (pp. 108–120). New York: Springer.

Section IV

Mobile Applications in Healthcare

Chapter XIII

M-Health: A New Paradigm For Mobilizing Healthcare Delivery

Nilmini Wickramasinghe[1],
Stuart Graduate School of Business, Illinois Institute of Technology, USA

Steve Goldberg,
INET International Inc., Canada

Abstract

Medical science has made revolutionary changes in the past decades. Contemporaneously, however, healthcare has made incremental changes at best. The growing discrepancy between the revolutionary changes in medicine and the minimal changes in healthcare processes is leading to inefficient and ineffective healthcare delivery and one if not the significant contributor to the exponentially increasing costs plaguing healthcare globally. Healthcare organizations can respond to these challenges by focusing on three key solution strategies, namely, (1) access – caring for anyone, anytime, anywhere; (2) quality – offering world-class care and establishing integrated information repositories; and (3) value – providing effective and efficient healthcare delivery. These three components are interconnected such that they continually impact the other and all are necessary to meet the key challenges facing healthcare organizations today. The application of mobile commerce to healthcare, namely, m-health, appears to offer a way for healthcare delivery to revolutionize itself. This chapter serves to outline an example of adopting mobile commerce within the healthcare industry, namely, in the area of a wireless medical record. In particular, it discusses an appropriate, feasible mobile solution to enable hospitals operate effectively and efficiently in today's competitive and costly healthcare environment as well as meet all the necessary regulatory requirements. The lessons learnt from these case study data should be of interest to both practitioners and researchers since they will outline realistic and feasible solutions to enable hospitals to incorporate a wireless/m-commerce solution as well as highlighting key areas for further research in this important area of high-quality, effective, and efficient healthcare management.

Introduction

Currently the healthcare industry in the United States as well as globally is contending with relentless pressures to lower costs while maintaining and increasing the quality of service in a challenging environment (Pallarito, 1996, pp. 42–44; Wickramasinghe & Silvers, 2003, pp. 75–86). It is useful to think of the major challenges facing today's healthcare organizations in terms of the categories of demographics, technology, and finance. Demographic challenges are reflected by longer life expectancy and an aging population; technology challenges include incorporating advances that keep people younger and healthier; and finance challenges are exacerbated by the escalating costs of treating everyone with the latest technologies. Healthcare organizations can respond to these challenges by focusing on three key solution strategies, namely, (1) access – caring for anyone, anytime, anywhere; (2) quality – offering world-class care and establishing integrated information repositories; and (3) value – providing effective and efficient healthcare delivery. These three components are interconnected such that they continually impact the other and all are necessary to meet the key challenges facing healthcare organizations today.

In short then, the healthcare industry is finding itself in a state of turbulence and flux (Wickramasinghe & Mills, 2001, pp. 406–423). Such an environment, we believe, is definitely well suited for a paradigm shift with respect to healthcare delivery. Therefore, in this chapter we address the issue of wireless solutions for healthcare delivery and management.

First, we discuss the findings from INET's study on mobile Internet (wireless) technology in healthcare by Ontario Hospitals in Canada. We use these findings as a launching place to review a rigorous way to accelerate healthcare delivery improvements. Next, we outline some preliminary evidence for using a standardized mobile Internet (wireless) environment in healthcare. For example, INET International is advocating the use of a wireless healthcare portal to validate the possible reduction in IT infrastructure costs. A portal may reside on a wireless PDA device as single point of contact for clinicians to obtain immediate patient data (radiology reports, lab results, and clinical findings). This wireless portal may also improve patient care outcomes with access to the best available clinical evidence at the point of care. We shall also describe the current status of a standardized mobile Internet (wireless) environment in terms of technology requirements, security readiness, and IT management practices. In addition, we will also outline some of the key challenges that a hospital's IT department, medical units, administration, and clinicians will face regarding a wireless project and provide some reasonable solutions to these challenges. Finally, we shall outline the key steps necessary for a hospital to transition from proprietary information systems to a three-tier Web-based architecture.

INETS Study in Mobile Internet Technology in Healthcare

INET International Inc. delivers rigorous e-business acceleration projects in large corporate, government, and healthcare organizations. The organization focuses on custom Mobile, Internet, Intranet, Extranet (INET), and wireless solutions. It was founded by Steve Goldberg in 1998 and it leads a Wireless Technology Consortium (WTC) to collect evidence on the best way to use wireless technology to accelerate healthcare delivery improvements. These applications are designed to improve patient care, reduce costs, increase healthcare quality, and enhance teaching and research.

Over a period of 2 years INET has been conducting research that has been directed at how to apply mobile Internet wireless technologies' low-cost advantages to evolve a wireless healthcare portal. A portal is a single point of contact for healthcare providers and handheld technology applications (HTA) to access and process various data pertaining to patients such as: (1) patient-specific data (i.e., patient ID, radiology reports, lab results, clinical findings, and research data); (2) medical knowledge (primarily from evidence-based medicine training and journals); (3) clinical guidelines (i.e., association guidelines such as the Association of Radiologists clinical practice publications); and (4) reimbursement rules and data (i.e., Ontario Health Insurance Plan, known as OHIP). This research has shown that mobile/wireless solutions for healthcare can achieve four critical goals of (1) improving patient care, (2) reducing transaction costs, (3) increasing healthcare quality, and (4) enhancing teaching and research.

Improving Patient Care and Increasing Healthcare Quality

In the final report compiled by the Committee on the Quality of Healthcare in America (AIM, 2001), it was noted that improving patient care is integrally linked to providing high-quality healthcare. Furthermore, in order to achieve a high quality of heathcare the committee identified six key aims, namely, (1) healthcare should be safe – avoiding injuries to patients from the care that is intended to help them; (2) effective – providing services based on scientific knowledge to all who could benefit and refraining from providing services to those who will not benefit (i.e., avoiding under use and overuse); (3) patient centered – providing care that is respectful of and responsive to individual patient preferences, needs, and values, and ensuring that patient values guide all clinical decisions; (4) timely – reducing waiting and sometimes harmful delays for both those receiving care and those who give care; (5) efficient – avoiding waste; and (6) equitable – providing care that does not vary in quality based on personal characteristics.

Most of the poor quality connected with healthcare is related to a highly fragmented delivery system that lacks even rudimentary clinical information capabilities resulting in poorly designed care processes characterized by unnecessary duplication of services and long waiting times and delays (AIM, 2001). The development and application of sophisticated information systems is essential to address these quality issues and improve efficiency, yet healthcare delivery has been relatively untouched by the revolution of information technology that has transformed so many areas of business today (Wickramasinghe & Mills, 2001, pp. 406–423; Wickramasinghe & Silvers, 2003, pp. 75–86). This, then, certainly justifies the need for e-business solutions for healthcare delivery; however, from various applied research scenarios (Goldberg et al., 2002a, 2002b, 2002c, 2002d, 2002e), we can see why mobile solutions appear to be superior since they enable even better care for patients with the added advantage of significant cost savings. Physicians themselves are excited by the possibilities offered by mobile solutions as the following quotation discussing benefits to radiologists exemplifies:

"The demands on radiologists in today's fast-paced and cash-strapped hospitals are tremendous," says Dr. Brian Yemen, radiologist at Hamilton Health Sciences. "The possibility of transmitting radiology reports between hospitals and physicians on-demand is exciting. It will mean less time and travel for busy doctors, but more importantly, it will mean faster turnaround and quicker results for patients."

A mobile Internet wireless solution uses a personal digital assistant (PDA) device at the point of patient care. Such handheld technology applications are simple to use and require little, if any, training. Furthermore, they can significantly reduce change management costs which are usually dominated by technology education. Typically, the PDA devices and application require very little training unlike most PC-based e-health alternatives where even simple skills like typing prove to be challenging hurdles for physicians and other users to overcome. This means that not only do physicians have the information they need in a timely fashion but they also are not restricted by typing or other activities that are not directly going to benefit the patient's encounter. In Canada, most manufacturers and IT national resellers have found no need to set up educational practices to support over 1 million PDA users. Currently, Canadian retail stores provide the bulk of the customer service requirements, such as, product pricing, availability, promotions, and product warranty. Usually, the product warranty process replaces defective PDA devices with a new or refurbished unit within hours or the next business day with minimal inconvenience to users.

Cost Reductions: Wireless Technology May Reduce IT Infrastructure Costs by 84%

- Integral to the set up of such a mobile solution is the wireless healthcare portal. This portal can dramatically reduce IT infrastructure project costs. For example, the government of Ontario, Ministry of Health and Long-Term Care is planning to spend $150 million to develop a turn-key IT solution for 6,400 physicians. By using

Table 1. Cost estimates

Component	Total Cost	Comment
Mobile Internet (wireless) Infrastructure Support	$60	Annual
Mobile Internet (Wireless) Infrastructure Upgrade	$100	Amortize Over 3 Years
PDA Device	$133	Amortized Over 3 Years
Cell Phone/Data Cable/Modem	$66	Amortized Over 3 Years
Wireless Communication Costs	$420	$35/Month – transmits 25 to 50 patient records/day
Physician Wireless Portal	$480	$40/Month – assume four handheld technology applications

wireless technology it is possible to reduce this IT infrastructure cost to $24 million by leveraging hospital IT infrastructure investments. The preliminary cost estimate is $1,259/year/physician over 3 years. Table 1 outlines the cost estimates/year/physician for each component.

Thus, the findings summarized above from the research conducted by INET demonstrate not only does the mobile/wireless solution support improving patient care, but it also is a cost-effective solution. Furthermore, not only can cost savings be enjoyed by leveraging off existing infrastructure, but the wireless solution also lends itself to rapid healthcare delivery improvements as well as the ability to deploy healthcare improvements within short time cycles as discussed below.

Achieve Rapid Healthcare Delivery Improvements

By adopting a mobile/wireless healthcare delivery solution, it is possible to achieve rapid healthcare delivery improvements, which impacts both the costs and the quality of healthcare delivery. This is achieved by using an e-business acceleration project which provides hospitals a way to achieve desired results within a standardized mobile Internet (wireless) environment as shown in Figure 1.

Integral to such an accelerated project is the ability to build on the existing infrastructure of the hospital. This then leads to what we call the three-tier Web-based architecture. In such an environment, tier 1 is essentially the presentation layer, which contains the Web browser but no patient data is stored within this layer, and thereby ensuring compliance with international security standards/policies like HIPAA. Tier 2 then provides the business logic, including but not limited to, lab, radiology, and clinical transcription applications; messaging of HL7, XML, DICOM, and other data protocols; and interface engines to a Hospital Information Systems (HIS,) Lab Information Systems (LIS), Radiology Information Systems (RIS), as well as external messaging systems such as Smart Systems for Health (an Ontario Healthcare IT infrastructure project). Finally there is the tier 3 architecture which consists of the back-end databases such as Oracle or Sybase.

Figure 1. E-business accelerated project

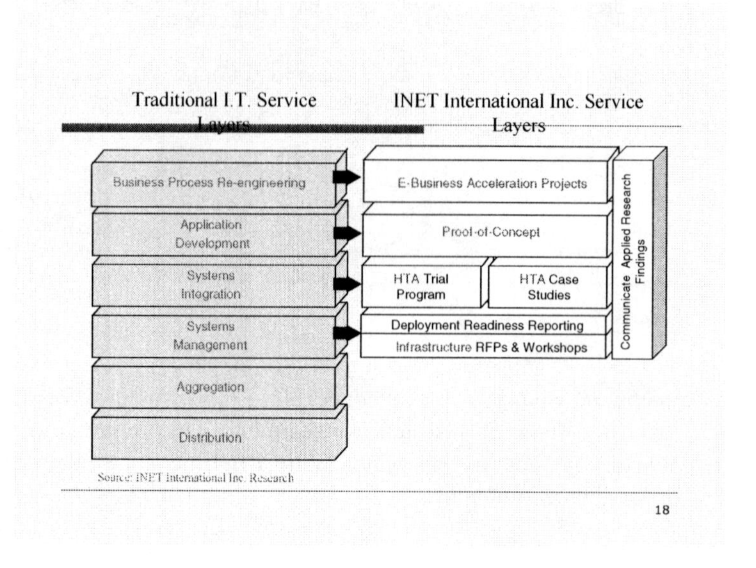

Deploying Healthcare Improvements Within Short Time Cycles

A mobile Internet (wireless) infrastructure may achieve rapid technology deployments to individual practices, small clinics, and healthcare networks at a very low cost. For example, for $35/month a physician can have wireless access to a Radiology Information System (RIS) to retrieve diagnostic imaging (DI) reports. With a minimum cost these RIS systems may easily and securely access reports from the hospital, multiple hospitals (using Smart Systems for Health) or interface into other diagnostic imaging (DI) clinics as a single point of contact. The objective is to improve patient care by getting DI results faster to patients (through referring physicians). Once the solution achieves approval for general release, it has the possibility of being deployed to 3,000 to 4,000 physicians within short time cycles so that, for example, clinicians using a mobile Internet (wireless) connection are able to hook into an existing hospital infrastructure as in the case of the Ontario hospitals where they use the Hamilton and Toronto downtown hospitals' IT infrastructures.

Teaching and Research

The provision of the best available clinical evidence at the point of care is playing a pivotal role in medical research and training (www.cochrane.org). Today, a physician-led research center (www.cebm.utoronto.ca/projects/index.htm) is conducting an applied research project on the impact on mobile Internet (wireless) technology to help clinicians with the following:

- Consistently translate high-quality clinical evidence into practice.
- Obtain the information they need from the right resources.
- Engage in a process to replace the need to read 17 articles per day.

The challenges of using mobile Internet technology have been documented by this center. This includes the following:

- What kinds of information are most useful to clinicians?
- What is the most effective way of querying evidence-based resources?
- How do we format answers?

INET International Inc. applied research on a standardized mobile Internet (wireless) environment may provide some unique solutions. The solutions may include the following:

- Clinicians can participate in a rigorous e-business acceleration project to narrow information requirements into a few useful requirements. For instance, radiologists have identified and confirmed the importance of referring physicians getting imaging reports and results to patients faster.
- By harnessing a hospital's information systems (HIS) and standardized mobile Internet (wireless) environment a physician can gain access to patient-specific data (i.e., imaging reports). This may provide a new opportunity to evaluate how physicians can use current patient data to determine the best ways to query evidence-based resources. Also, researchers may extend the use of patient-specific data to nurse practitioners, home care providers, and patients to help develop the most effective way of querying evidence-based resources within a consumer-centric model.
- Finally, INET International Inc. research on a standardized mobile Internet (wireless) environment may help physician-led research centers select and aggregate Internet search forms and Internet "answer" formats from multiple resources into a single wireless personal digital assistant (PDA) presentation. Additionally, researchers can make rapid changes to Internet "answer" formats to accommodate user feedback. They work within this standardized environment to control iterative prototyping for the delivery of small and frequent proof of concepts and randomized trials. The use of release management and quality assurance tools enable the scaling of successful trials into wireless production systems for tens of thousands of healthcare providers and patients within a very short time cycle.

To summarize, based on the findings from the various research endeavors conducted by INET Inc. (Goldberg et al., 2002a, 2002b, 2002c, 2002d, 2002e), using a standardized mobile

Internet (wireless) environment appears to be the simplest way to access the best-available clinical evidence at the point of care for everyone, anywhere, anytime at a very low cost. The achievement of superior quality cost-effective healthcare is uppermost on the agendas of most countries' healthcare initiatives. We believe this can only be realized by adopting some type of the wireless solution.

The Mobile Vision

Having described the key benefits of a mobile/wireless solution, let us now turn to focus on how to set about actualizing such a mobile vision. We illustrate the possibilities by discussing INET's strategies and approaches.

Research Goal and Approach

Simply stated, the research goal is to use a standardized mobile Internet (wireless) environment to improve patient outcomes with immediate access to patient data and provide the best-available clinical evidence at the point of care. To do this, INET International Inc. research starts with a 30-day e-business acceleration project in collaboration with many key actors in hospitals such as clinicians, medical units, administration, and IT departments. Together they follow a rigorous procedure that refocuses the traditional 1–5 year systems development cycle into concurrent, 30-day projects to accelerate healthcare delivery improvements. Figure 2 highlights the key success factors required in such an approach.

The completion of an e-business acceleration project delivers a scope document to develop a handheld technology application (HTA) proof of concept specific to the unique needs of the particular environment. The proof of concept is a virtual lab case scenario. A virtual lab operates within a mobile Internet (wireless) environment by working with hospitals and technology vendors. The final step is the collection of additional data with clinical HTA trials consisting of 2-week hospital evaluations.

Figure 2. Key success factors

Mobile Internet (wireless) Technology Key Success Factors For Health Care	
Physician-led	Widespread physician acceptance of handheld technology applications.
Simple to Use	Personal Digital Assistance, Bluetooth, Java Cell Phones, Digital communicators.
Low Cost	Convergence of wireless IP WAN and wireless IP LANs.

E-Business Acceleration Project Outcomes

The first e-business acceleration project for healthcare was conducted for Hamilton Health Sciences (HHS) Diagnostic Imaging (DI) Department.

They followed a procedure to divide an HHS enterprise DI process into smaller mini-processes. This provided a new way for radiologist to enhance DI delivery in small manageable pieces and minimize risk to patients. Hence we can see that a critical step in the e-business accelerated project is often to reengineer existing processes to ensure they are as effective and efficient as possible.

During the first e-business acceleration project a DI sign-off report process was selected and a handheld technology application (HTA) proof of concept was developed.

The next step, HHS is planning an HTA trial to evaluate a low-cost, secure, and simple way for radiologists and referring physicians to access critical healthcare information at the point of patient care. The radiologist and referring physicians could retrieve and review a patient's DI reports using a wireless PDA device—getting diagnostic results faster to patients and thus enhancing quality as well as effective and efficient healthcare delivery.

A series of 2-week evaluations are planned for hospitals in Ontario and United States. This trial program is presented in Figure 3. Each hospital provides a different environment

Figure 3. A wireless handheld technology applications (HTA) trial program

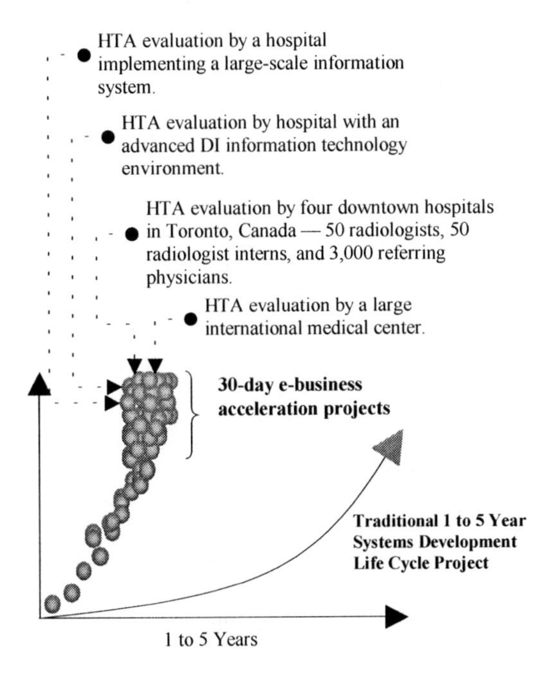

to evaluate the scalability of the DI sign-off report HTA. The aggregation of such 2-week evaluation outcomes provides hospitals evidence on key areas such as:

- Showing clearly how reduction in large-scale IT infrastructure project costs can occur.
- Demonstrating that the PDA is simple to use, requiring little or no training of clinicians.
- Clearly outlining how to achieve rapid healthcare delivery improvements.
- Demonstrating how to scale and deploy healthcare delivery improvement to clinicians within short time cycles.

Applied Research Challenges

Engaging in such a project, naturally, has several challenges. Initially technology vendors saw the immediate opportunities to engage in INET International Inc. Virtual Lab to work with Hamilton Health Sciences (HHS). The lab is a low-cost way for vendors to connect demonstration facilities and resources to conduct proof of concepts and clinical trials. However, growing the lab to 10 hospitals involves the formalization of a hospital procurement process that includes applied research, that is, a proof of concept and a trial program.

Traditionally IT operates within 1- to 5-year systems development life cycle to meet hundreds of end-user requirements. Whereas the e-business acceleration project process selects one or two requirements to deliver improvements in concurrent short time cycles. To expand the applied research of a rigorous approach to accelerate healthcare delivery improvements requires additional investment by hospitals, something that is clearly challenging.

Possibly the most significant impact of the virtual lab is the resolution of mobile Internet (wireless) technology challenges. The HHS applied research findings (lesson learned) provide a way to help wireless projects achieve the same results in a much shorter time cycle. HSS is an academic health sciences center whose primary role is healthcare delivery. It is one of the largest hospitals in Ontario, operating across four sites with approximately 8,000 employees and 1,000 physicians and thus the research findings from HHS are detailed and extensive and a useful resource to help any other wireless project in this area.

A Standardized Mobile Internet (Wireless) Environment

INET virtual lab also encounters unknown challenges when working with key actors in hospitals such as clinicians, IT departments, medical units, and administration. This is to be expected because each situation is unique and it typically happens during consensus building on HTA proof of concepts and clinical HTA trial programs. The

Figure 4. A standardized mobile Internet (wireless) environment

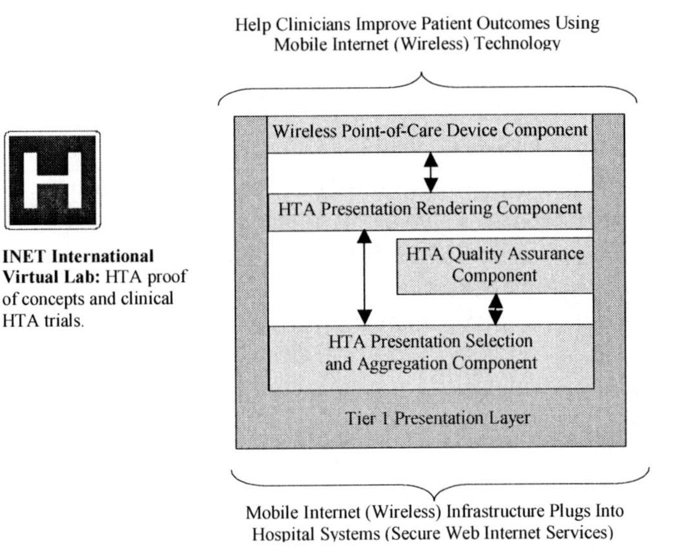

virtual lab uses these challenges to identify, prioritize, and select the requirements for a standardized mobile Internet environment.

Ontario's Management Board Secretariat (MBS) is taking the lead in the use of a Three-Tier Web-Based Architecture, a standardized application environment to implement and scale applications for the Ontario Public Sector. Used as the starting point, this is the foundation to architect a standardized mobile Internet (wireless) environment for an Ontario hospital.

By design, the mobile Internet (wireless) environment is a very thin layer on top of a standardized application environment and represents the extension of the tier 1 presentation layer. The purpose is to prevent time-consuming and costly tier 2 (business logic) and tier 3 (back-end database) development (see Figure 4 on this reading from the bottom up).

INET International Inc. started a wireless technology consortium in 2001 to collect evidence on the best way to define a standardized mobile Internet (wireless) environment. Every member works in collaboration to build a consensus on the IT management practices and IT professional services related to each component. The member's competency is validated through applied research on using mobile Internet (wireless) technology in collaboration with an Ontario hospital. The virtual lab is the basis of INET's applied research. INET uses case scenarios by running selected handheld technology application proof of concepts and trial evaluations. All unpublished data (Goldberg et al., 2002a, 2002b, 2002c, 2002d, 2002e) is available to INET International Inc. collaborations with business and physician-led research centers.

In 2001 the WTC members' roles and responsibilities are listed in Table 2, while Appendix 1 provides the necessary background information on all participating members.

The WTC acceptance process is currently under review. The process may include the following steps:

1. Identify vendor's IT management practice investments.
2. Map IT management practices and IT professional services to a mobile Internet (wireless) environment component(s).
3. INET assigns WTC responsibilities: WTC leaders, complete management qualifications, IT service transition plans and workshops. Build a consensus on a mobile Internet (wireless) technology certification program. Assist IT associations in training.
4. Each member meets or sets international management practice standards (based on ITIL Information Technology Infrastructure Library http://www.itil.co.uk/index.html). They participate in academic conference panels, present papers, and collaborate on applied research.

Table 2. Wireless Technology Consortium virtual lab roles and responsibilities

<table>
<tr><th rowspan="2">Wireless Technology Consortium Member</th><th colspan="4">Virtual Lab Roles and Responsibilities</th></tr>
<tr><th>Management Responsibilities</th><th>Provide IT Professional Services</th><th colspan="2">Validate Mobile Internet Technology</th></tr>
<tr><td>INET International Inc. [1]</td><td>Project management</td><td>Collaboration, consensus building, and coordination services to deliver a handheld technology application (HTA) within a wireless healthcare portal</td><td colspan="2">Standardized mobile Internet (wireless) environment</td></tr>
<tr><td rowspan="2">TELUS Mobility [2]</td><td rowspan="2">Service-level management</td><td>Wireless WAN help-desk services</td><td>Wireless IP WAN solution</td><td rowspan="3">Wireless point-of-care device component</td></tr>
<tr><td>Connector help-desk services</td><td>Cable/Card solution</td></tr>
<tr><td rowspan="3">Palm Canada Inc. [3]</td><td rowspan="3">Incident management</td><td>PDA help-desk services</td><td>Device solution</td></tr>
<tr><td>Browser help-desk services</td><td colspan="2" rowspan="2">HTA presentation rendering component</td></tr>
<tr><td>HTA help-desk services</td></tr>
<tr><td rowspan="3">Compuware Corporation [4]</td><td rowspan="2">Release management</td><td>Mobile Internet (wireless) infrastructure deployment services</td><td colspan="2" rowspan="3">HTA quality assurance component</td></tr>
<tr><td>HTA production release services</td></tr>
<tr><td>Availability management</td><td>Application management services</td></tr>
<tr><td rowspan="3">NetManage Inc. [5]</td><td rowspan="2">Change management</td><td>HTA development</td><td colspan="2" rowspan="3">HTA presentation selection and rendering component</td></tr>
<tr><td>HTA enhancement</td></tr>
<tr><td>Problem management</td><td>HTA defect recovery service</td></tr>
</table>

5. Participate in INET virtual lab proof of concepts and trial evaluations to validate a standardized mobile Internet (wireless) environment.
6. At any time INET International Inc. may reassign a WTC role, request a new member to apply to the WTC, or ask WTC to reengage the WTC acceptance process in step 1.

Key Challenges

The introduction of wireless into healthcare naturally brings with it many unique challenges which must also be addressed if success is to ensue in these initiatives. One of these challenges is concerned with security and privacy of highly sensitive patient data. In a healthcare setting, in the case of the mobile electronic patient record, typically the physician is accessing patient data from the hospital's Web services. The order information is transmitted is from the mobile device to a base wireless station, and from there, through the mobile communication infrastructure, to the wireless application gateway of the hospital. There are naturally many valid business reasons why all concerned parties—the patient, healthcare institution, and physician, to name a few—should expect the information exchanged during such a mobile transaction process to be secure. Specifically, parties should expect the data associated with a transaction to be *confidential* and delivered without violation of *integrity* or *authenticity*. Further, all parties should similarly expect *nonrepudiability*; neither the client nor the hospital should be able to deny the completion of the transaction if such a transaction, in fact, occurred.

Possible security vulnerabilities can occur at several points including: (1) between the physician and the mobile device; (2) between the mobile device and the mobile infrastructure operator; (3) between the mobile infrastructure operator and the wireless application gateway of the merchant; and (4) between the wireless application gateway and the Web services of the hospital. There are similarities and differences between a mobile and a fixed (those not using a mobile device) transaction with respect to security concerns. As in the case of a fixed transaction, *content* of a mobile transaction must be protected from unauthorized access and alteration whether it is in storage or being transmitted. However, protecting data that move is a bit more difficult than data that do not move since one has to worry about parameters that do not come into play in a fixed network, such as (a) location management, (b) mobility management, and (c) radio resource management.

Security and privacy are important considerations. In the United States, all healthcare institutions must be HIPAA compliant in this regard. Yet, while HIPAA details what is required, it does not discuss how this should be addressed. We believe that a sound approach to achieve a high degree of end-to-end security in a mobile/wireless environment is to develop a suitable mobile trust model using symmetric keys (Goldberg et al., 2002e). This will not only ensure HIPAA compliancy but will also ensure at least as secure an environment as its wired counterpart. We are confident that these security challenges can be addressed and then healthcare organizations can enjoy the full power and potential of their mobile/wireless solutions.

Figure 5. (Source: Wickramasinghe & Goldberg, 2004, pp. 140–156)

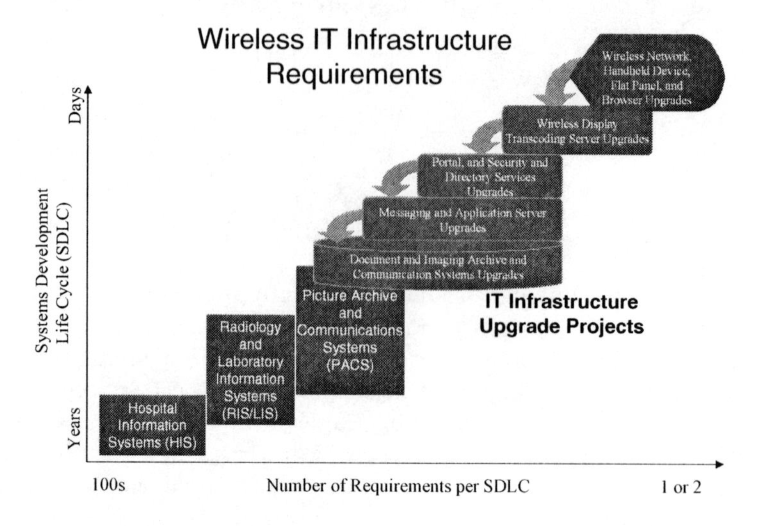

Security is not the only challenge, however. Other important challenges include compatibility issues and the ability to share information throughout the existing hospital networks. We believe this is less problematic because of the advantages inherent in constructing the healthcare portal. Furthermore, in the designing of the wireless, infrastructure directly builds on top of existing hospital infrastructure as can be seen in Figure 5.

INET's studies (Goldberg et al., 2002a, 2002b, 2002c, 2002d, 2002e) have shown that this approach appears to circumvent several problems pertaining to information sharing. Clearly though incorporating techniques such as information management and knowledge management would serve to enable even more efficient and effective use of all necessary information (Wickramasinghe & Goldberg, 2004). We also suggest that some business process reengineering would be required to ensure that processes are as effective and efficient as possible and make full use of the potentials of the wireless technologies.

Discussion

"Today's hospitals demand zero tolerance for software defects and downtime. That's where Compuware comes into play," says Cathy Lippert, Distributed Systems Product Management Director, Compuware Corporation, Farmington Hills, Michigan. "Our tools can set a new pace for accelerating the testing and quality assurance processes for handheld-technology applications based on mobile Internet standards. This means lower costs, reduced risk, a shorter learning curve, and the confidence that applications can be scaled from one user to thousands of users."

In an effort to improve medical information systems within 30-day cycles and help establish international mobile Internet standards, the WTC is creating a virtual research lab in Ontario. The virtual lab will play a key role in setting new standards for the delivery of high-impact handheld technology applications to clinicians. "The WTC's establishment of a virtual lab designed to help establish global mobile Internet standards should rapidly accelerate the development of handheld applications across a broad spectrum of healthcare industries," says Matthew Hickey, director, enterprise solutions for Palm Canada. "Already, Palm handhelds are at the forefront of this revolution and we expect to continue pushing the envelope in terms of innovation and collaboration." "Our 2.5G iDEN technology allows the seamless transfer of information from wireless handheld devices to hospital servers," says Jawad Shah, manager of Wireless Data, TELUS Mobility, Scarborough, Ontario. "The virtual lab can examine the proper implementation of firewalls, authentication servers, biometrics, intrusion detection, virus protection, and virtual private networks (VPN) to assure that patient-sensitive data can be transmitted securely over the entire TELUS Mobility iDEN network." The preceding quotations from various qualified people in the field are presented to illustrate some key points. First, wireless is here to stay and this technology offers us many advantages. Second, healthcare requires a new way to manage current challenges and thereby provide cost-effective quality care to us all. Finally, the combination of wireless in healthcare appears to be a real solution. Clearly we are technically capable of delivering wireless solutions to healthcare institutions as we have outlined in this chapter. Now appears to be the opportune time to do so. We strongly believe that hospitals of the 21st century should be wireless to some extent at least and a mobile healthcare delivery solution will facilitate the necessary revolutionary changes required by healthcare to contend with today's challenging environment.

Conclusions

Wireless technology advancements have made it possible for physicians to use handheld computers to produce diagnostic imaging reports on demand and within seconds. Along with improving access to medical information and the overall speed of healthcare delivery, mobile Internet wireless technologies can play a significant role in lowering IT infrastructure costs for hospitals and preventing disruption to healthcare delivery due to technology problems. The preceding has served to demonstrate that (a) we have the capabilities to offer wireless solutions and (b) wireless solutions will enable healthcare institutions to enjoy many advantages. By detailing the studies conducted by INET, we have shown how doable wireless solutions are in healthcare. Furthermore, we have highlighted the many benefits including cost savings and the ability to deliver a higher level of quality care. We have also endeavored to highlight the key challenges that any wireless initiative must address so that the full power and potential of such an iterative can be realized and thereby truly making mobile healthcare delivery a necessary component of any healthcare arsenal.

We close with a strong message about the importance of thinking of wireless as a necessity in healthcare delivery for the 21st century and we urge for more research in this area as well as for healthcare institutions globally to move forward in this direction. Finally, we also wish to underscore that the lessons learned from INET Inc.'s accelerated project approach have broader implications above and beyond the healthcare domain. Specifically, we believe that the e-business accelerated project techniques are useful to many wireless endeavors and we encourage researchers and practitioners in the field to explore these possibilities.

References

Crossing the quality chasm. (2001). *A New Health System for the 21st Century*. Committee on Quality of Healthcare in America Institute of Medicine. Washington, DC: National Academy Press.

Goldberg, S., et al. (2002a). Building the evidence for a standardized mobile Internet (wireless) environment in Ontario, Canada. January 2002 update. Unpublished data.

Goldberg, S., et al. (2002b). *HTA presentational selection and aggregation component summary*. Unpublished data.

Goldberg, S., et al. (2002c). *Wireless POC device component summary*. Unpublished data.

Goldberg, S. et al. (2002d). *HTA presentation rendering component summary*. Unpublished data.

Goldberg, S. et al. (2002e). *HTA quality assurance component summary*. Unpublished data.

Pallarito, K. (1996). Virtual healthcare. *Modern Healthcare, March*, 42–4.

Wickramasinghe, N., & Goldberg, S. (2004). How M=EC2 in Healthcare. *International Journal of Mobile Communications, 2*(2), 140–156.

Wickramasinghe, N., & Mills, G. (2001). MARS: The electronic medical record system: The core of the Kaiser Galaxy. *International Journal Healthcare Technology Management, 3*(5/6), 406–423.

Wickramasinghe, N., & Silvers, J.B. (2003). IS/IT the prescription to enable medical group practices to manage managed care. *Health Care Management Science, 6*, 75–86.

Endnotes

- This chapter has been adapted from an earlier version of a paper by Wickramasinghe, N., & Goldberg, S. (2004). How M=EC^2. *International Journal of Mobile Communications*, *2*(2), 140–156.

[1] Corresponding author.

Appendix I: Background Information on Consortium Members

1. **About INET International Inc.**

   INET International Inc. is a Canadian technology management consulting firm. Currently the firm is focusing on how to use wireless technology to accelerate healthcare delivery improvements and leads a wireless technology consortium. INET International is not on the Web. If you need additional information, please do not hesitate in contacting Steve Goldberg at sgoldberg@sprint.ca, or t: 905 889-2704.

2. **About TELUS Mobility**

   TELUS Mobility, combining the national wireless operations of Clearnet, QuébecTel Mobilité, and TELUS, provides a full suite of wireless services to more than 2.3 million clients across Canada. For more information, please visit us at www.telusmobility.com.

   TELUS Corporation (TSE: T, T.A; NYSE: TU) is one of Canada's leading telecommunications companies, providing a full range of telecommunications products and services that connect Canadians to the world. The company is the leading service provider in Western Canada and provides data, Internet Protocol, voice, and wireless services to Central and Eastern Canada. For more information about TELUS, visit www.telus.com.

3. **About Palm, Inc.**

   Palm, Inc. is a pioneer in the field of mobile and mobile and wireless Internet solutions and a leading provider of handheld computers, according to IDC (December 2000). Based on the Palm OS(R) platform, Palm's handheld solutions allow people to carry and access their most critical information wherever they go. Palm™ handhelds address the needs of individuals, enterprises, and educational institutions through thousands of application solutions. The Palm OS platform is also the foundation for products from Palm's licensees and strategic partners, such as Franklin Covey, Handspring, IBM, Kyocera, Sony, Symbol Technologies, and HandEra (formerly TRG). The Palm Economy is a growing global community of industry-leading licensees, world-class OEM customers, and approximately 170,000 innovative developers and solution providers that have registered to develop solutions based on the Palm OS platform. Palm went public on March 2, 2000. Its stock is traded on the NASDAQ national market under the symbol PALM. More information is available at http://www.palm.com.

4. **About Compuware Corporation**

   With fiscal 2001 revenues of more than $2 billion, Compuware Corporation provides business value through software and professional services that optimize productivity and reduce costs across the application life cycle. Meeting the rapidly changing needs of businesses of all sizes, Compuware's market-leading solutions improve the quality, ease the integration, and enhance the performance of distributed, e-business, and enterprise software. Compuware employs approximately

13,000 information technology professionals worldwide. For more information about Compuware, please contact the corporate offices at 800-521-9353. You may also visit Compuware on the World Wide Web at http://www.compuware.com.

5. **About NetManage**

Founded in 1990, NetManage, Inc. (NASDAQ: NETM) delivers information access, publishing, integration, and support software and services that maximize a company's investment in existing information systems and applications. It provides an instant bridge to e-commerce. NetManage offers a significantly broader range of application integration software, host access software, centralized management, and live interactive support solutions than competitors. Only NetManage instantly transforms corporate information assets into powerful e-business solutions. NetManage sells and services it products worldwide through its direct sales force, international subsidiaries, and authorized channel partners. For more information, visit www.netmanage.com, send e-mail to pr@netmanage.com, or call 408.973.7171 (Pacific Time).

Chapter XIV

A Prehospital Database System For Emergency Medical Services

Nada Hashmi, 10Blade, Inc., USA

Mark Gaynor, Boston University School of Management, USA

Marissa Pepe, Boston University School of Management, USA

Matt Welsh, Harvard University, USA

William W. Tollefsen, Boston University School of Medicine, USA

Steven Moulton, Boston University School of Medicine, USA

Dan Myung, 10Blade, Inc., USA

Abstract

Emergency Medical Services (EMS) are not only responsible for providing prompt and efficient medical care to many different types emergencies, but also for fully documenting each and every event. Unfortunately, the vast majority of EMS events are still documented by hand. The documents are then further processed and entered manually into various billing, research, and other databases. Hence, such a process is expensive, labor intensive, and error prone. There is a dire need for more research in this area and for faster, efficient solutions. We present a solution for this problem: Prehospital Patient Care Record (PCR) for emergency medical field usage with a system called iRevive that functions as a mobile database application. iRevive is a mobile database application that is designed to facilitate the collection and management of prehospital data. It allows point-of-care data capture in an electronic format and is equipped with individual patient sensors to automatically capture vital sign data. Patient information from the field is wirelessly transmitted to a back-end server, which uses Web service standards to promote interoperability with disparate hospital information systems, various billing agencies, and a wide variety of research applications. In this chapter, we describe the current state of EMS, the iRevive application, a mini-trial deploying iRevive in real scenarios, the results, and a future direction for our solution.

Introduction

There are times when an individual's life may depend on the quick reaction and competent care of emergency medical technicians (EMTs). These highly trained, prehospital healthcare providers are dispatched by 911 operators to incidents as varied as motor vehicle crashes, heart attacks, near-drowning events, childbirth, and gunshot wounds. Their first priority is to stabilize a patient's cardiopulmonary status. They must then determine the nature and severity of the patient's condition and whether the patient has any preexisting medical problems. EMTs follow strict rules and guidelines in their provision of emergency care and often use special equipment such as backboards, defibrillators, airway adjuncts, and various medications before placing patients on stretchers and securing them in an ambulance for transport. At a medical facility, EMTs transfer the care of their patients to emergency department personnel by reporting their observations and actions to staff.

Equally important is EMS personnel documenting the care they provide. They do so in the form of a prehospital record, which must be completed for each patient who is treated or transported by them. The prehospital record is a medical and legal document used by emergency medical technicians to record a variety of data concerning a patient's current illness or injury, past medical history, treatment rendered, and subsequent improvement or worsening of the patient's condition (Mann, 2002). This type of prehospital documentation is used to support the actions of the crew, the transfer of care, and to justify reimbursement from various insurance companies; it is also used for quality improvement programs and research. Unfortunately, the vast majority of EMS events are still documented manually by hand on paper. This leads to an extensive amount of manual data processing as the often illegible handwritten data must sometimes be deciphered, then manually entered into various billing, research, and other databases. The whole process is expensive, labor intensive, and error prone.

The rest of this chapter is sectioned as follows: first, an overview of the current state of EMS workflow, documentation methods, and research is provided. This section emphasizes the National Highway Traffic Safety Association's goals for EMS in the future, including the call for a national EMS database and improved information systems, so that prehospital information can be linked with the hospital record. The next section is a description of one solution called iRevive, a mobile database for EMS professionals that streamlines data capture, communication, reimbursement processing, quality assurance, and research. It takes advantage of tiny wireless sensors to automatically record vital sign data. It permits multilevel decision support; the local EMT over his/her patient, the regional commander over a selected vicinity, and the central level of control over all the events occurring at a particular time. Actual deployment of iRevive, for live field-testing by Professional Emergency Services of Cambridge, Massachusetts, is examined and critiqued. This trial version was conducted without sensors or multilevel decision support. Finally, a future vision of iRevive is described, including the addition of many different types of sensors such as chemical sensors and GPS devices for location information. All exchange of data will be interfaced through Web services and conform to standards such as HL-7 to help the increase of data exchange and interoperability.

Background

Ambulances of the early 1900s were regarded as a means of transportation for the sick and injured from homes, work site, and public places to hospitals, where real treatment could begin. It was not until the advent of cardiopulmonary resuscitation (CPR) and the 1966 publication of a National Academy of Sciences paper entitled, "Accidental Death and Disability: The Neglected Disease of Modern Society," that modern EMS systems came into being (Callahan, 1997). Later, with the introduction of cardiac defibrillation by trained crewmembers and more extensive airway training, the back of ambulances became the sites of true life-saving treatments. While emergency medical services have grown rapidly over the past 30 years, the scope of EMS research has not. Most EMS research focuses on a single intervention or health problem, and it rarely addresses the inherent complexities of EMS systems (Delbridge, Bailey, & Chew, 1998).

It is estimated that EMS systems treat and transport up to 30 million patients per year (NHTSA, 2001) and it is assumed that EMS intervention positively affects patient outcomes, but this is difficult to quantify. Studies have shown that early defibrillation and administration of certain drugs save lives, but other interventions including certain instances of intubations in the field may in fact cause more harm than good (Adnet, Lapostolle, Ricard-Hibon, Carli, & Goldstein, 2001; Vahedi, Ayuyao, & Parsa, 1995). The fact that so few therapies have been examined in outcome studies illustrates a lack of evidence regarding the benefits of many prehospital interventions.

The lack of EMS systems data can be attributed in part to the healthcare industry's delay in utilizing technology; it is one of the last industries to transition to the use of computers for daily operation (Cheung et al., 2001; Foxlee, 1993; Mikkelsen & Aasly, 2001; Tello, Tuck, & Cosentino, 1995). Although some elements of the system are automated (e.g., computer automated dispatch), most EMS personnel record clinical information and other run data using paper and pen. Data collection is therefore limited and highly inefficient. In addition, the patient care report (PCR) that is completed after each EMS transport does not contain data regarding overall patient outcome. The reason for this is that outcome data is held by several different entities, sometimes including other ground and air transport services, hospitals, rehabilitation centers, and physician offices. These various healthcare entities may be affiliated with each other, but seldom are they officially linked and rarely do they exchange prehospital patient information. This lack of information exchange is further hampered by patient privacy laws, incompatible (proprietary) systems, limited data mining methods, and little impetus to form a continuous patient care record. The resultant lack of outcome data severely limits the type and amount of EMS research that can be carried out (Dunford, 2002). Compounding the overall problem is the recognition that serious medical errors can arise in the setting of incomplete data (Foxlee, 1993; Tello et al., 1995). These errors in the handing off of patient care can range from duplicative or delayed therapy to complete lack or inappropriate therapy.

Current Methods

EMS personnel usually work in teams of two and divide the workload at a particular event. While one provider tends to the patient, the other interviews family members or bystanders, sizes up scene conditions, and searches for medications, identification cards, and insurance cards. The NHTSA currently mandates a data set of 40 items to be collected for each patient for each event, including such items as incident location, crewmember identification numbers, patient's social security number, and physical exam findings. Additional insurance and billing information is required by ambulance services so that patients can be billed, while Medicare calls for waivers and prescription forms. It is when the patient's condition is more critical that EMS team members must give their full attention to patient care, forgoing any attempts at data capture or documentation. It is data from these types of events, however, that are of greatest interest to emergency department personnel, researchers, and system administrators. And as EMS systems evolve to offer more advanced care, more time is needed for hands-on patient care and, in turn, more information must be documented (Mears, Ornato, & Dawson, 2002). To overcome these obstacles to better patient care, EMS systems must adopt information systems that streamline the recording, storage, retrieval, and application of quality information.

New methods are being developed to quantify and organize the plethora of data. In 1996, the NHTSA published an article entitled, "EMS Agenda for the Future: Implementation Guide". This article stressed the need for a standardized EMS information system, based on uniform data elements and uniform definitions (NHTSA, 2001). In order to accurately draw conclusions there must be more information regarding care in the field, transportation, emergency department care, hospital care, and final patient disposition. To achieve these goals, EMS information systems must develop new ways to store and retrieve patient data so that patient information is always available. Data must be pooled from a communications center, ambulance personnel, emergency department staff, and finally, other agencies including fire departments, police departments, and medical examiners. Only then can there be a complete database containing all of the information necessary to describe an entire EMS event and facilitate continuous EMS system evaluation and research across multiple systems and to support patient care and EMS-related research (Delbridge, 1998). Our system, iRevive, attempts to address these problems and provides a novel solution in streamlining the data collection. The next section describes the iRevive system built to generate prehospital patient care record.

iRevive: The Mobile Database System

In working toward this vision of compatible EMS database systems that support healthcare data integration, changes must start with regard to how data is originally collected. iRevive is a mobile prehospital database system that allows point-of-care data capture in an electronic format. It consists of a network of wireless, handheld computers running the iRevive application. Wireless vital sign sensors, called VitalDust motes, automatically capture and integrate patient vital sign data directly into each developing patient care record. All of this information can be sent wirelessly from the field to a server that stores and relays selected patient information to receiving hospitals. The stored data is accessible to authorized users for billing and research purposes using Web services (Figure 1).

iRevive explores improved documentation of EMS events by streamlining data capture and providing essential prehospital information for subsequent integration with the developing hospital record. Data entry is designed to be logical and intuitive. iRevive can "walk" an EMT through an assessment and remind him/her, for example, to evaluate a patient's neurological response, an essential step that could easily be left out during the high-paced transport of a critically ill patient. Real-time sensors, which collect heart rate and blood oxygen saturation, provide real-time monitoring and enrich the data collection process. During transport or soon after arrival at a hospital, the EMT may choose to complete the PCR using a preset narrative template on the handheld computer. Alternatively, he/she can enter his/her own narrative using either the handheld computer's pen pad or a computer terminal in the receiving hospital's emergency department. Once the PCR is complete and electronically signed, it can be saved (in the central database) and printed (at the hospital) so that it can be included in the patient's paper-based

Figure 1. Current iRevive System Architecture

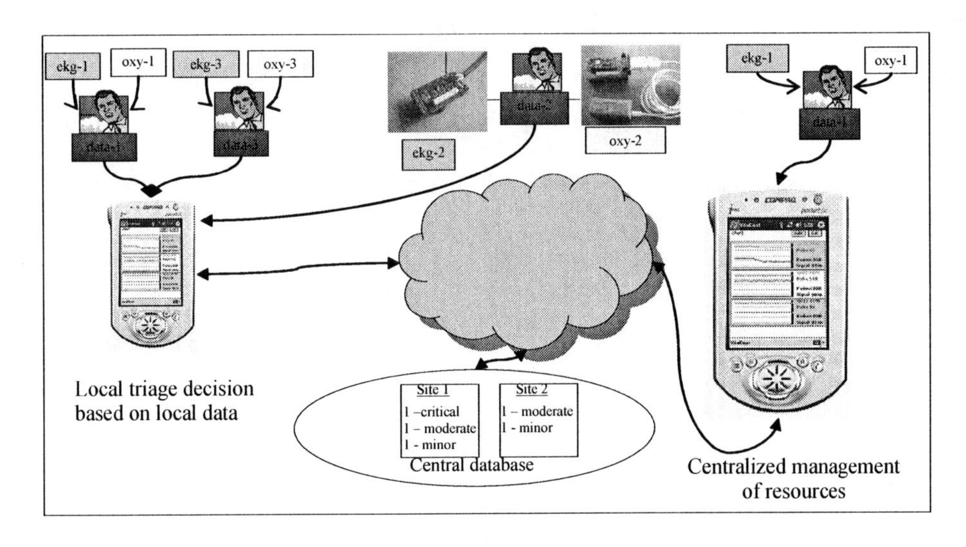

hospital record. The administrative back end aggregates the raw data into summary reports and statistics for operational analysis, performance measurement, and effective managerial decision making.

Prehospital Patient Care Records

The iRevive prehospital patient care record consists of and captures three types of data. The first type of data is the information captured by the EMT which contains the preexisting conditions and the current conditions. Specially, it contains demographic information, a history of the patient's illness or injury (HPI), past medical history (PMH, including medications and allergies), procedural information (e.g., IV access, splinting, and endotracheal intubations), and disposition information (Figure 2). The HPI section includes standardized narratives for more than 30 common complaints, which range from allergic reactions to "dead on scene." The use of standardized narratives allows EMTs to quickly describe each incident using a series of drop-down option boxes, rather than writing an entire incident out in prose—a difficult task on a PDA. The addition of pertinent negatives and additional information in the event of special circumstances is also allowed. These narratives enrich the database and, in the future, will aid in the development of new knowledge-based treatment algorithms.

The second type of data captured is the documentation of critical examinations. It is devoted to an extensive physical exam including head, neck, chest, abdomen, pelvis, extremities, and nervous system. From the physical exam main menu, users may select and describe any part of the anatomy that has been found to be abnormal (Figure 3a). Selecting an "abnormal" button brings the user to additional pages where abnormal physical findings may be documented in detail using a series of pull-down lists (Figure 3b).

Figure 2. iRevive patient care report sections

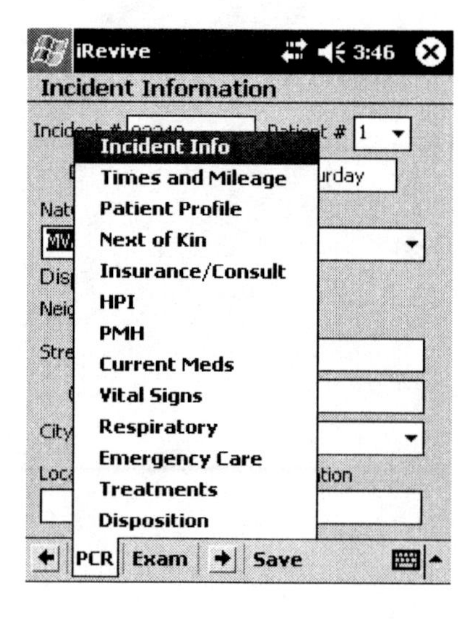

Figure 3a. iRevive physical exam main menu. If a normal button is selected, a statement is entered in the PCR indicating a normal exam. If an "abnormal" button is selected, then a detailed exam screen opens (CNS = central nervous system; PNS = peripheral nervous system; Cranial Nn = cranial nerves).

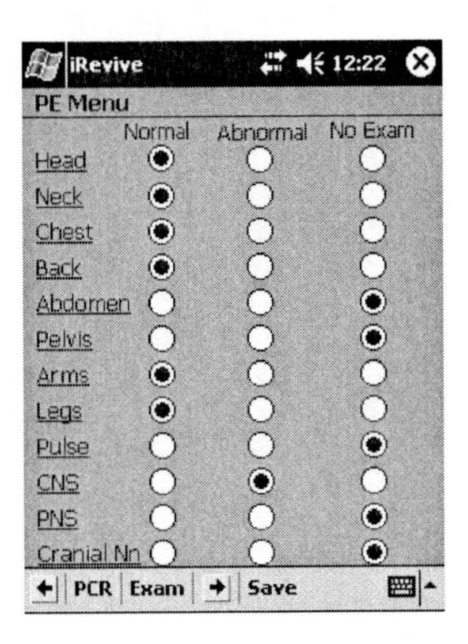

Figure 3b. A detailed exam screen for the central nervous system (CNS) is shown (GCS = Glascow Coma Score, React. = papillary reactivity).

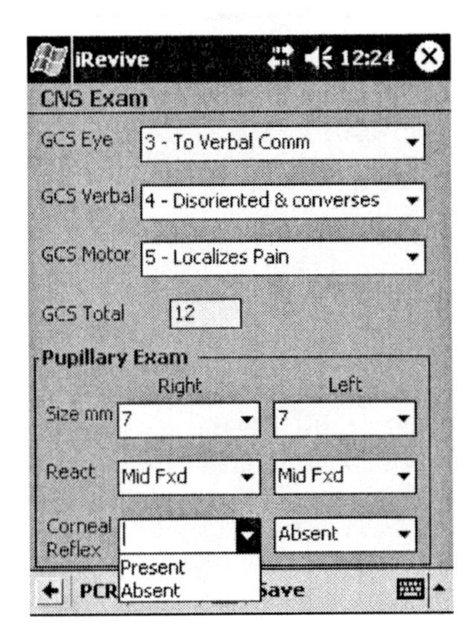

Figure 4. Sensor Data information transmitted wirelessly to the PDA.

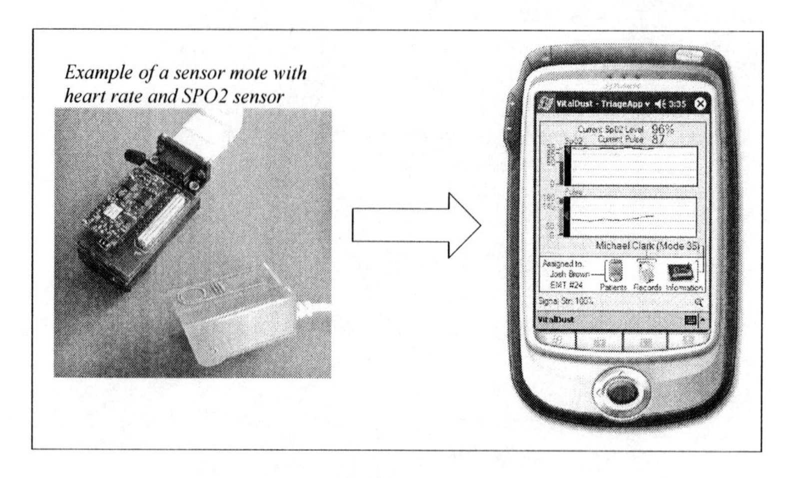

Finally, the third type of data captured and recorded is the sensor data from motes such as the pulse-Oxometer and Vital Dust (Figure 4). The vital signs, such as the pulse rate and SP02 levels, from the patient are automatically recorded and sent continuously from the sensor to the PDA. Critical changes can be immediately noted by the EMT. The EMT no longer has to physically monitor the patient's vital at regular intervals. He/she is automatically provided with more readings which results in a more accurate state of the patient at any particular time.

Current Application Architecture

The iRevive application is written in C# under the .NET environment and therefore built to run on many types of mobile devices including PDAs, laptops, and wearable computers (Figure 5). iRevive provides a graphical user interface that users can navigate with a stylus. The application is primarily menu driven with customizable drop-down menus that increase efficiency by allowing for quick navigation and data collection. Once the data have been uploaded to a central database, a Web-based interface can be used to edit the PCR on any Internet-enabled PC.

Once field data have been synchronized with the iRevive database, EMS and hospital personnel can instantaneously track current patients and a complete patient care report can be generated and printed. Specific providers have access to the entire record base for billing, supply tracking, and continuous quality improvement (CQI) applications. Other users have the ability to access deidentified data that are nonconfidential for use in overall emergency service research and systems management. All data transfer is Health Insurance Portability and Accountability Act (HIPAA) compliant and the security is end-to-end (using standard Transport Layer Security). Users must be authenticated to use the system and only authorized users can view specific types of data.

Figure 5. The Future iRevive System Architecture

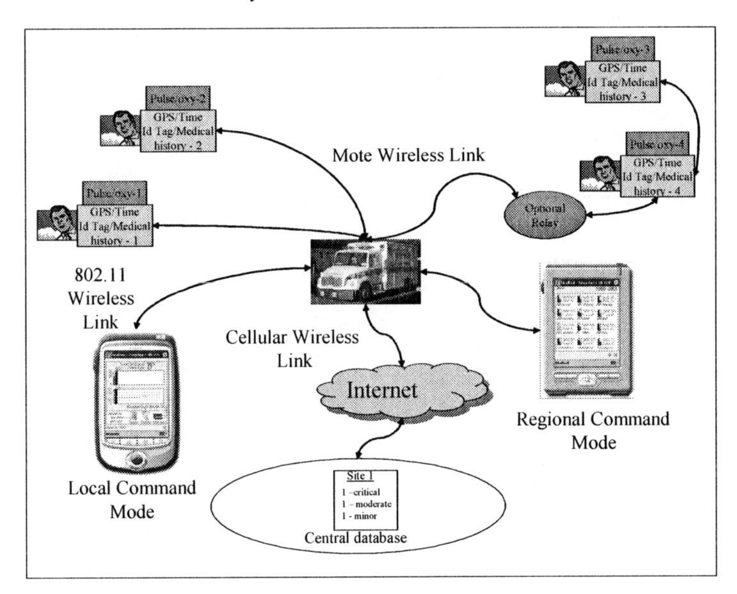

To encourage interoperability, the iRevive system is built around emerging standards such as the National Highway and Traffic Safety Board Uniform Prehospital EMS Data Set (NHTSA-UPDS) and the Data Elements for Emergency Departments Standard (DEEDS), which is the Centers for Disease Control and Prevention's (CDC's) specification for emergency department data that was created to assist data integration across information resources.

Field Trial Analysis Without Sensors

One ambulance from Professional Emergency Services (Pro) of Cambridge, Massachusetts, was selected to field test iRevive. Pro is a moderate-sized private ambulance company with eight ambulances and 50 emergency field providers who respond to over 16,000 emergency calls per year. iRevive was used in parallel with the service's existing handwritten documentation methods during emergency responses. Data were collected using a Hewlett Packard iPAQ Pocket PC, which was synchronized with a laptop satellite station before being uploaded to the server at the end of each shift. In an attempt to steer further product development, users focused on how the iRevive application could be improved. The effectiveness of iRevive in integrating and streamlining data capture was studied, along with its ability to merge with the current workflow of EMS professionals. Factors examined included ease of use, documentation completion, and content. The use of iRevive in the field was observed by a proficient user and ambulance crew chief. This was further examined via interviews with other ambulance crew members. Printouts of iRevive PCRs were then compared to handwritten reports from the same events.

A Case Study

In one real-life scenario, a call concerning chest pain and difficulty breathing was dispatched to an ambulance. While en route to the location, one EMT began to enter data into iRevive. Information included the incident location, type of response, and other pertinent dispatch information known at the time; the time of response was recorded in the times and mileage page. Upon arrival at the scene, time was recorded again while dispatch was notified of arrival via the radio. The time of arrival at the patient's location in the third-floor apartment was also recorded. The first EMT began an initial assessment, obtaining vital signs, attaching a heart monitor, administering oxygen, and prepping an IV site. At the same time, the second EMT interviewed the patient and bystanders in order to gain a detailed account of the patient's medical history, allergies, medications, and a history of the present illness. All pertinent information was recorded using pull-down menus within the iRevive program.

After recording placement of an IV catheter, oxygen delivery rate, and the type of oxygen device used, the medics determined that the patient should receive one dose each of nitroglycerine and aspirin. This was also recorded in iRevive. Once the patient was placed in the ambulance, a second set of vital signs indicated the need for a second dose of sublingual nitroglycerine, which was recorded in iRevive with a time stamp. En route to the hospital, the first EMT took over use of iRevive to record the time of transport and to complete additional sections as necessary. At the hospital, a verbal report was given to the emergency department staff as the patient's care was transferred to an emergency physician. The iRevive PCR was then completed by an EMT with additional supplemental data pertaining to the transport and final patient disposition. A standardized narrative was completed to finish the event documentation. The PCR was then saved on the handheld and uploaded to the 10Blade server allowing immediate access to the PCR for emergency department, ambulance service, and billing.

Results

iRevive was used in conjunction with the standard method of EMS documentation at Professional Ambulance for 16 emergency transports. Of these transports, 12 were calls for medical help, 3 for trauma, and 1 for an assist. iRevive was used by 12 of the 50 field providers at Pro.

It was noted that both the methods captured the same type of data. However, as the iRevive data were electronic, once captured, they were easily ported to the necessary storage and are now available for any further research or analysis. There were difficulties encountered while capturing the data using iRevive which are presented in the next subsection. These difficulties were taken into consideration for the next version of iRevive.

Difficulties

The main goal behind deploying iRevive in the field was to obtain feedback and find out what problems should be addressed for the next release. Problems experienced can be categorized into two main categories: usage problems and need for more data points (Tollefsen, 2003).

Usage Problems:

- History of present illness page had to be navigated prior to discovering key information. This page would be better placed at the end of the report.
- Placement of times and mileage page at the beginning of the PCR required jumping back from pages in use to record times during procedures.
- Dispatch, scene arrival, scene departure, and hospital arrival times are recorded by the dispatch center; this information could be automatically synchronized with the 10Blade server in place of manually recording each time in iRevive.
- Program did not allow a PCR to be saved unless certain fields were completed.
- Inability to return to saved PCR until synchronized with server caused inability to continue to update report.
- Lengthy pull-down menus for certain items took too long to navigate; these could be improved by reordering or starting in a more appropriate range.

Need for More Data Points:

- Lack of appropriate values in some pull-down menus required manual entry of these values.
- Insufficient space to record multiple allergies.
- Inability to draw a picture of patient or scene. Addition of human figure in order to point to areas injured is recommended. Integration with a digital camera was suggested.

Discussion

Overall, the ability of capturing data in an electronic manner proved to be a huge success. Having electronic data available automatically and as soon as the data are captured is a huge gain from the current state of the art. However, in this domain, the granularity for the type of data and the data content are very important. Even though the current version of iRevive contains all 40 points of the Massachusetts Office of EMS prehospital data set, the need for additional data points was recognized. In order to capture as many data points as possible and make iRevive more attractive to a broader number of EMS providers, several additions have been proposed, including the following:

- Type of response: lights and sirens vs. with traffic flow.
- Location type: to explain difficulties or irregularities (e.g., if the incident occurred at a school, parents would not be present).
- Additional patient information, such as estimated weight.
- Signs and symptoms of chief complaint, such as headache or syncope associated with a cut on a patient's head.
- EMT's impression of the event and patient state (e.g., psychiatric issues).
- Cause of trauma and additional information about the cause, such as severity of automobile damage, and whether the patient was wearing a seat belt.
- Reason for transport (e.g., generalized weakness).
- Additional information about the pain a patient is experiencing during the physical exam.
- Condition of the patient's skin.
- Additional transport information, including position of patient and traffic delays.
- Signature page for capture of patient receipt.

Future Version of iRevive

The future version of iRevive aims to increase the usability and efficiency of the complete system. This will be done by the addition of more devices and functionality and increasing the role the ambulance plays (Figure 5).

It was observed the ambulance can play a key role in helping communications and transactions that take place during a response. In the future version of iRevive, the ambulance will play the role of a hub bridging real-time communication from the PDA to the hospital housing the central command. The ambulance will contain a wireless network connection (satellite, cell) as well a global positioning system (GPS) tracking its location and time. Information from the PDA to the central command will be exchanged in real time; the EMTs will be able to receive the patient's information immediately as well as be able to send the condition right away allowing for the hospital to make important decisions such as which hospital can facilitate to the patient's needs. This real-time data exchange will help routing of the ambulances efficiently. Furthermore, the PDAs as well as the sensor motes will also be equipped with GPS and wireless connections. EMTs will be required to match the PDA with the sensor mote of the patient to acquire the vital sign information.

The New Architecture

We plan to acquire and implement the following components (devices):

1. Sensors

   We plan to integrate additional customizable sensors, medical and nonmedical, that are capable of patient monitoring, data filtration, and ad-hoc networking. An example of such sensors would be chemical sensors to detect levels of toxicity and the GPS sensors to record time and location.

2. Web Services

   In order to ease construction of exchange partnerships and increase flexibility of integration with hospital legacy systems and yet-to-be-released communication protocols, the exchange of data will be transitioned into using Health Level-7 (HL-7) Web services. HL-7 is a widely used data definition and delivery standard that has been recommended by the National Committee on Vital and Health Statistics. The HL-7 data format, a Web service based on open standards that are being accepted by the medical community at large, allows any client to access iRevive's Web service as long as he/she has permission and follows the standards set forth in calling the data. A server node will be the manager of sensor streams. A default is a manager of sensor streams on ambulance which manages data for connected PDAs and GPS sensors. Other modes for the server node may include:

   - Operating with attached databases for further storage.
   - Acting as a forwarder to another server that is running with an attached database.
   - Act as an aggregator of multiple server nodes running on multiple ambulances.

An example of such a server node would be a daemon running on the ambulance listening for PDAs and hospital databases.

3. Client Device

   A PDA will subscribe to an ambulance to receive sensor streams as well as send PCR data for storage/transmission. Laptops may also be employed at different locations that receive the complete information of which ambulance and sensors are in which locations, and the type of conditions that are being treated. This will help create a FedEx routing of the ambulances.

Conclusion

The current methods employed by the EMS for gathering and recording data are outdated, time consuming, and error prone. New technologies and solutions must be employed to improve the system. iRevive provides a solution to overcome many of the problems. EMTs can save time and efficiently capture the necessary information. Furthermore, the data are in electronic format readily available for exchange and analysis with many different systems. A wireless environment for the exchange of data saves time and provides up-to-date real-time information. Also, iRevive can be employed on a variety of different platforms and devices, including but not limited to PDAs, laptops, tablets, and so forth. It was tested in the field and many improvements were made. More innovative changes are in the process: addition of many more emerging sensor and sensor network technology, embracing of industry standards and accepted data formats, a new level of prehospital care is on the horizon. In time, broad deployment of prehospital applications such as iRevive will allow healthcare providers to monitor and document in real time how various procedures and types of therapy affect patient status and outcome. As more data are accrued, new algorithms will be developed to accurately guide and control all types of medical care. Eventually, automated acute care algorithms will be developed to enable sensor-based, computer-controlled patient care.

Acknowledgments

This material is based upon work support by the National Institute of Health. Any opinions, finding, and conclusions or recommendations expressed in this material are those of the author(s) and do not necessarily reflect the views of the National Institute of Health.

References

Adnet, F., Lapostolle, F., Ricard-Hibon, A., Carli, P., & Goldstein, P. (2001). Intubating trauma patients before reaching hospital—revisited. *Critical Care*, *5*(6), 290–291.

Ananthataman, V., & Han, L.S. (2001). Hospital and emergency ambulance link: Using IT to enhance emergency pre-hospital care. *International Journal of Medical Informatics*, 61(2–3), 147–161.

Callahan, M. (1997). Quantifying the scanty science of pre-hospital emergency care. *Annals of Emergency Medicine*, *30*, 6.

Cheung, N.T., Fung, K.W., Wong, K.C., et al. (2001). Medical informatics—the state of the art in the hospital authority. *International Journal of Medical Informatics*, *63*(2–3), 113–119.

Delbridge, T., Bailey, B., Chew, J., et al. (1998). EMS agenda for the future: Where we are ...where we want to be. *Pre-hospital Emergency Care, 2*(1), 1–12.

Dunford, J. (2002). Performance measurements in emergency medical services. *Pre-Hospital Emergency Care, 6*(1), 92–98.

Foxlee, R.H. (1993). Computer-aided documentation. Quality, productivity, coding, and enhanced reimbursement. *American Journal of Clinical Oncology, 16*(5), 455–458.

Hill, J., Szewczyk, R., Woo, A., Hollar, S., Culler, D., & Pister, K. (2000, November). System architecture directions for networked sensors. *Proceedings of Architectural Support for Programming Languages and Operating Systems (ASPLOS)*, Cambridge, MA.

Mann, G.E. (2002). *Data capture in the United States healthcare system—the pre-hospital phase of patient care.* Unpublished master's thesis, Boston University School of Medicine–Division of Graduate Medical Sciences.

Mears, G., Ornato, J., & Dawson, D. (2002). Emergency medical services information systems and a future EMS national database. *Pre-hospital Emergency Care, 6*, 123–130.

Mikkelsen, G., & Aasly, J. (2001). Concordance of information in parallel electronic and paper based patient records. *International Journal of Medical Informatics, 63*, 123–131.

National Highway Traffic Safety Administration (NHTSA). (2001). National research agenda. Retrieved from *www.nhtsa.dot.gov/people/injury/ems/ems-agenda*

Silver, M.S. (1990). Decision support systems: Directed and non-directed change. *Information System Research, 1*(1), 47–70.

Silver, M.S. (1998). User perceptions of decision support system restrictiveness: An experiment. *Journal of Management Information Systems, 5*(1), 51–65.

Tello, R., Tuck, D., & Cosentino, A. (1995). A system for automated procedure documentation. *Computational Biology, 25*(5), 463–470.

Tollefsen, W. (2003). *The history, development and field trials of iRevive, a handheld mobile database for use in the pre-hospital setting.* Unpublished masters thesis, Boston University School of Medicine–Division of Graduate Medical Sciences.

Vahedi, M., Ayuyao, A., Parsa, M., et al. (1995). Pneumatic anti-shock garment-associated compartment syndrome in uninjured lower extremities. *The Journal of Trauma, 38*(4), 616–618.

Welsh, M., Myung, D., Gaynor, M., & Moulton, S. (2003). Resuscitation monitoring with a wireless sensor network. *Circulation, 108*(Supplement IV), 1037.

Chapter XV

Adoption of Mobile E-Health Service: A Professional Medical SMS News Service in Finland

Shengnan Han, Åbo Akademi University, Finland

Pekka Mustonen, The Finnish Medical Society Duodecim, Finland

Matti Seppänen, The Finnish Medical Society Duodecim, Finland

Markku Kallio, The Finnish Medical Society Duodecim, Finland

Abstract

This study investigates physicians' willingness to adopt a professional medical SMS news service in the Finnish healthcare sector. A concise survey using SMS mobile technology was conducted on March 5, 2003. Two hundred and fifty-nine out of 685 responded within 24 hours, and 90% of these answers were received within 6 hours after the survey was sent out. The response rate was 38%. Findings from this simple SMS survey showed that physicians had positive perceptions of the SMS news service. Nearly 60% of the respondents have used it. Some of the answers included spontaneous feedback about the SMS news service, which revealed valuable comments and suggestions regarding further improvements to it. The SMS survey as a new data collection technique needs academic attention. Implications and future research are briefly discussed.

Introduction

Mobile commerce (m-commerce) has been an important focus of research in recent years. Generally, m-commerce is defined as the extension of electronic commerce (e-commerce) from wired to wireless computers and telecommunications, and from fixed locations to any time, anywhere, and anyone (Keen & Mackintosh, 2001), that is, the use of mobile technologies and devices to provide, sell, and buy convenient, personalized, and location-based services. Many healthcare organizations are turning to m-commerce or wireless solutions in order to achieve better, more effective, and efficient practice management (Wickramasinghe & Misra, 2004). A number of companies are extending their Internet services for physicians for use with personal digital assistants (PDAs) or other mobile devices. For example, the use of PDAs among doctors is rising, and had reached 27% by 2001 in the United States (Harris Interactive, 2001). In Europe the leaders, in terms of the percentage of general practitioners who used PDAs in their practices, were The Netherlands (31%), the United Kingdom (18%), Spain (17%), France (11%), and Germany (10%) (Harris Interactive, 2002). Mobile e-health services might offer an answer to healthcare challenges in the 21st century (Goldberg & Wickramasinghe, 2003).

Text-based technology or short messaging service (SMS) is one of the underlying technology platforms for m-commerce. Compared with wireless Web-based technologies, that is WAP, another m-commerce platform, SMS has a simple user interface and is supported by most mobile phones. Recent years have seen the adoption of SMS worldwide in many sectors of commerce, for example, news, weather forecasting, retail, entertainment, and so forth.

The rapid diffusion of SMS has also inspired some applications in the healthcare industry, for example, a professional medical SMS news service. Medical knowledge is changing constantly. It is not easy for physicians to keep their knowledge and information up to date to help in their patient care and patient management efficiently on the one hand, and to maintain the level of their professional competence on the other (Jousimaa, 2001). A professional medical SMS news service might help physicians keep their knowledge up to date and provide information about recent medical development trends and new discoveries. The aim of this chapter is to investigate how ready physicians are to adopt an available professional medical SMS news service that is currently implemented in the Finnish healthcare sector. First, we present the theoretical background underlying the study. In the following, we present the study context and the survey administration. The method we used for data collection is described in Section 4. In section 5, the results are reported. Discussions and conclusions are at the end.

Physicians' Adoption of Technology

Users' perceptions of and intentions to adopt an information system (IS) and the rate of diffusion and penetration of technology within and across organizations are two important foci in IS research (e.g., Straub, Limayem, & Karahanna-Evaristo, 1995). They are understood to constitute an essential aspect, property or, value of information technology (Orlikowski & Iacono, 2001). It is generally accepted that the usage of information systems at work could increase employees' productivity in their work, and improve individual and organization performance. System usage is an important dimension for measuring IS success (DeLone & McLean, 1992, 2003). In particular, physicians' adoption of IS in healthcare is aimed to improve the health quality of human beings. In the past few decades, the conclusions of many studies based on different theoretical approaches and research methods have proved and confirmed that usefulness of a supportive medical IS is very important in influencing a physician's decision to use it. For example, Chau and Hu (2002a, 2002b) based their study on classical IS adoption models (e.g., Davis, Bagozzi, & Warshaw, 1989; Moore & Benbasat, 1991) and collected data by means of a questionnaire to investigate physicians' behavior towards adopting new technology. They found that usefulness was one factor determined a physician's adoption of telemedicine technology. Jayasuriya (1998) conducted his research in a similar style. His study indicated that health professionals were willing to use technology in their jobs when they perceived it to be useful for their performance. Many studies have also been conducted using qualitative methodologies. For example, Mayer and Piterman (1999) studied the attitudes of Australian general practitioners towards evidence-based medicine by using a focus group method. They concluded that physicians' readiness to accept technology was influenced by its relevance (usefulness) to general practice, as well as the local contextual and patient factors. Using a prototype simulation study for 1 week in a hospital in Germany, Ammenwerth et al. (2000) successfully found that in order to meet the diverse requirements of different professional groups, the developer was needed to design "multi-device mobile computer architecture". Then, the usefulness of the mobile information and communication system in clinical routines was more likely to be improved.

Obviously, results from different studies by different research methods seem to confirm that usefulness is a fundamental determinant of physicians' readiness to adopt different technologies. The selection of different research methods varies according to the individual IS researcher and largely depends on the problem he/she aims to solve. As suggested by Moody and Buist (1999), the real question is not whether the research method is appropriate per se, but whether it is appropriate in order to answer the question being asked.

Study Context and the Survey Administration

The Finnish Medical Society Duodecim is a leading provider of medical knowledge and information in Finland. It has put much effort into improving the quality of medical knowledge. It has also adapted new technologies to distribute knowledge to physicians, for instance, CD-ROM, intranet, the Internet, and wireless technology (e.g., Han, Harkke, Mustonen, Seppänen, & Kallio, 2004a, 2004b, 2004c; Jousimaa, 2001).

Duodecim Publishers Ltd., owned by the Finnish Medical Society Duodecim, has developed and provided the SMS professional medical news service (SMS news service for short) for physicians currently employed in the Finnish healthcare sector, including hospitals, healthcare centers, private doctors, and so forth. The news team at Duodecim Publishers Ltd. maintains the service. The physicians subscribe to a specific number of services. After the service has been set up, the physicians receive one item of news a day. Currently, they do not need to pay for the service because of support from Pfizer Finland Ltd. Limitations of SMS technology mean that the service does not include any pictures. With the rapid diffusion of multimedia message (MMS) technology, pictures will be included as well. The SMS news service began in 2002.

After the SMS news service had been implemented for about 1 year, the developer, Duodecim Publisher Ltd. and its supporter Pfizer Finland Ltd., conducted a survey to investigate the physicians' acceptance of the SMS news service and their feedback. Time is of the essence for the service providers in order for them to respond to physicians' changing needs effectively and efficiently. Traditional methods of conducting surveys, for example, mailed paper questionnaires, usually take weeks or months to complete. Therefore, the use of a new medium to distribute the survey was needed. The practitioners commended SMS technology. The SMS survey, thus, was conducted on March 5, 2003. It was sent to the target group (685 physicians; among those, 259 subscribers to the SMS news service) as an SMS message, and the answers were gathered via the same medium.

Concerning the limitations of SMS technology, the survey had only one question, that is, "How do you perceive the SMS medical news service by Duodecim?" with three predefined answer alternatives, which were:

a. "The service is really good and useful."
b. "It is OK, but needs to be improved."
c. "I haven't subscribed to the service."

The survey was administered by a company that has experience of mobile marketing and is located in Helsinki. The question was sent out on March 5, 2003, at 12:52 p.m. The whole time span for data collection was 24 hours. On March 6, 2003, 12:00 p.m. sharp, the procedure was finished.

Data Collection

In Figure 1, we present the time span and numbers of answers collected between March 5, 2003, 11:00 a.m. and March 6, 2003, 11:00 a.m. The x-axis indicates the time series, categorized by hour. The left y-axis shows the numbers of responses received each hour and the y-axis on the right illustrates the cumulative sums of all received answers. Altogether 259 answers (out of 685) were received and the response rate was 38%. Two

Figure 1. Data collection procedure

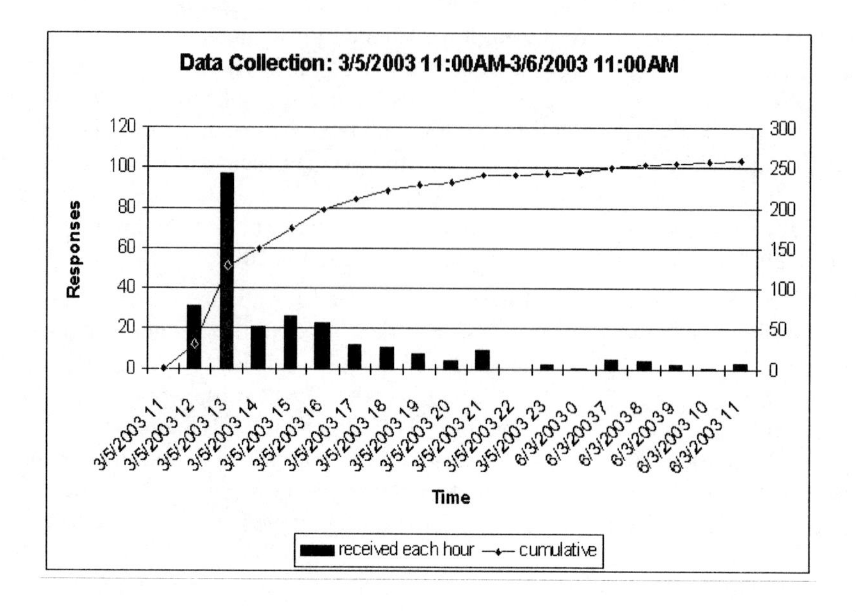

responses contained empty replies. Among the 259 who subscribed to the service, the response rate was extremely good, over 50%. Some of the answers included spontaneous feedback about the service, which was very valuable.

The majority of answers (>85%, n = 221) were gathered within 6 hours (before March 5, 2003, 19:00) of sending out the question, over 50% (n = 149) within 2 hours (before March 5, 2003, 14:00). Before midnight March 5, we received 243 answers, nearly 95% of the responses; only 16 were sent back on March 6. We soon noticed that the answers received each hour reached a peak between 1 and 2 p.m. on March 5, 2003. The number of responses during this time span was close to 38% (n = 97) of the total.

Results

A basic summary report of our quantitative data is shown in Figure 2. Of 257 usable answers, 31% (n = 79) perceived that the SMS news service was really good and useful. Seventy-three (28%) thought some improvement was necessary. Thirty-two percent of the physicians had not yet subscribed to it. We received 22 answers that did not select any of the predefined alternatives, and we therefore assigned them to D.

We analyzed the content of the feedback provided spontaneously with the three alternatives and other comments categorized as D.

Figure 2. Survey results

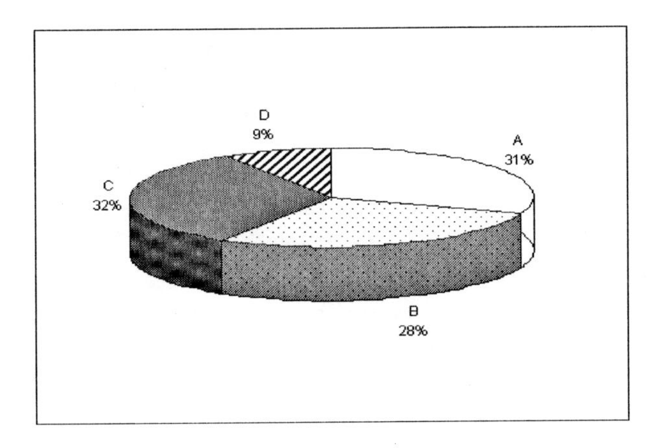

- **Arguments with A**

  Physicians who replied with the A option really liked the service and thought it was very useful for their work. They also showed a positive future intention towards the service. As one physician wrote:

  "[The service] is very useful and I would like to receive the messages in the future. ..."

  Although acknowledging the usefulness of the SMS news service, physicians still expected information from traditional channels. A physician replied:

  "The messages are excellent for me and I would be happy to receive other newsletters as well."

- **Arguments with B**

  Most of those who selected B did not specify very much about how to develop the system. Several physicians did give suggestions, however. One physician mentioned that

  "Occasionally one gets news from a short message. [It is] a different picture of the matter compared with the original text [e.g., in newspapers]."

  This message indicated that, as an emerging information distribution channel, mobile SMS was expected to be compatible with other traditional media, such as the texts published in newspapers or on the Internet. On the other hand, physicians also welcomed the SMS news:

"I have during weekdays several times had the time to read the same news on the Internet, but it's good to have repetitions."

Three physicians suggested that the SMS news might include more extensive sources or specialized fields.

"... [the news might come] along with Internet references or more extensive sources."

"... Develop the messages for specialized fields, i.e., information to the correct address...."

"... More accurate sources of references could make it easier to find the article (e.g., about the disadvantage of epilepsy medication)."

Those suggestions shed light on how to improve the content of the SMS news service. Reference sources that help physicians go further to check the latest medical information are quite demanding. Physicians also need information from their specialized fields that are better suited to meet their personal requirements as to the specific knowledge they want to acquire.

Some physicians pointed out that the SMS medical news should be longer; the shortness of SMS was considered a drawback. Comments were received such as:

"The messages could be even longer."

"Apparently the shortness of SMS causes limitations."

- **Arguments with C**

  The feedback came with option C provided several reasons why physicians did not subscribe to the services. The first and obvious reason was they did not need the SMS news service, for example:

  "Not used and not needed."

  "Not used and I don't want it."

  The second reason was some physicians still stick to other distribution channels, such as newspapers or the Internet.

"I haven't followed the text message news. I read newspapers."

"It is easier to read from the Internet."

The third reason was that technical problems with the SMS news service and physicians' lack of relevant technical knowledge hindered their usage of the SMS news service. Complaints included:

"I have not received any news even though I have subscribed."

"I have tried to get the service without success...."

"Not used since I don't know how...."

The fourth reason was the small screen of the mobile device: *"I would prefer to read them as e-mails due to the small screen of my mobile phone ... probably better with the communicator."* The size of the screen is not comparable with PC-based media apparently.

Some physicians also lacked the time to subscribe to the service: *"I'm in principle interested, but haven't had time to pick up the instructions."* Physicians are usually very busy. Lack of time is frequently cited as a reason for not adopting new information systems (Berg, 1999).

Besides those negative comments, we also found positive intentions from some nonadopters:

"I will activate."

"I would like to have in use..."

"Reading news as SMS can in principle be feasible."

- **Arguments collected from D**

  A content analysis of the feedback we classified as D was conducted as well. We found that there were three main suggestions from the physicians:

1. The service still had some bugs; problems with receiving and opening the service were common complaints. For example:

"None of the text messages that I ordered have been activated, so I have not received any news at present."

"I have not received any news even though I got confirmation of my order."

"For some reason, I have not received news for a long time."

"I acted according to instruction but didn't receive any messages."

"How is it activated?"

These messages indicated that the developer and the service content provider have to pay more attention to the technical problems associated with the SMS news service. The growing number of complaints of the difficulties in activating and receiving it might be a severe barrier for physicians to continue using it.

2. The current length of the message (160–200 characters) was considered both a limitation and an advantage. We got freely formulated feedback of the kind:

"160 characters is quite a limited amount. But on the other hand, one doesn't bother to read too long stories."

"During office hours a short and concise [message] is better...."

Obviously the limited length means that the news has to be concise. This means that they can then save time when reading the news during a busy working day.

3. The news service is quite general; more categories of specialties and more in-depth news are required. For example:

"Seems like a good idea. Personally, I would like to choose those fields that interest me."

"I would wish for more information about urology...."

These messages revealed that physicians demand more services that meet their "personal" needs. The current service is targeted at all physicians as one uniform group. The developer has to take "personalization" into consideration as a further improvement to the SMS news service.

Discussions

On the Survey Question

The findings from this simple SMS survey showed that physicians had positive perceptions of the service. Those responses for options A and B indicated that, generally, the ones (31%) who were happy with the service really liked it, and an almost equal number (28%) thought the service should be improved. The total answers in categories A and B reached 59%.

Arguments came with the predefined alternatives and those categorized as D overlapped in some cases. In brief, they highlighted several important issues concerning how to improve the service in the future. First, technical problems have to be solved immediately. It constitutes a barrier to physicians' usage of the service continuously, or hinders their subscriptions. Possible guidance on how to open and use the service should be improved as well. Ease of use is always considered a crucial driver for the success of an information system. Mobile technology is no exception (Han, Mustonen, Seppänen, & Kallio, 2004d). Second, the length limitation of SMS technology means not only that the news has to be concise, which was considered a time-saving way to read it, but also made it impossible to include more information in one message. With the development and diffusion of MMS, technically the news items could easily be made longer. Third, the physicians demand "personalization." The fact that the current service included only general topics did not exploit to the full advantages of mobile technology, that is, personalized services to mobile users. Therefore, it might undermine the value added by the SMS medical news service. A personalized service for physicians' news needs would enhance their knowledge, which would be highly relevant to their specialties, and provide the information that they desire most. Fourth, as an emerging technology, SMS competes with traditional channels of distributing medical news. The traditional channels, for example, newspapers and the Internet, have dominated information delivery for a long time. Currently, SMS is a complementary channel, but far from the dominating one. A good thing is that it is warmly welcomed by most of the physicians. It has strong growth potential for the future. It adds comparatively more value to physicians' work in terms of flexibility and personalization or localization.

Nearly one third of the respondents had not yet subscribed to the SMS professional medical news service. Attention has to be paid to the reasons why they did not subscribe to the services in order to increase the speed of its diffusion. Compared with the important issues for improving the service referred to above, we find that strong efforts to improve the service might remove the negative "reasons" that impede some physicians from subscribing to it.

On the Data Collection Method

The survey was supported by SMS mobile technology, namely, an SMS survey. We received a 38% response rate within 24 hours, and 90% of responses came within 6 hours after the survey was sent out. We acknowledge the pros and cons of the SMS survey.

Pros

- It is a time-efficient method. The response time is very short compared to other media. Most of the respondents answered within 6 hours after the question was sent out. Within 24 hours, or 1 day, the procedure was completed.
- It is a method that attracts more responses. Compared to traditional survey methods, which usually receive response rates lower than 20%, the survey received

a 38% response rate, which was very encouraging. It is also worth highlighting that we received such a high response in 1 day. This is not possible via other media.

- It is a cost-effective method. By using SMS technology, the practitioners have saved a lot of money compared with surveys conducted by other media, for example, paper mail. In Finland, an SMS message costs less than 0.16 Euro.
- It is a method that makes a survey focus on the most important questions. So far, the limitation of the SMS technology, only allowing 160–200 characters, does not permit inclusion of many questions.
- It is a method supported by new mobile technology. With the development of new technologies, IS researchers try to adapt them to conducting specific studies. For example, tape recording, video recording, Internet, and so forth. The adoption of SMS mobile technology is a trend.
- It is a method containing both quantitative and qualitative data. In the survey, we obtained answers to our predefined alternatives as well as freely formulated feedback. This feedback provided us with fruitful and interesting suggestions, for example, how to improve the SMS medical news service, or why someone did not subscribe to the service.

Cons

- It is a method restricted by technological limitations. The length limitation, 160–200 characters, made it impossible to include more questions that might cover more aspects of a research focus.

Beside those pros and cons of the SMS survey, conducting the survey itself to investigate physicians' acceptance of the SMS medical news service was a very successful experience. Time is crucial for the practitioners to adjust their business models according to the changing demand of customers. Apparently, adoption of mobile technology to support this research is more efficient than other media.

Conclusions

Users are a key driver for the success of mobile commerce as well as mobile e-health services. It has become increasingly important for IS researchers and practitioners alike to investigate the important factors triggering users' adoption behavior in the new technological context. This chapter is intended to investigate physicians' acceptance and opinions regarding the SMS professional medial news service, a simple mobile e-health service. Physicians' opinions were evaluated using data collected from 257 respondents practicing in the Finnish healthcare sector. Some implications can be drawn from the findings of the study.

An important further development of the mobile e-health service would be to enhance "usefulness" by "personalizing" it. User adoption theories assert that usefulness is the main driver for an individual to adopt a specific information technology. The perceptions of usefulness of the SMS medical news by physicians demonstrate strong possibilities for continuing usage in the future. Providing personalized service is a better way to enhance the "usefulness" of the service. A successful mobile service should be flexible, add value, and take full advantage of mobile technology (e.g., Carlsson & Walden 2001, 2002a, 2002b). It is quite demanding to personalize the SMS professional medical news service according to physicians' different information needs. System developers and contents providers alike should move beyond the limits of current possibilities and devise a service that physicians must have in their working and daily lives. Physicians' work is characterized as busy, pragmatic, and lacking in time (Berg, 1999). A personalized SMS medical news service, on the other hand, fits their work practice. Concise and desirable personalized SMS medical news will both save their time in a busy working day and provide them with new medical knowledge at any time and in any place.

From a managerial standpoint, the findings of this study reveal that, in order to facilitate physicians' motivations to use a technology, it is important to have proper training programs or information sessions to help overcome potential technical difficulties, and also to focus on how the medical news service can help keep their professional knowledge updated. It is better to encourage and cultivate positive intentions towards using the SMS medical service in the work practice. We have to be aware that it is the users and their use, not advanced mobile technology that will drive m-commerce growth to a new level (Jarvenpaa, Lang, Takeda, & Tuunainen, 2003). Obviously, it is equally important for any mobile e-health service as well.

Using SMS to do market research, so-called mobile marketing (Haig, 2002), is not new. The survey reported in this chapter has revealed the benefits of such research. It is a new data collection technique that is time efficient and cost effective. It provided answers in a very short time. The medical news service developer and content providers received "instant" results that help them investigate the problems and find solutions almost at the same time. The SMS survey is not like traditional surveys that usually take months or years; it is a new attempt to do the survey in hours or days with the aid of new mobile technology. It also reduces the cost of the survey. Mail surveys are usually very expensive, as are Internet surveys, which charge a monthly fee. The main weakness of the SMS survey is obviously its length limitation; it excludes the possibility of many questions in the survey. However, it means concentrating on the most important message that practitioners want to know immediately. Such instant feedback might represent high business value for practitioners.

From the management perspective, the SMS-based data collection technique is a very important form of two-way communication and an interaction mechanism for contacting the users—physicians—in the study directly. It is a much more powerful way to speak with physicians and learn their experiences in a relatively short time. It would be very beneficial for service providers to improve the SMS medical news service according to physicians' real needs. The responses from nonadopters revealed that service providers also could adapt the SMS text-messaging survey to inspire awareness of this new mobile medical service, which might lead to mass adoption quickly (Rogers, 1995).

Future research efforts are needed to address these questions rigorously. We might define a complete set of alternative answers for the question (from positive to neutral to negative). We also could ask permission from more physicians to participate in such surveys in order to receive more information both for service development and management purposes. The technical limitations of SMS made it impossible to investigate more relevant questions. Therefore, possible integration with other data collection methods might be necessary.

The research reported here could be seen as a preliminary study reflecting physicians' reactions to a mobile e-health service, especially an SMS professional medical news service. Since the vast majority of the respondents had probably used the SMS professional medical news service in their daily work, their appreciation of the service may be considered as a solid basis for future improvement of the mobile service in healthcare. As far as the external validity of our results is concerned, we have to be cautious since the findings describe the situation in Finland, a country characterized by a very high penetration of mobile phones, and a strong national "mobile fever." Physicians in the Finnish healthcare sector seem to be no exception to this general rule.

Acknowledgment

Our warmest thanks go to Pfizer Finland Ltd. for its support for the survey. The first author would also like to thank Ms. Stina Störling-Sarkkila for her support in translating the Finnish data into English.

References

Ammenwerth, E., Buchauer, A., Bladau, B., & Haux, R. (2000). Mobile information and communication tools in the hospital. *International Journal of Medical Informatics, 57*, 21–40.

Berg, M. (1999). Patient care information systems and health care work: A sociotechnical approach. *International Journal of Medical Informatics, 55*, 87–101.

Carlsson, C., & Walden, P. (2001, August 6–12). The mobile commerce quest for value-added products and services. *Proceedings of SSGRR 2001*, L'Aquila, Italy.

Carlsson, C., & Walden, P. (2002a, June 17–19). Mobile commerce: A summary of quests for value-added products and services. *Proceedings of the 15th Bled Electronic Commerce Conference, e-Reality: Constructing the e-Economy,* Bled, Slovenia.

Carlsson, C., & Walden, P. (2002b, July 4–7). Extended quests for value-added products and services in mobile commerce. *Proceedings of the International Conference on Decision Making and Decision Support in the Internet Age,* Cork, Ireland.

Chau, P.Y.K., & Hu, P.J. (2002a). Investigating healthcare professionals' decisions to accept telemedicine technology: An empirical test of competing theories. *Information & Management, 39*(4), 297–311.

Chau, P.Y.K., & Hu, P.J. (2002b). Examining a model of information technology acceptance by individual professionals: An exploratory study. *Journal of Management Information Systems, 18*(4), 191–229.

Davis, F.D., Bagozzi, R.P., & Warshaw, P.R. (1989). User acceptance of computer technology: A comparison of two theoretical models. *Management Science, 35*(8), 982–1003.

DeLone, W., & McLean, E. (1992). Information systems success: The quest for the dependent variable. *Information Systems Research, 3*(1), 60–95.

DeLone, W., & McLean, E. (2003). The DeLone and McLean model of information systems success: A ten-year update. *Journal of Management Information Systems, 19*(4), 9–30.

Goldberg, S., & Wickramasinghe, N. (2003, January 6–9). 21st century healthcare—the wireless panacea. *Proceedings of the 36th HICSS*, Big Island, Hawaii.

Haig, M. (2002). *Mobile marketing: The message revolution*. London: Kogan Page.

Han, S., Harkke, V., Mustonen, P., Seppänen, M., & Kallio, M. (2004a, May 23–26). Physicians' perceptions and intentions regarding a mobile medical information system: Some basic findings. *Proceedings of the 15th IRMA international conference*, New Orleans, LA.

Han, S., Harkke, V., Mustonen, P., Seppänen, M., & Kallio, M. (2004b, June 13–15). Professional mobile tool: A survey of physicians' perceptions of and attitude towards a mobile information system. *Proceedings of the 5th GITM conference,* San Diego, CA.

Han, S., Harkke, V., Mustonen, P., Seppänen, M., & Kallio, M. (2004c, June 13–15). Mobilizing medical information and knowledge: Some insights from a survey. *Proceedings of ECIS 2004*, Turku, Finland.

Han, S., Mustonen, P., Seppänen, M., & Kallio, M. (2004d, August 5–8). Physicians' behavior intentions regarding a mobile medical information system: An exploratory study. *Proceedings of AMCIS2004,* New York.

Harris Interactive. (2001, August 15). Physicians' use of handheld personal computing devices increases from 15% in 1999 to 26% in 2001. *Health Care News, 1*(25).

Harris Interactive. (2002, August 8). European physicians especially in Sweden, Netherlands and Denmark, lead U.S. in use of electronic medical records. *Health Care News, 2*(16).

Jarvenpaa, S.K., Lang, K.R., Takeda, Y., & Tuunainen, V.K. (2003). Mobile commerce at crossroads. *Communications of the ACM, 46*(12), 41–44.

Jayasuriya, R. (1998). Determinants of microcomputer technology use: Implications for education and training of health staff. *International Journal of Medical Informatics, 50*, 187–194.

Jousimaa, J. (2001). *The clinical use of computerised primary care guidelines*. Published doctoral dissertation, University of Kuopio, Finland.

Keen, P., & Mackintosh, R. (2001). *The freedom economy: Gaining the m-commerce edge in the era of the wireless Internet.* Berkeley, CA: Osborne/McGraw-Hill.

Mayer, J., & Piterman, L. (1999). The attitudes of Australian GPs to evidence-based medicine: A focus group study. *Family Practice, 16*, 627–632.

Moody, D.L., & Buist, A. (1999). Improving links between information systems research and practice—lessons from the medical profession. *Proceedings of the 10th ACIS* (pp. 645–659).

Moore, G.C., & Benbasat, I. (1991). Development of an instrument to measure the perception of adopting and information technology innovation. *Information Systems Research, 2*(3), 192–223.

Orlikowski, W.J., & Iacono, C.S. (2001). Research commentary: Desperately seeking the "IT" in IT research—a call to theorizing the IT artefact. *Information Systems Research, 12*(2), 121–134.

Rogers, E.M. (1995). *Diffusion of innovations* (4th ed.). New York: Free Press.

Straub, D.W., Limayem, M., & Karahanna-Evaristo, E. (1995). Measuring system usage: Implications for IS theory testing. *Management Science, 41*(8), 1328–1342.

Wickramasinghe, N., & Misra, S.K. (2004). A wireless trust model for healthcare. International *Journal of Electronic Healthcare, 1*(1), 60–70.

Note

An earlier and short version of the paper is published in the proceedings of the 4th International Conference on Electronic Business (ICEB2004), December 5–9, 2004, Beijing, China.

Chapter XVI

Outcomes of Introducing a Mobile Computing Application in a Healthcare Setting

Liz Burley, Swinburne University of Technology, Australia

Helana Scheepers, Monash University, Australia

Barbara Haddon, Intelligent Data Pty Ltd, Australia

Abstract

Interest in mobile computing applications has been increasing over the past few years. The healthcare sector has begun recognizing the potential for providing at point-of-care access to applications through mobile devices. However, there are challenges for the successful implementation of mobile computing applications. This chapter explores the implementation of a mobile computing solution in two Australian residential aged care facilities. The chapter compares the results of the implementation with previous studies and outlines a hierarchy of three levels of impact within the two organisations. The chapter furthermore describes the challenges, impacts, and outcomes. Finally it lists some strategies for alleviating some of the difficulties with mobile computing solutions.

Introduction

Interest in mobile computing applications has been increasing over the past few years. One indication of this is that by 2003, Microsoft had registered 11,000 applications, and now has more than 380,000 professional Windows mobile developers worldwide (Smith, 2004). The mobile computing applications of most interest to corporations are e-mail, calendars, sales force automation (SFA), customer relationship management (CRM), and field force automation (Smith, 2004). The healthcare sector has also begun recognizing the potential for providing at point-of-care access to applications through mobile devices, for the healthcare professional (Burley & Scheepers, 2004; McCreadie, Stevenson, Sweet, & Kramer, 2002; Rothschild, Lee, Bae, & Bates, 2002).

However, there are challenges for the successful implementation of mobile computing applications.

First, organisations need to plan carefully for the implementation of the mobile computing initiative. What are the benefits they hope to achieve by introducing the mobile computing initiative? What strategies will they put in place to ensure that they achieve those benefits? In a survey of 100 companies conducted by Optimize Research only 10% of the respondents had fully realized a return on investment (ROI) from their mobile-technology investment. It was suggested that the reason for this could well be lack of clear goals or coordinated strategy (Violino, 2003). One of the key findings in a recent study on achieving business value from implementing mobile computing was the importance of having a clear business objective and a "willingness to make business changes to embrace the transformation to core business processes which are driven by the mobile technologies" (Scheepers & McKay, 2004).

Second, there needs to be careful consideration of the business processes involved, the type of information accessed by the mobile application, and the required availability of that information. What information architecture will be deployed? Not all information needs to be in real time. As Evans (2003) notes, "the important thing about 'real time' is that it's not always the right thing to do. Not all business processes can be improved by speeding them up.... So in effect, business processes should run at 'right time' not 'real time' for your specific situation" (Evans, 2003). The ideal mobile computing application allows a balance between synchronised and real-time data (Haywood, 2002, cited in Synchrologic, 2003).

Third, organisations need to anticipate possible outcomes of the use of the mobile computing application for end users (Scheepers & McKay, 2004). What effect will the mobile computing application have on the day-to-day work practices of the end user? This chapter explores the planning, implementation and usage outcomes of a mobile computing solution in two residential aged care facilities. The chapter compares the results of the implementation with previous studies in the pharmacy sector and outlines a hierarchy of three levels of impact within the two organisations. The chapter furthermore describes the challenges, impacts and outcomes. Finally it lists some strategies for alleviating some of the difficulties with mobile computing solutions.

Literature Review

In 1998, the British National Health Service (NHS) issued a report outlining a strategy for making health information shareable between health providers and allowing individuals to view their own health record (NHSIA, 1998). A major part of the strategy was to provide a lifelong electronic health record for every person in the country by 2005. A report, *Building the Information Core,* issued in 2002, addressed the steps required to bring the 1998 vision into reality (NHSIA, 2002a). A further report, *Share With Care*, also issued in 2002, researched patients' and the public's attitudes to patient consent and confidentiality.

As noted in the *Share With Care* report:

> *Without access to appropriate information, a health system is, at best, inefficient and frustrating and, at worst dangerous. Modern healthcare services cannot function without those involved having the information they need to provide and receive care. (NHSIA, 2002b, p. 5)*

Healthcare is an information-intensive industry. As noted by Cho and Choi (2003), the healthcare industry is facing constant challenges to provide healthcare professionals access to patient information wherever and whenever it is required. They say this access can be achieved through mobile computing (Cho & Choi, 2003). Several recent studies have discussed the use of personal digital assistants (PDAs) to document healthcare services at the point of care (Brody, Camano, & Malony, 2001; Clark & Klauck, 2003; Lau, Balen, & Lam, 2001; Lynx, Brockmiller, Connely, & Crawford, 2003; Paradiso-Hardy, Seto, Ong, Bucci, & Madorin, 2003; Reilly, Wallace & Campbell, 2001; Scheepers & McKay, 2004; Silva, Tataronis, & Maas, 2003).

Table 1 lists some of these studies and summarises their findings. An intervention is any planned or spontaneous action taken to manage, modify, or assist with a condition or situation related to the recipient's needs.

With the exception of Brody et al. (2001), the studies found that documenting healthcare interventions on PDAs provided an advantage over more traditional (i.e., paper-based) documentation methods. The advantages listed were:

- more complete and standardized documentation (Clark & Klauck, 2003; Paradiso-Hardy et al., 2003; Scheepers & McKay, 2004)
- greater efficiency (Clark & Klauck, 2003; Scheepers & McKay, 2004)
- more interventions recorded (Clark & Klauck, 2003; Silva et al., 2003)
- greater user satisfaction (Clark & Klauck, 2003)
- increased visibility and recognition of work done by staff (Lynx et al., 2003; Paradiso-Hardy et al., 2003)

Table 1. Studies of pharmacy PDA applications

Author	Type of Application	Number of Users	Purpose of Study	Finding
Clark & Klauck (2003)	PDA system to document pharmacists' interventions. Developed using Satellite Forms.	Not stated in study.	"To determine whether PDAs allow pharmacists to document interventions more completely and efficiently and in greater number than interventions documented on an identical paper form" (Clark & Klauck, 2003, p. 1772).	There were more interventions per new order recorded over a 4-week period using the PDA (i.e., 919 per 13,184 new orders written) compared with interventions per new order recorded using the paper-based forms (697 interventions per 15,979 new orders written). "Data collected using a PDA were consistently more complete than data collected on paper" (Clark & Klauck, 2003, p. 1773).
Lynx et al. (2003)	PDA system to document pharmacists' interventions. Developed using Pendragon Forms v 3.2.	14 staff pharmacists, 17 pharmacy technicians	"To develop a systematic and uniform approach to documenting pharmacists' interventions that could be used to generate reports for the medical staff regarding the type and quantity of interventions or recommendations made to each physician" (Lynx et al., 2003, p. 2342).	"The initial report provided to the medical staff was met with astonishment. Both the number and magnitude of interventions and recommendations were far more than were expected by the medical staff" (Lynx et al., 2003, p. 2344). The report covered a 3-month period. "The PDA is an exceptional tool, at a reasonable cost, that can be used to improve and support the clinical interventions and documentation aspects of pharmaceutical care" (Lynx et al., 2003, p. 2344).

The time taken to record an intervention was found to be about the same for both PDA and paper-based systems (Clark & Klauck, 2003). An additional advantage was the ability to generate reports on the data collected. Paradiso-Hardy et al. (2003) stated that "unlike the paper-based system, the PDA-based data collection sheet standardizes documentation and generates reports that are comprehensive and consistent" (Paradiso-Hardy et al., 2003, p. 1944). A limitation noted by Paradiso-Hardy et al. (2003), however, was the limited view of data on the PDA.

However, there are challenges to achieving the above benefits in healthcare institutions. As Dickerson (2003) states, the implementation of IS systems "reflect a larger non-technical business-process change, but IT is often the front-line messenger of such change." Even if the system produces productivity improvement according to objective

Table 1. (cont.)

Author	Type of Application	Number of Users	Purpose of Study	Finding
Paradiso-Hardy et al. (2003)	PDA system to document the progress of heart valve replacement patients treated with warfarin. Developed using Pendragon Forms, v 3.2.	Not stated in study.	Not explicitly stated in study.	"Pharmacists have found the PDA to be a valuable, convenient, user-friendly patient-monitoring tool" (Paradiso-Hardy et al., 2003, p. 1944). "The various subforms serve as task lists to ensure that essential patient-related activities are completed as specified by hospital guidelines. Unlike the paper-based system, the PDA-based data collection sheet standardizes documentation and generates reports that are comprehensive and consistent" (Paradiso-Hardy, 2003, p. 1944).
Silva et al. (2003)	PDA system to document pharmacists' cognitive services. Developed using HanDBase relational database.	4 pharmacy residents, 4 clinical pharmacists	To investigate the use of PDAs to document "pharmacist cognitive services and estimate potential reimbursement" (Silva et al., 2003, p. 911).	There were more interventions recorded over a 6-month period using the PDA (i.e., 7,319) compared with interventions recorded over a previous 6-month period using paper-based forms (i.e., 5,028).

measurements, the end users will often complain that it slows them down. Dickerson (2003) suggests that the best way to alleviate this is to ensure that end users receive the training they need for the new system and also basic IT training.

Furthermore, it helps to have a healthcare professional system "champion" providing encouragement, training, and support. Wolf (2003) outlines the reasons for a successful implementation of a computerized physician order entry (CPOE) system in a U.S. hospital. They were detailed planning, executive commitment, a dedicated physician leading the effort, early-adopter physicians providing training to others, financial resources past implementation, and ongoing user support (Wolf, 2003).

Research Method

Case study research lends itself to the exploration of new areas of research (Eisenhardt, 1989) such as mobile computing. The research strategy allows for in-depth description of the relationships in context (Benbasat, Goldstein, & Mead, 1987; Galliers, 1993). The case research strategy was chosen here owing to the novelty of mobile technology applications within organisations (Yin, 1994).

Four in-depth, 1-hour interviews were conducted. One with the CEO of the aged care facility A, another with the director of Nursing at aged care facility B, and one of the directors of SmartHealth was interviewed twice. The interviews were relatively unstructured and the participants were free to discuss the main issues/advantages of the mobile computing initiative from their perspective. A qualitative approach was chosen to allow in-depth analysis of the perceptions of the participants in the study.

Background: Residential Aged Care in Australia

Government funded care for the elderly in Australia is provided either through community aged care packages (CACPs) for those living in their own homes or through residential aged care facilities for "live-in" residents. A report published in 2003 by the Australian Institute of Health and Welfare states (as of June 2002) there were 2,961 residential aged care facilities in Australian providing a total of 144,695 places (AIHW, 2003). In June 2003, the average residential care occupancy rate was around 96% and just over half the residents were 85 years or older. The majority of the residents were female (72%) and most of the residents received a government pension (77%).

Funding for Residential Aged Care

Residential aged care funding is paid by the Commonwealth government through the Department of Health and Ageing. The funding for residential aged care for 2001–2002 was AUD$4.0 billion. Each residential aged care facility is paid a subsidy through the health and aging payment system. For the year 2001–2002, the subsidy for a high-care resident was AUD$38,685 and for a low-care resident was AUD$13,380 (Commonwealth of Australia, 2002a). According to the *Aged Care in Australia* report, a high-care resident "usually involves 24-hour care. Nursing care is combined with accommodation, support services (cleaning, laundry, and meals), personal care services (help with dressing, eating, toileting, bathing, and moving around), and allied health services (physiotherapy, occupational therapy, recreational therapy, and podiatry). Low-level care focuses on personal services ... and provides nursing care when required" (Commonwealth of Australia, 2002b, p. 7).

Resident Classification Scale (RCS)

When a person is admitted to an Australian residential aged care facility their dependency level is assessed, as indicated by the Resident Classification Scale (RCS). There are eight categories, which represent eight levels of care in descending order of need from 1 to 8. Funding from the Commonwealth government (the Commonwealth care subsidy) is based on the level of care indicated by each RCS category (AIHW, 2003). The RCS is a funding tool that allocates funding dollars per resident per day based on the assessed care needs of the resident. The higher the care need, the higher the category and the higher the funding (Commonwealth Attorney-General's Department, 1998).

Levels 1–4 are classified as High Care, and levels 5–8 as Low Care. As of June 30, 2002, 64% of residents fell in the High Care category (19% – level 1, 25% – level 2, and 15% – level 3). Only 2% of the residents were classified at level 8. This is probably due to most of those residents being able to continue to stay at home and be part of the CACPs scheme (AIHW, 2003).

A care plan is prepared for each resident admitted, based on the level of care required. The nurses and carers for the resident follow the care plan providing services as indicated by the plan. If an additional service is required then this is noted as an exception to the plan. The nurse/supervisor assesses the exceptions and if it has been found that the resident is requiring additional care, the care plan is updated to reflect the higher level of care required.

The review of resident classifications, introduced as part of the Commonwealth Government Aged Care Act 1997, has had a major impact on residential aged care facilities. This is particularly evident in increasing the importance of maintaining adequate documentation of the type and level of care provided to each resident on a day-to-day basis. During 2001–2002, 11,200 reviews of RCS classifications were conducted and 36% of these resulted in reductions of funding, of which 4% were appealed (Commonwealth of Australia, 2002a).

The challenge for nurses and carers in the residential aged care facility is to ensure that every intervention they provide their patients is documented at the end of their shift. The amount of documentation required soon becomes very onerous and, if not carefully done, can result in the resident's classification being downgraded. This results in a decrease in funding to the residential aged care facility from the Commonwealth Care subsidy scheme.

The Vendor and the Aged Care Mobile Computing Application

SmartHealth, a small IS vendor in Melbourne, Australia, recognised the need for a computing application which assisted the staff in documenting all the care interventions conducted by the staff throughout their day-to-day tasks with the residents. They were in an ideal position to develop the software because one of their directors had previously been a nurse in the aged care sector and is very familiar with the processes required.

The OurCare mobile computing application was developed by SmartHealth to not only allow carers to document their day-to-day tasks with the residents, but also to allow domestic staff to enter observations, kitchen staff to be able to view dietary changes, and maintenance staff to view their tasks for the day and special needs for the residents. (Note: Both "OurCare" and "SmartHealth" are pseudonyms.)

Case A. AgedCare A

AgedCare A has 80 staff members consisting of office staff, division 1 nurses, personal care assistants (PCAs), domestic staff, a handyman, and kitchen staff. They currently only take low-care residents at registration stage and at the time of writing have 102 residents of whom 10% are now high care. All staff, except the office staff, use the handheld device in their day-to-day activities. The staff at AgedCare A began using the mobile computing application in February 2003.

Case B. AgedCare B

AgedCare B caters for high-care residents. They have 74 staff members consisting of office staff, division 1 nurses, PCAs, domestic staff, a handyman, and kitchen staff. At the time of writing they have 74 residents with an average age of late 80s to early 90s. The staff at AgedCare B began using the mobile computing application in October 2003.

Planning

The main objective therefore for the adoption and implementation of the OurCare mobile computing application was to improve the data capture of the day-to-day care interventions by the staff at both AgedCare A and AgedCare B. To ensure successful implementation of the application, SmartHealth worked with each residential aged care facility to prepare an implementation plan which suited their requirements. They also provided training on change management. Most often SmartHealth uses a phased implementation approach with only a few modules introduced at each stage.

Implementation: Business Processes and Information Architecture

As noted in the introduction of this chapter, there needs to be careful consideration of the business processes involved, the type of information accessed by the mobile application and the required availability of that information. The residential aged care facilities do not require real-time access to the data. It was considered sufficient to simply wait for breaks in the shift and at the end of the shift to upload the care interventions conducted during that shift. The data is therefore held locally on the PDA device until the staff member synchronises the PDA to upload work during morning tea, lunchtime, and at the end of their shift.

The OurCare system in both aged care facilities uses the off-line application model (Synchrologic, 2003). Staff log in to OurCare and synchronise the mobile device at the beginning of their shift to download resident interventions for their shift. At regular intervals during their shift the staff member logs off the OurCare system and synchronises the mobile device to upload work completed to that point (SmartHealth, 2004).

Table 2. Comparison of pharmacy PDA and AgedCare A and B findings

Findings From Previous Pharmaceutical Case Studies	AgedCare A and B
More complete and standardized documentation (Clark & Klauck, 2003; Paradiso-Hardy et al., 2003; Scheepers & McKay, 2004)	Yes
Greater efficiency (Clark & Klauck, 2003; Scheepers & McKay, 2004)	Yes
More interventions recorded (Clark & Klauck, 2003; Silva et al., 2003)	Yes
Greater user satisfaction (Clark & Klauck, 2003)	No evidence
Increased visibility and recognition of work done by staff (Lynx et al., 2003; Paradiso-Hardy et al., 2003)	Yes
The time taken to record an intervention was found to be about the same for both PDA and paper-based systems (Clark & Klauck, 2003)	Yes
Ability to generate reports on the data collected (Paradiso-Hardy et al., 2003)	Yes
Limited view of the data on the mobile device (Paradiso-Hardy et al., 2003)	Yes
Complaint that the system slows the user down (Dickerson, 2003)	Yes
	Greater continuity of care due to the data stored on the mobile system
	Provided training and upskilling to low- skilled staff

Both residential aged care facilities find that this works well. However, initially both experienced problems with staff being impatient and removing the mobile device prior to completion of synchronisation via the USB cradle.

The key decision making/reporting is done at the consoles (i.e., desktop PCs) by division 1 nurses. The console operators are responsible for monitoring the care plan for each resident, scheduling interventions, reviewing exceptions, generating action items, and updating resident care plans accordingly (SmartHealth, 2004).

Use of Mobile Computing Application

SmartHealth recognized the importance of ensuring that the residential aged care staff are comfortable using the OurCare system. They therefore placed a heavy emphasis on training the staff. At both aged care facilities, staff were surveyed to determine their level of expertise with computers. They found at AgedCare A, that most of the staff had poor IT skills with 40% who had never touched a computer. Initially SmartHealth ran a 2-day course on how to transfer data, then 4 weeks later, it ran a 2-day course on how to use the console. AgedCare A found this unsatisfactory and SmartHealth has now changed this to a 4-day off-site train-the-trainer course. AgedCare B, who implemented the OurCare system after AgedCare A, found that the revised training strategy worked well.

Comparison of Pharmaceutical Case Studies and AgedCare A and B Findings

The following table compares the findings of the implementation of the mobile system at AgedCare A and B with the findings from the mobile systems in the pharmaceutical industry.

Discussion

As noted by Scheepers and McKay (2004), the marketing of mobile devices has concentrated on the individual use of mobile devices and very little focus has been placed on the value of mobile computing at an organisational level. Clarke (2001) listed four value propositions which may be derived from mobile technology: ubiquity, convenience, localisation, and personalisation (Clarke, 2001). However, as Scheepers and McKay (2004) argue, these all come from the interaction between the organisation and the customer who is using the mobile device. This view is therefore limited in understanding other areas of potential impact from mobile technologies. Scheepers and McKay (2004) introduced another value proposition —that of internal value for the organisation as well as external value. The internal value proposition involves improving the effectiveness and efficiency of the staff within the organisation through mobile computing. The external value is derived from the interaction of the staff using mobile computing and their clients (see Figure 1). Based on these value propositions three levels of outcomes can be identified: managerial, mobile staff, and resident. On each level a number of outcomes can be identified. Following is a short discussion of the outcomes.

Figure 1. Mobile technologies enabling internal and external value propositions (adapted from Scheepers & McKay, 2003)

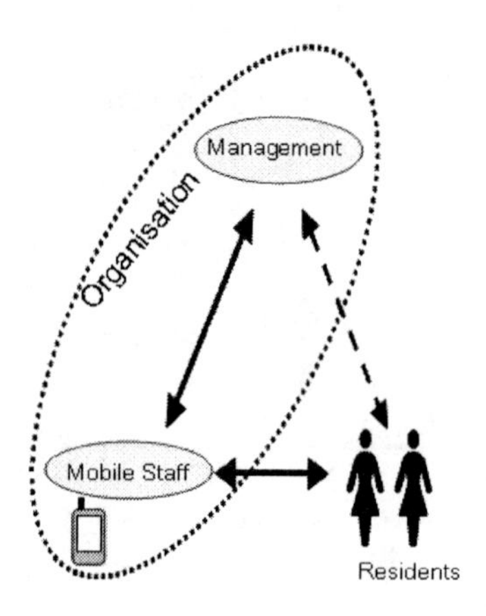

Managerial Outcomes

With the increased sophistication and complexity of care, and the increase in the number of healthcare professionals partnering in the care of the patient, it has become even more important to ensure continuity of care. To do this, each healthcare participant needs access to information describing past care interventions. This is one of the aims of the consolidated healthcare record. Even within a healthcare organisation, such as a hospital or residential aged care facility, it is important to have ready access to a central electronic health record to ensure continuity of care between shifts.

One of the major outcomes found after using the system for 6 months has been the increased availability and quality of management information regarding the number and type of care interventions performed by staff. As the director of AgedCare B said after 6 months of using the SmartHealth system, "It has allowed us to have more control over information and allows us to tie the loop on care requirements." She also mentioned how it stops things "falling through the cracks." According to the director of SmartHealth, the number of staff interventions per high-care resident has also been much more than expected. The managerial outcomes noted above are consistent with that documented in other studies regarding documentation of healthcare interventions (Clark & Klauck, 2003; Paradiso-Hardy et al., 2003; Silva et al., 2003).

Staff Outcomes

There is anecdotal evidence that the quality of the information recorded by staff is improving over time as they use the mobile device for recording interventions. The information is more consistent with greater detail provided in each intervention record. It has also raised the profile of some of the staff, particularly PCAs, domestic staff, and handymen as indicated by the following scenarios.

The director of AgedCare A, mentioned the following scenarios:

> *"One of our domestic staff was stripping the bed linen as usual in the morning for one of our male aged care residents when she noticed blood on the sheets. She entered this into her mobile device and then went on with her work. The device was synchronised at the end of her shift and the console operator entered an intervention – 'Call the doctor' for the aged care resident. The doctor came the next day and after the results of tests it was found that the aged care resident had bowel cancer."*

> *"Our handyman sometimes places a bet on the TAB for the residents and this is now put through the system as a form of assistance for that resident."*

Since the PCAs, the domestic staff and even the kitchen staff have access to the SmartHealth system through their mobile devices, they are now able to participate in the care of the residents in a more direct and visible way. In a sense, it has empowered these workers. This increased visibility and recognition of work done by staff is also consistent with other studies (Lynx et al., 2003; Paradiso-Hardy et al., 2003).

Resident Outcomes

There is also anecdotal evidence that the care has now become more resident-focused rather than staff-routine focused. The preferences of the resident are documented in their care plan and now that this has become more accessible and visible, via the mobile device, the staff are accommodating those preferences in the day-to-day care of the residents. This accommodation of resident preferences is also consistent with a study of a handheld computer-based system used by nurses for preference-based care planning (Ruland, 2002). The care plan is now updated more frequently and the care for residents is upgraded as their needs increase.

Challenges

On the whole, though, the challenge is to ensure the staff feel that the day-to-day entry of their care interventions is valuable. The director of AgedCare B has expressed concern that the staff do not see the value to them for the daily grind of data entry. Because of the size of the device it is difficult for them to get feedback. This is consistent with the limitation noted by Paradiso et al. (2003). One strategy to alleviate this has been proposed by the director of the AgedCare B. She has proposed that the staff should be allowed to view the console on the desktop PCs to see the amount of information that is visible to the console operator and how that information is acted on.

It has also been found that documenting the intervention on the PDA does not save time. The time for recording an intervention on the PDA is the same as the time for recording the intervention on the paper-based form. This is consistent with the findings of Clark and Klauck (2003). It is, however, important to note that the recording of transactions is done at the point of care and not at the end of the shift. This helps to ensure that information is not lost and is not forgotten. It also ensures that all staff are actively involved in the documenting process where previously it had been left to a few. The mobile system provides easy access to recording interventions, and also provides the carer with the personalized care plan for the resident.

Evaluation of Investment

Another way to evaluate the implementation of the SmartHealth mobile system is to evaluate the benefits and value that an organisation derives from such an implementation. In previous research a framework was used to evaluate the return on investment of mobile systems that is based on three broad types of benefits (AvantGo, 2003; Scheepers & McKay, 2004). Type 1 benefits are those associated with encouraging and motivating

Table 3. Evaluation of investment of AgedCare A and B

Type 1 Benefits: Improve adoption	**AgedCare A and B**	**Mobile impact**	**Revenue impact**
Data exchange quality	Carers get more detailed information about interventions required by residents.	D	I
	Management get increased accuracy in reporting of interventions provided to resident.	D	D
Data exchange frequency	None perceived. The exchange of data still provided on a daily basis.	-	-
User satisfaction	New system is easy to use for all staff. PCAs do not perceive the data entry to be less time consuming than the previous manual process. More visible contributions of other staff, such as domestic staff to the care provided to residents.	D	I
Outcomes	*Ready adoption*		

Type 2 Benefits: Reduce Operating Costs	**AgedCare A and B**	**Mobile impact**	**Revenue impact**
Care management	Increase the ability to control and manage the care plan of a resident and provide the Division 1 nurses with the ability to manage care provided to residents	D	D
Management of corporate information	Decrease paper documentation	D	I
	Less manual copying of information (fewer errors)	D	D
	Increased accuracy of records about residents and interventions performed	D	D
Management of Operations	Infrastructure costs increased	D	D
Outcomes	*Uncertainty in terms of net position*		

Type 3 Benefits: Increase organisational effectiveness	**AgedCare A and B**	**Mobile impact**	**Revenue impact**
Staff effectiveness	No change in capabilities of Carers	-	-
	More integrated view of residents care plan is available	D	I
Increased effectiveness in management processes	Division 1 nurses are able to access information about residents care plan more effectively which leads to faster re-evaluation and updating of care plans.	D	I
	Improved management of core business processes	D	I
Client value	Care more resident-focused rather than staff-routine focused	D	I
Outcomes	*Some definite effectiveness gains*		
Overall	*Overall gain. Provided AgedCare A and B with the ability to provide documentation to support their funding from government and to provide resident focused care to residents.*		

(Legend: D = direct, I = indirect)

the mobile staff to adopt and use the technology, a fundamental hurdle to be overcome if organisational benefits are to be derived. Type 2 benefits have to do with efficiency gains, in most instances associated with reduced costs or decreased time to complete a task. Type 3 benefits are associated with effectiveness gains that resulted from the mobile technology. For each identified benefit, an attempt was made by the researchers to decide whether this outcome was a direct or indirect result of the implementation of the mobile technologies (mobile impact), and whether the benefit would have a direct or indirect impact on costs or revenues (revenue impact). Table 3 outlines the evaluation of the findings for AgedCare A and B, followed by a short discussion of the findings.

Type 1 benefits relate to the adoption of the mobile system by staff. The staff of AgedCare A and B readily adopted OurCare. Challenges exist to ensure that all staff understand how their data entry contributes to the care provided to residents. Type 2 benefits relating to the reduction of operating cost did not indicate a direct reduction in cost of operation. The system did, however, have an impact on the organisational effectiveness (type 3 benefits) as a more integrated view of the care provided to residents is available. The overall evaluation of the mobile system is that it provides the aged care facilities with the ability to produce records of interventions to support their funding provided by the government. The mobile system also encourages staff to move from staff-routine care to resident-focused care.

Conclusions

This chapter compared the results of the implementation of OurCare into two residential aged care facilities with the previous studies in the pharmacy sector and outlines a hierarchy of three levels of impact within the two organisations. It was found that, on the whole, a number of the outcomes are similar to previously published case studies. It was, however, found that in the case of the residential aged care facilities, where novice computer users were using the system, some differences exist. The most significant difference was that the use of the mobile system resulted in training and upskilling of low-skilled staff. They also found that the small size of the mobile device inhibited the understanding of the users. The aged care facilities worked around this problem by providing all staff with read access to all of the information.

The findings from the previous pharmaceutical case studies were compared with the findings from the AgedCare A and B case studies and they were consistent except for greater user satisfaction where there was no evidence in the AgedCare A and B findings. There were, however, two additional advantages found in the AgedCare A and B case studies: greater continuity of care due to the data stored on the mobile system and the provision of training and upskilling to low-skilled staff.

The internal value proposition of the implementation of the mobile system was also evaluated and it was found that the system was readily adopted by staff and that it contributed to the organisational effectiveness.

A longitudinal evaluation of the findings in these cases is necessary to ascertain whether the value derived from the mobile system changes over time. The evaluation of the mobile system was done from an organisational perspective and a more in-depth study of individual user perceptions is required to gain an understanding of how the mobile system impacted the work practices of individual users.

References

Australian Institute of Health and Welfare (AIHW). (2003). Residential aged care in Australia 2001-02: A statistical overview, Australian Institute of Health and Welfare - AIHW cat. no. AGE 29. Retrieved 14 Sept, 2003, from *www.aihw.gov.au/publications/age/racsa01-02/racsa01-02.pdf*

AvantGo. (2003). Know your ROI ... Building the business case for mobile SFA (AvantGo White paper). Retrieved from *www.avantgo.com*

Benbasat, I., Goldstein, D., & Mead, M. (1987). The case research strategy in studies of information systems. *MIS Quarterly, 11*(3), 368–386.

Brody, J., Camano, J., & Malony, M. (2001). Implementing a personal digital assistant to document clinical interventions by pharmacy residents. *American Journal of Health-System Pharmacy, 58*, 1520–1522.

Burley, L., & Scheepers, H. (2004). Emerging trends in mobile technology development: From healthcare professional to system developer. *International Journal of Healthcare Technology and Management, 6*(1/2), 179–193.

Cho, H., & Choi, J. (2003). Ubiquitous computing in healthcare. *Business Briefing: Global Healthcare, 2003*. Retrieved September 13, 2004, from *www.bbriefings.com/pdf/28/gh031_p_CHO.pdf*

Clark, J., & Klauck, J. (2003). Recording pharmacists' interventions with a personal digital assistant. *American Journal of Health-System Pharmacy, 60*, 1772–1774.

Clarke, I. (2001). Emerging value proposition for m-commerce. *Journal of Business Strategies, 18*(2), 133–146.

Commonwealth Attorney-General's Department. (1998). *Classification principles 1997 made under subsection 96-1 (1) of the Aged Care Act 1997*. Prepared by the Office of Legislative Drafting, Attorney-General's Department, Canberra, Australia.

Commonwealth of Australia. (2002a). Report on the operation of the Aged Care Act 1997. Retrieved September 26, 2003, from *www.ageing.health.gov.au/reports/acareps/rep2002.pdf*

Commonwealth of Australia. (2002b). Aged care in Australia. Retrieved July 30, 2004, from *www.health.gov.au/acc/about/agedaust/agedcare.pdf*

Dickerson, C. (2003). Sometimes, IT can't win. *InfoWorld, 25*(41), 26.

Eisenhardt, K. (1989). Building theories from case study research. *Academy of Management Review, 14*(4).

Evans, N. (2003). Is it the right time for "real time"? Business needs must dictate. *InternetWeek*. Retrieved April 11, 2004, from *www.internetweek.com/showArticle.jhtml?articleID=9400108*

Galliers, R. (1993). Choosing information systems research approaches. In *Information systems research: Issues, methods and practical guidelines* (pp. 144–162). Oxford: Blackwell Scientific.

Haywood, S. (2002). *Gartner's Adaptive Application Architecture. Gartner*, 19 July.

Lau, A., Balen, R., & Lam, R. (2001). Using a personal digital assistant to document clinical pharmacy services in an intensive care unit. *American Journal of Health-System Pharmacy, 58*, pp 1229-1232.

Lynx, D., Brockmiller, H., Connelly, R., & Crawford, S. (2003). Use of a PDA-based pharmacist intervention system. *American Journal of Health-System Pharmacy, 60*, 2341-2344.

McCreadie, S., Stevenson, J., Sweet, B., & Kramer, M. (2002). Using personal digital assistants to access drug information. *American Journal of Health-System Pharmacy, 59*, 1340-1343.

NHSIA. (1998). *Information for Health*. Retrieved April 12, 2004, from http://www.nhsia.nhs.uk/def/pages/info4health/contents.asp

NHSIA. (2002a). *Building the Information Core: Implementing the NHS Plan*. Retrieved April 12, 2004, from http://www.nhsia.nhs.uk/def/pages/info_core/contents.asp

NHSIA. (2002b). *Share with Care – People's Views on Consent and Confidentiality of Patient Information, Final Report 2002, Document reference 2002-IA-1099*. Retrieved April 12, 2004, from http://www.nhsia.nhs.uk/confidentiality/pages/docs/swc.pdf

Paradiso-Hardy, F., Seto, A., Ong, S., Bucci, C., & Madorin, P. (2003). Use of a personal digital assistant in a pharmacy-directed warfarin dosing program. *American Journal of Health-System Pharmacy, 60*, pp 1943-1946.

Reilly, J., Wallace, M., & Campbell, M. (2001). Tracking Pharmacist interventions with a hand-held computer. *American Journal of Health-System Pharmacy, 58*, pp 158-161.

Rothschild, J., Lee, T., Bae, T., & Bates, D. (2002). Clinician Use of a Palmtop Drug Reference Guide. *Journal of the American Medical Informatics Association, 9*(3), 223-229.

Ruland, C. (2002). Handheld Technology to Improve Patient Care. *Journal of the American Medical Informatics Association, 9*(2), 192-201.

Scheepers, H., & McKay, J. (2003). Delivering business value from mobile technologies: an empirical assessment of implementation outcomes. Paper presented at the *Proceedings of the Second Workshop on e-Business (WeB 2003)*, December 13-14, 2003, Seattle, USA.

Scheepers, H., & McKay, J. (2004, June 14-16). *An empirical assessment of the Business Value derived from Implementing Mobile Technology: A Case Study of two Organisations*. Paper presented at the ECIS 2004, Turku.

Silva, M., Tataronis, G., & Maas, B. (2003). Using personal digital assistants to document pharmacist cognitive services and estimate potential reimbursement. *American Journal of Health-System Pharmacy, 60*, 911-915.

SmartHealth. (2004). *OurCare Training Resource Book, Version 2.4.6, March 2004 (note SmartHealth and OurCare are pseudonyms)*.Unpublished manuscript.

Smith, B. (2004). Business Apps: Going for the Tried and True. *Wireless Week, 10*(1), 22.

Synchrologic. (2003). *A CIO's Guide to Mobilizing Enterprise Applications – Best Practices for developing a successful mobile architecture in the enterprise.* Retrieved April 11, 2004, from http://www.synchrologic.com/images/whitepapers/mobile_apps.pdf

Violino, B. (2003). *A Method to the Mobile Madness*. Retrieved September 10, 2004, from http://www.optimizemag.com/issue/026/gap.htm?_loopback=1, December 2003, Optimize Magazine, Issue 22.

Wolf, E. J. (2003). Critical Success Factors for implementing CPOE. *Healthcare Executive, 18*(5), 14.

Yin, R. K. (1994). *Case Study Research — Design and Methods* (2nd ed.). Thousand Oaks, CA: Sage Publications.

Section V

Mobile Technologies in International Markets

Chapter XVII

Market Configuration and the Success of Mobile Services: Lessons from Japan and Finland

Jarkko Vesa, Helsinki School of Economics, Finland

Abstract

This chapter introduces a novel analytical framework called Mobile Services Matrix (MOSIM), which is used as the basis of a comparative analysis between the Japanese and the Finnish mobile services markets. The results indicate that as the mobile industry shifts from highly standardized voice services towards more complex mobile data services, the vertical/integrated market configuration (i.e., the Japanese model) appears to be more successful than the horizontal/modular configuration (i.e., the Finnish model). A brief overview of the UK market shows that the leading UK mobile network operators are transforming the industry towards a more vertical, operator-driven market configuration. The role of national regulatory framework in this industry evolution process is discussed.

Introduction

There is a paradox in the mobile services industry in Europe today: Even though the industry will be experiencing a major transformation during the next few years, for the time being the business is going too well for the senior management of many telecom operators to take action to redefine their business models and strategies in order to be well positioned in the new era of mobile multimedia services (aka mobile Internet, mobile data services, or nonvoice services).

Take Finland, for instance. Once the Internet bubble burst and business risks that were built into the growth strategies of the leading telecom operators were realized, the growth-oriented senior management was replaced with new management teams with a strong focus in cost cutting and downsizing. Financial markets expected operators to clean up their balance sheets after notorious UMTS licence auctions and other unsuccessful attempts to become serious players in the international mobile market. As a result of this development, Finnish operators have delayed their investments in new services and networks (for instance, the first UMTS service for commercial use was launched by TeliaSonera in October 2004, even though Finland was one of the first countries in Europe to allocate spectrum for 3G!) and sold those parts of their operations that are not considered to be their core business. Against this background it is easy to understand why the leading Finnish network operators opt for the current situation where they keep on making nice profits in the saturated market instead of actively seeking to change the competitive arena. Although this approach is understandable from individual companies' point of view, recently there has been much discussion as to whether this kind of risk-avoiding strategy will turn Finland into a yesterday's hero when it comes to actively building the brave new world of mobile services. Operators can rest on their laurels for a few more years, but the longer they neglect developing their business models for the competition of the future, the more painful process lies ahead of them. Some people within the industry have realized the destructive nature of the current price-driven competition in the Finnish market: a representative of TeliaSonera expressed his concern that many of the players in the mobile market are there to cannibalize the market with their aggressive pricing schemes, not to develop the market in order to secure healthy business also in the coming years (Tietoviikko, 2004). The industry seems to ignore the fact that the worst is yet to come, as new disruptive technologies such as voice-over-WLAN, WiMax, and free Internet telephone services such as the Skype service become increasingly popular in the coming years, especially now as eBay acquired Skype in order to enhance their online auction platform.

Unfortunately mobile operators are not the only ones sticking to old voice-centric business paradigm and earnings logic. Even the national regulatory authorities fail to see the need to adjust the regulation of mobile markets to the changes in technology and in the business environment in general. While the mobile phone usage has slightly increased (i.e., the minutes of use), in mature markets like Finland, the decrease in call tariffs has led to a situation where average revenue per user has remained flat or even decreased. Although 2004 was regarded as highly exceptional due to the introduction of mobile number portability in Finland, there is a widely shared view that call tariffs will continue to fall at a rate of 20–30%.

The current development in the traditional mobile voice market has made the leading operators to turn their eyes on nonvoice services. However, so far the European operators have not managed to turn mobile multimedia services into a similar success story as their Japanese counterparts. This raises the question why, despite the similarity in services and content offered, have nonvoice mobile services not taken off as expected? Based on a comparison of two very different mobile markets, namely Japan and Finland, this chapter argues that the lack of success of mobile Internet services in Europe is more a result of wrong business models and industry structure than it is about quality of individual services or products. As mobile services evolve from highly standardized and commodized voice-based communication services (i.e., person-to-person communication) towards the personalized and complex world of digital content and services (i.e., mobile multimedia or person-to-content type services), new challenges emerge also for the creation and delivery of mobile services. I argue that mobile data services represent a "complex good," which Mitchell and Singh (1996) define as "an applied system with components that have multiple interactions and constitute a nondecomsable whole." In a complex system like this, the overall performance depends on component performance, as a chain is only as strong as the weakest link. In the closely integrated and interrelated world of mobile Internet services, not only must all components meet users' requirements, but all the elements of the service offering (i.e., networks, handsets, services, and content) must also work seamlessly together. As the analysis of the two mobile markets presented in this chapter will demonstrate, this is where Europe has performed much worse than during the previous paradigm shift in the mobile telephony industry, which was the transition to digital mobile networks in the beginning of the 1990s.

During the past 5 years, both Japanese and European mobile markets have been studied extensively. However, the focus of these studies is often technical (e.g., how various technologies used in handsets and networks have evolved during different generations of mobile telephony, or how technical standards have been adopted in various markets), or they have focused on macroeconomic issues (e.g., mobile phone penetration rates or how well competition works in different markets). In this chapter I will present a different approach that raises the level of analysis above the discussion of superiority of competing technical standards, which all too often overshadows more important issues. In fact, I argue that the European way of introducing new technologies to the market suffers from two mind-sets that are not optimal in the context of mobile services. The first one could be described as business reductionism. As Albert-Lászlo Barabási points out in his book *Linked*, there appears to be a widely accepted belief that "once we understand the parts, it will be easy to grasp the whole" (Barabási, 2003). By following the ideals of reductionism, some researchers have tried to understand the differences in the success of mobile markets by comparing various technologies used: What kind of markup language has been used? Are the mobile networks circuit or packet switched? How many pixels do cameraphones support? Unfortunately it looks like the dynamics of complex networked industries, such as the mobile services industry, cannot be explained simply by analyzing all the pieces of the puzzle separately. In all fairness I have to admit that this was the way I was thinking when I sat on a plane on my way to Tokyo to meet some of the key players of the Japanese mobile industry in October 2002. However, it did not take long to realize that there was something bigger than just technologies, protocols, and standards behind the success of mobile services in Japan, as we will find out later in this chapter.

The other ideal causing problems for the Finnish mobile market is that open standards seem to be the only goal worth pursuing, regardless the maturity of products and services offered. A good example of the "open-market thinking" that prevails in Europe was a conversation between a representative of J-Phone and some of my colleagues at the Helsinki School of Economics. When our Japanese guest presented the J-Phone business model where the operator controls all the key components of mobile services, several people in the audience raised the question whether this kind of model is acceptable. Their argument was that as the customers of J-Phone are not free to use other operators' networks, services, or content, this kind of model is not as elegant as the European open-market approach. Our Japanese guest asked why we Europeans always emphasize so much this question of openness and the freedom to move between every possible service provider's offering? Is it not possible that a consumer or business customer in some cases would prefer to deal with only one service provider, if that service provider offered high-quality services and content delivered in a seamless, easy-to-use way at a reasonable price? While being aware of the strong argumentation in favor of open markets and standards in economic theories such as network externalities, at least I could not help finding the Japanese concept attractive. Perhaps one reason for this reaction was the fact that I was still suffering from mental trauma caused by unsuccessful use of wireless application protocol (WAP) services. Without going into details in what went wrong in the launch of WAP in Europe, one could argue that the key players of the industry in Europe had probably too much faith in the power of open standards, while in reality the products and services that were built on the WAP protocol were everything but standardized.

By combining the two assumptions that lie behind the widely accepted notion of what would be an optimal mobile market from the European perspective, we end up with something that could be described as "standards-based technical reductionism," which aims at free and efficient competition. It is easy to see that this kind of market would look very much like today's PC business (albeit there we are talking about de facto standards, thanks to the dominance of Intel and Microsoft). Although there are many excellent qualities in the market structure and product architecture of the PC industry, the analysis of mobile services market will demonstrate several reasons why simply copying the PC business model does not work for mobile services—at least not in the current phase of evolution. One of the key findings of this analysis of the mobile industry is that the optimal industry structure and product architecture is a function of time, that is, one should not make the mistake of presuming that a product/industry configuration that has turned out to be successful in some industries at some point of time in history would necessarily be the right alternative for another industry or another point of time, despite the fact that they may share some similar characteristics.

As the previous discussion indicates, the objective of this chapter is to offer an alternate approach into the analysis of the structure and dynamics of the mobile services markets. I will first present a new analytical framework that has been derived from the Double Helix model by Charles Fine (1998). This framework will then be used in the analysis of two very different markets—Japan and Finland. As the reasons for choosing these two markets will be explained in detail later in this chapter, it is simply stated here that Japan and Finland represent the two extremes of the continuum of contemporary mobile services business models. However, in order to better understand the dynamics of mobile market

the UK market, which represents an
·emes. The analysis shows that the UK
integrated market configuration as in
and is stuck with a horizontal business
dicate that in a regulated industry such
llowed to implement business models
nore, recent development in Finland
nobile services market are reluctant to
trategic behavior. It is argued here that
e powerful vision of the kind of market

ything the key players in a given market
rds a more favorable one? What are the
oduct architecture for service providers
ing various markets if they all are very
mobile telephony, size of the market)?
me of these questions—or even better,
/een mobile service development and
he mobile industry within which these

ll begin by presenting a novel model for
3, I will apply the model to the Japanese
mobile market is analyzed by using the
son between the current status and future
tion 6 discusses the impact of market
/ices in a given market, and section 7

s Industry Matrix

hods have been used in the analysis of the
ireless industries. Various types of value
els, and business ecosystems approaches
ve followed academic research or market
nodels have certain good qualities, when
ndustry they seem to be deficient in one
y provide a snapshot of an industry without
pment one can expect to see in the industry.
urrently taking place in the mobile industry
complex mobile multimedia services, and
ks to open packed-based IP-networks), new
approa of this fast-moving industry.

One attempt to apply a new kind of approach to the research of the mobile industry was presented by Vesa (2003, 2004a) who applied the Double Helix model developed by Charles Fine (1998) in the context of the mobile services industry. Although the Double Helix model appeared to capture nicely the current trends in the mobile market, the model itself was criticized for lacking solid theoretical foundations. Some critics have even pointed out that the name "double helix" is not correct—a more appropriate name for the model would be the "Double Donut" or the "Pretzel Model"! Nevertheless, because the use of the Double Helix model in the context of mobile services has received also very much positive feedback, the matrix model presented in this chapter (see Figure 1) builds similar constructs as Fine is using in his model. However, in the Mobile Services Industry Matrix (MOSIM) the Product Architecture dimension has been replaced with Service Architecture in order to emphasize that despite of all the exciting technologies, the mobile services industry really is what its name claims it to be, a SERVICE industry. This line of reasoning is supported by recent statements within the service research discipline emphasizing the "nested relationship between service and goods" (Vargo & Lusch, 2004).

Let us take a closer look at the MOSIM presented in Figure 1. The matrix consists of two dimensions that are called Service Architecture and Industry Structure. The two extremes of the Service Architecture dimensions are defined as integrated service architecture and modular services architecture. Likewise, the Industry Structure of a given market represents either horizontal industry structure or vertical industry structure—or something in between. This last point highlights the fact that in both dimensions there can be, and very often are, various types of hybrid or intermediate industry structures or product structures (see Vesa, 2005, for an in-depth analysis of factors influencing product architecture and industry structure).

The matrix identifies four possible service architecture/industry structure configurations: (i) *Horizontal Modular* (HM), (ii) *Vertical Integrated* (VI), (iii) *Vertical Modular* (VM), and (iv) *Horizontal Integrated* (HI). The first two configurations, the Horizontal Modular and the Vertical Integrated, are the two extremes between which an industry

Figure 1. The Mobile Services Industry Matrix (MOSIM)

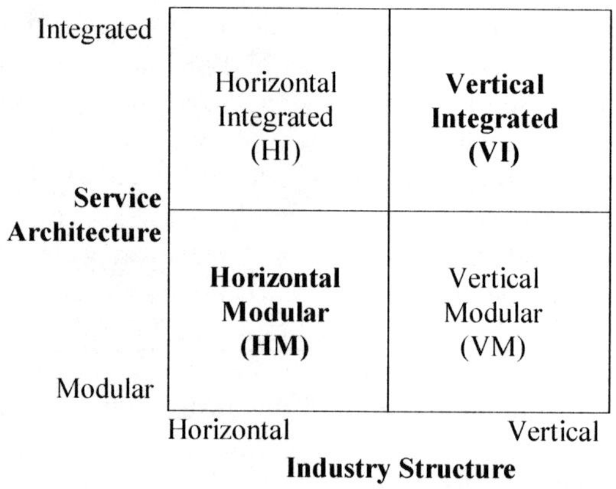

Service Architecture	Horizontal	Vertical
Integrated	Horizontal Integrated (HI)	**Vertical Integrated (VI)**
Modular	**Horizontal Modular (HM)**	Vertical Modular (VM)

Industry Structure

cycles in "an infinite double loop" (Fine, 1998, p. 43). However, the MOSIM allows us to identify also two other possible combinations, that is, Horizontal Integrated and Vertical Modular configurations. In the following section, the MOSIM will be used in the analysis of the Japanese mobile services industry.

The Mobile Services Industry in Japan

There is a huge amount of academic and business research covering the various aspects of the Japanese mobile market (e.g., Baker & Megler, 2001; Funk, 2004; Kodama, 2002; Matsunaga, 2001; Natsuno, 2003; Vesa, 2003, 2005). Japan is considered to be one of the leading markets in the field of mobile Internet services: approximately 86% of the mobile phone subscribers (as of December 2003) also subscribe to the mobile Internet services, and over 20% of mobile operators average revenue per user (ARPU) comes from data access fees. For NTT DoCoMo the figures are even more impressive: almost 90% of DoCoMo's over 47 million subscribers are using the i-mode service. Furthermore, for FOMA user (the 3G service of DoCoMO), data ARPU is about 32%, which translates into over US$30 worth of data services per user each month (NTT DoCoMo, 2004). In addition the data access fees, the operators take 9–12 % of the revenue generated by content and services offerings by third parties.

In Japan, the industry structure is vertically integrated (see Figure 2) which means that mobile operators control directly or indirectly all different levels of value chain. Carriers act as wireless Internet service providers, access providers, mobile phone providers, retailers, and content aggregators (Baker & Megler, 2001). The Japanese mobile industry

Figure 2. Japan: Vertical integration with integrated product architecture

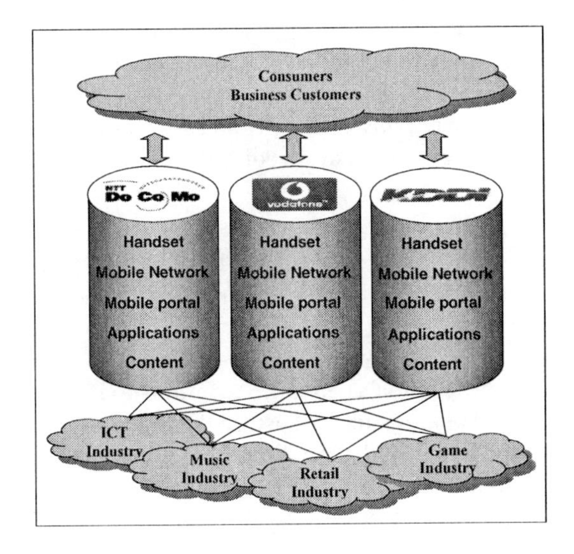

structure resembles the computer industry in 1975–1985 when the three largest companies (i.e., IBM, DEC, and HP) were highly integrated vertically (Fine, 1998). There is, however, one major difference: big part of the components used in the total service offering of the Japanese mobile operators comes from their partners.

In the Japanese mobile industry the competition is between business networks or ecosystems where the focal companies are the mobile operators (see, e.g., Natsuno, 2003; Vesa, 2003, 2005). An excellent depiction of the Japanese mobile industry is the NTT DoCoMo case study by Kodama (2002), which describes the creation of a broad business network around DoCoMo's highly successful i-mode service. As Figure 2 demonstrates, the Japanese mobile operators are linking together several different industries such as the music industry and the game industry.

There seems to be an excellent match between the requirements of the next-generation mobile multimedia services and the traditional Japanese way of doing business. According to Hoshi, Kashyap, and Scharfstein (1991), one key component of the Japanese business environment is the concept of *keiretsu*, which is an industrial group that "coordinates the activities of member firms and finances much of their investment activity" (p. 34). Berger, Sturgeon, Kurz, Voskamp, and Wittke (1999) have named the Japanese business networks "captive value networks" that rely on dominant lead firms, i.e. suppliers of various elements of the mobile services (e.g., handsets, network technology, applications, content) are highly dependent on one or a few key customer firms. Lead firms often "urge affiliated suppliers to adopt specific technologies" (Berger et al., 1999), as is the case in the Japanese operator-specific mobile networks and handset specifications.

The service architecture of mobile services is highly integrated in Japan. Mobile service offering consists of a handset, mobile network, and mobile portal (i.e., "a window" or "a door" to different kinds of mobile content and services) that are all closely integrated. Mobile phones are sold under an operator's brand and each phone model is designed to be used only in that operator's network. All three operators have their own mobile portal, which plays an important role in the mobile Internet business. Due to the limited screen size of a mobile phone and because of the huge amount of content available in mobile portals, it is very important how the links to various services are presented to the users when they access mobile Internet services. The last components of the Japanese mobile services business model are applications and content. Mizukoshi, Okino, & Tardy (2001) note that operators do not buy content from content owners or aggregators, nor do they create their own content, but they do control the content business through their certification process and billing service.

There are some disadvantages in the vertically integrated industry structure and integrated service architecture in Japan that have been identified in previous research. Perhaps the biggest disadvantage is that operators have to subsidize the price of the mobile phone—and the subsidy may be as high as 90% of the end-user price. Furthermore, mobile phone manufacturers are forced to develop and manufacture mobile phones that can only be used in one operator's network. According to Baker and Megler (2001), this increases the R&D and manufacturing costs in the Japanese system. It is important to keep in mind, however, that even the European "open and standardized" GSM world has been criticized for unfairness in the distribution of benefits resulting from standard-

Figure 3. Positioning the Japanese market in the Mobile Services Industry Matrix

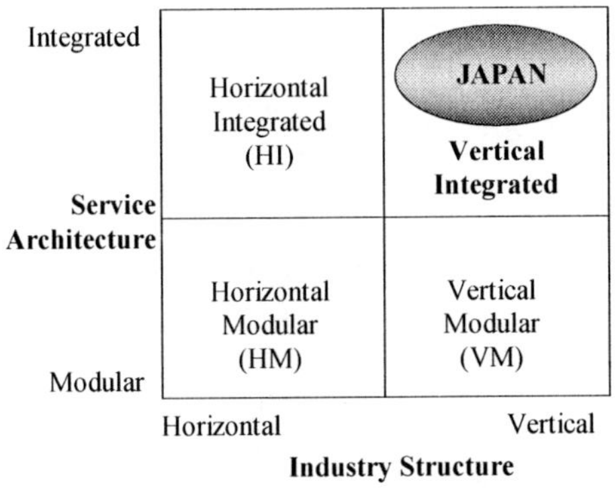

Service Architecture	Horizontal	Vertical
Integrated	Horizontal Integrated (HI)	JAPAN **Vertical Integrated**
Modular	Horizontal Modular (HM)	Vertical Modular (VM)

Industry Structure

ization. As the CEO of a leading European mobile operator pointed out in his presentation in the 3GSM World conference in Cannes in February 2004, the only one to benefit from the fact Nokia manufactures over 100 million phones per year is the company itself—the benefits of standardization remain within the walls of the mobile phone giant!

Due to the large domestic market of over 80 million mobile phone users, the Japanese operators have managed to reach a critical mass of users for their services, despite the lack of or limitations in the interoperability of the three commercial mobile networks in Japan (even though one could argue that at least from the technical point of view there are more than three mobile or cellular networks in Japan).

As Figure 3 demonstrates, the Japanese mobile services industry is positioned in the "Vertical/Integrated" quadrant of the MOSIM. This concludes our brief overview of the Japanese mobile services market. Next we will examine what the Finnish mobile market looks like.

The Finnish Mobile Market

Since the time Nokia introduced their first analog mobile phones and the state-owned telecom operator Telecom Finland (the predecessor of Sonera, which later merged with Swedish telecom operator Telia) opened their NMT (Nordic Mobile Telephony) network, Finland has been one of the world leaders of mobile telephony. The role of Finnish mobile market was also important during the launch of the digital GSM networks. Against this background it is easy to understand why the Finnish mobile market (despite the population of only 5 million people) is still considered to be an interesting subject for the research of the structure and dynamics of the mobile industry. The regulatory framework of the Finnish mobile market in particular receives much attention in different parts of the world, for reasons we will discuss later in this section.

The Finnish mobile services market is almost the opposite of the Japanese market. The industry structure in Finland is horizontal: the competition is taking place on each of the horizontal layers of the market, that is, operators are competing against each other, mobile phone manufacturers are competing against each other, and so forth (see Figure 4). Finland is one of the few countries in Europe (along with Italy) that does not allow the use of so called SIM-lock, which prevents the users from switching to another operators network without paying penalties. In addition, the legislation in Finland prohibits mobile operators from bundling mobile phones and subscriptions—or more accurately, customers' decision to buy subscription must not affect the price of the mobile phone, in case subscription and mobile phone are sold at the same time. One of the implications of the contemporary regulatory framework is that mobile phones and mobile subscriptions are sold separately—and sometimes through different channels. An important element of the mobile business in Finland is so-called "finders fee," which operators pay to independent retail stores and chains (such as specialized stores selling only mobile phones, departments stores, and electronics and household appliance resellers) for each new subscriber. Therefore, it can be argued that the regulatory environment has a direct impact on the business models of the mobile services industry in Finland.

Handset subsidies and the bundling of mobile subscription and handset will be allowed for 3G handsets during a period of two years starting in the beginning of 2006. However, for the 2nd generation GSM handsets, bundling remains prohibited. The implications to the structure of the Finnish mobile services industry remain to be seen . The second largest mobile operator, Elisa (which has cooperative agreement with Vodafone), and handset manufacturer SonyEricsson would have preferred "the Central European" model that allows handset subsidies. The legislative authorities decided to maintain the existing legislation; in other words, operators are not allowed to subsidize the handset price. According to a representative of MINTC, authorities are prepared to reevaluate the situation if, for instance, 3G services, which were launched at the end of 2004, does not take off as expected.

Figure 4. The horizontal and modular structure of the Finnish mobile market

Handsets	Nokia, Samsung, Siemens, Sony Ericsson, Motorola
Open standards (GSM, GPRS, EDGE, UMTS)	
Network operators	TeliaSonera, Elisa, Finnet
Service operators	Sonera, Saunalahti, Elisa, Kolumbus, DNA, ACN, Tele Finland, MTV3, Cubio, Spinbox, Fujitsu
Open standards (WAP)	
Mobile Portal	Sonera MobilePlaza, Zed, MTV3, Helsingin Sanomat, Buumi.net
Open standards (Java, XML)	
Applications	Java games, Browser, Messaging, Location-based services
Content	Movie trailers, Weather, Music, News

Figure 5. Positioning the Finnish market in the Mobile Services Industry Matrix

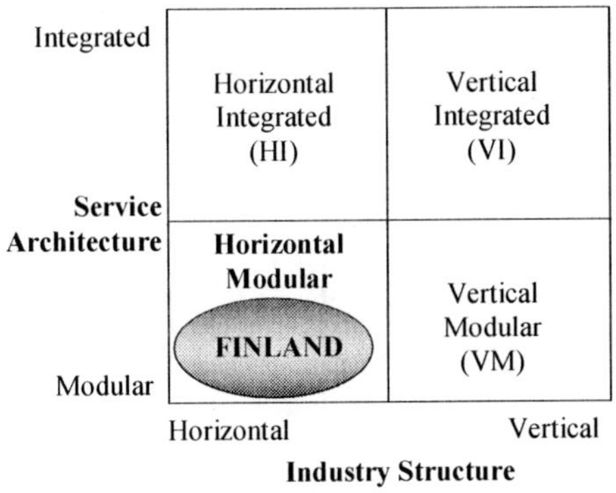

Integrated

Horizontal Integrated (HI)

Vertical Integrated (VI)

Service Architecture

Horizontal Modular

FINLAND

Vertical Modular (VM)

Modular

Horizontal

Vertical

Industry Structure

As Figure 5 demonstrates, the Finnish mobile services industry is positioned in the "Horizontal/Modular" quadrant of the MOSIM.

This concludes our brief overview of the Finnish mobile services market. In the following section we will focus on the differences between these two markets that represent the two extremes in the dichotomy introduced in the Double Helix model by Fine (1998) that has inspired the development of the MOSIM.

Differences Between the Two Mobile Markets

Next we will compare the structure and dynamics of the mobile services industries in Japan and Finland based on the MOSIM presented in section 2, and the analyses of the two markets presented in the previous two sections.

By now it has come very clear that the Japanese market (i.e., vertical/integrated configuration) and the Finnish mobile services market (i.e., horizontal/modular configuration) could not be farther apart from each other than what they are. This may not, however, be any permanent condition for neither of the markets. According to Fine (1998), neither vertical/integral nor horizontal/modular industry configurations are very stable as the forces of integration and disintegration are causing industry structures to oscillate between the two extremes in an infinite double loop. This led Fine to introduce the double helix of DNA as a metaphor when he wanted to describe the cyclic nature of industry evolution.

In Figure 6 the cycle between vertical/integrated and horizontal/modular is demonstrated by using the MOSIM. One of the benefits of the matrix approach when compared with

Figure 6. The cycle between vertical/integrated and horizontal/modular configuration

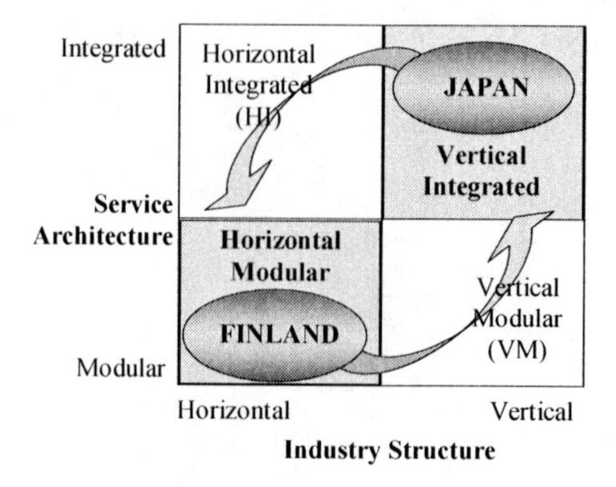

the Double Helix loop is that it identifies the two types of intermediate market or industry configurations between the two extremes. The mobile services industry in a given country may evolve from horizontal/modular (HM) to vertical/modular (VM), for instance, if an operator chooses to use open and standardized product and service components, but wants to achieve more control in the market by integrating vertically upstream or downstream in the industry value chain, or even to different but related industries. Likewise, after moving from proprietary technologies to open standards, an industry may enter a configuration where industry structure becomes horizontal, but at the same time operators try to maintain an integrated service architecture. Later in this chapter we will test the matrix model by adding a few more markets into the matrix.

Based on the analysis presented in this section, we can conclude that the industry structures of Japan and Finland are like day and night, or Yin and Yan. Next, we will try to see what is the impact of a given market configuration on the success of mobile services in the respective markets.

How Successful is a Given Market Configuration?

One of the objectives of this chapter is to study the relationship between an industry configuration (i.e., industry structure vs. service architecture), and the success of mobile services in a given market. In order to do this, we will next compare the key performance indicators (KPIs) of Japan and Finland. The assumptions behind this comparison are as follows:

Table 1. Comparison of the Japanese and Finnish mobile services markets

Key Performance Indicators	Japan	Finland
Mobile phone subscribers (1,000)	84,3131[1]	4,880[2]
Mobile network operators	DoCoMo, KDDI, Vodafone	TeliaSonera, Elisa, Finnet
Industry structure	Vertical	Horizontal
Service architecture	Integrated	Modular
% of users using mobile Internet (excl. SMS)	89%[1]	5%[7] (estimate)
ARPU US$/user/month	US$89[1]	US$48[3]
Data ARPU % (incl. SMS)	22–32%[4]	11–12%
Non-SMS data revenue (Person-to-content)	14.3%[5]	1%[6]
Minutes of use (MoU)	219	157[3]
Churn	1–1.5%[1]	20–30%[3]

[1] *Source: NTT DoCoMo, 2004.*

[2] *Source: Helsingin Sanomat, 2004.*

[3] *Source: Results of Q1–Q3 of FY2004, Elisa Oyj and TeliaSonera Oyj.*

[4] *Source: Vodafone, June 2004; NTT DoCoMo's October 2004 (FOMA service).*

[5] *Vodafone Japan, data revenue in quarter ended 30 June, 2004 (excl. messaging).*

[6] *Estimate by a representative of a leading telecom operator in November 2004.*

[7] *According to Statistics Central of Finland, 80% of Finnish mobile phone subscribers used SMS services in 2002.*

- If industry configuration has a role in the success of mobile services, then there ought to be a significant difference in the KPIs of Japan and Finland because the industry configurations are practically the opposite.
- Success of mobile services can be measured by using the KPIs of the markets and the leading operators in the markets.

Table 1 presents some of the key performance indicators of the two case markets. The values presented here are not exact values but estimates indicating the magnitude of a certain KPI (as all operators do not provide all the KPIs presented in this comparison).

What can we learn from the comparison of the KPIs of the Japanese and Finnish mobile markets presented in Table 1? Let us focus first on three items that are particularly interesting for the purposes of this chapter. As Table 1 shows, a huge majority of Japanese mobile phone subscribers use also mobile Internet services. For instance, over 89% of NTT DoCoMo's customers subscribe to the i-mode service (NTT DoCoMo, 2004). Unfortunately there is no such information available for the Finnish market, but according to some estimates the respective figure in Finland is less than 10%. What makes this comparison somewhat challenging is that in Finland SMS is used not only for person-to-person communication but also as a means of ordering ring tones, logos, wallpapers, and so forth, digital content to mobile phones. Premium rate SMS is also an important

Figure 7. The UK market in the Mobile Services Industry Matrix

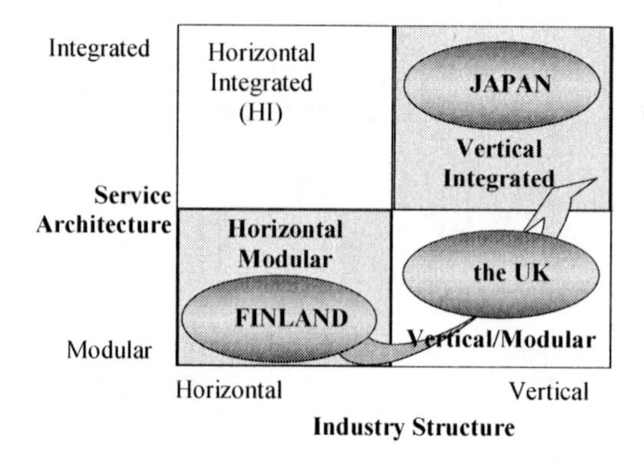

billing mechanism as operators do not offer any other micropayment services.

The second interesting item in Table 1 is the amount of data ARPU (average revenue per user) as a percentage of the total ARPU. In Japan data ARPU was 22% for Vodafone Japan's subscribers (mainly 2.5G users) and 32% for the subscribers of NTT DoCoMo's 3G services FOMA. In Finland data ARPU for the leading mobile operator Sonera was around 11–12% in Q3 2004. So despite the extensive use of SMS in Finland (for instance, the second largest operator's, Elisa's, subscribers sent on average 34 SMS messages per month; for Sonera this figure is below 30 messages per month), the Finnish operators are far behind the Japanese counterparts in making money with mobile data services.

The third item under scrutiny in this context is the amount of non-SMS data revenue, that is, the use of content over mobile networks (i.e., person-to-content services). Vodafone Japan reported in the end of the second quarter of 2004 that messaging represented 7.6% and data 14.3% of the total monthly ARPU. According to an estimate by a representative of a leading Finnish telecom operator, non-SMS data revenue in Finland is about 1% of the total ARPU.

Although this simple comparison of KPIs is not statistically relevant, it demonstrates the differences in the success of nonvoice mobile services in these two markets. Interestingly, both markets are considered to be advanced mobile markets—Japan because of the extensive use of mobile Internet services, and Finland due to high penetration rate of mobile phones and also because of the very low mobile phone call tariffs. It is also interesting that WAP services were introduced in the Finnish market roughly at the same time as NTT DoCoMo launched the i-mode service; however, while i-mode turned out to be a huge success story, WAP services failed miserably (although WAP technology is now starting to be an essential element of most mobile Internet services in Europe). Mobile networks in both countries were technically about on the same level although packet-based networks where first introduced in Japan. So if it was not about technology, timing, or people's willingness to adopt mobile technologies and services, why are mobile services much more successful in Japan and the gap seems to be growing? I argue that the vertical/integrated configuration of the Japanese mobile services industry has been more successful when the business paradigm is shifting from traditional voice-centric messaging-type services to content driven mobile Internet services.

Let us add one more market into our comparison. The UK market represents an interesting intermediary configuration between the Japanese and the Finnish markets. In the UK leading operators, such as Vodafone, mm02, T-Mobile, and Orange, have been building a new type of business model where mobile operators are orchestrating service delivery (i.e., moving from horizontal to more vertical industry structure) while maintaining more modular product architecture than the Japanese operators. If we add the UK market into the MOSIM, we can see that we have now occupied a new quadrant of the matrix.

What we can see in Figure 7 is that the UK market is moving towards the vertical/ integrated configuration, or the "Japanese model." This development is in line with the strategic statements by several leading European mobile operators who announced at the end of year 2002 that their goal was to increase operators' role in driving the mobile industry in the coming years. One of the reasons for this was the increasingly dominance of Nokia in the industry.

So how is the intermediate model of the UK doing, if we compare its success with Japan or Finland? The UK market is not even close to Japan but it has clearly advanced during the past 2 years. For instance, at Vodafone UK, data ARPU is close to 17% and the share of non-SMS data revenue is over 2%. If we compare these figures with the Finnish KPIs we can see that something has happened that has made the UK market more successful in the field of mobile services. Once again, I argue that this can be explained by comparing the differences in the market configurations.

Discussion and Conclusion

The objective of this chapter was to analyze the differences between the Japanese and Finnish mobile services markets by using the MOSIM. The matrix was derived from the Double Helix model developed by Fine (1998). The analysis revealed that the Japanese and the Finnish mobile industries represent very different kinds of market configurations: the Japanese market is a textbook example of vertically integrated industry structure, whereas the mobile services industry in Finland is currently in the horizontal and modular configuration. Our brief review of the UK market showed that there are also intermediate configurations between the two extremes. The transformation of the UK mobile services industry has been very fast, which indicates that the industry clockspeed (Fine, 1996; Mendelson & Pillai, 1999) is high in the mobile services business. What is particularly interesting from industry evolution perspective in this analysis is the role of regulation in the evolution of the mobile services industry. Both the Japanese market and the Finnish market have stuck to their existing regulatory frameworks. In Japan, the national regulatory authorities did not open up the market for competition, whereas in Finland the regulatory authorities did not allow the use of SIM-lock and the bundling of subscription and handset that are typical characteristics of a vertical/integral market configuration (Vesa, 2004b). At the same time, the leading UK operators are transforming their business model towards a more vertical and integrated operator-driven business model as the regulatory framework in the UK does not prevent them from doing this. This approach seems to increase the use of nonvoice mobile services in the UK market, as the key performance indicators of the leading mobile network operators indicate. However, year 2005 appears to be a turning point both for the Finnish and the Japanese markets. As discussed earlier in this chapter, the Finnish government is in the process of allowing 3G-handset subsidies for a period of two years. At the same time, the Japanese government is in the process of opening the Japanese market for two to three mobile operators. This development in the Finnish and the Japanese markets illustrates the kind of oscillation between different market configurations described in the Double Helix model by Fine (1998).

This paper took a very challenging approach by carrying out a "macro-level" analysis of a fast-moving service industry. The analysis of two different markets that are almost the opposite of each other gave us the opportunity to identify the key characteristics of these two extremes of the continuum between horizontal/modular and vertical/integrated configurations. Even though this approach illustrates the structure and dynamics of the mobile services industry, it suffers from a well-known limitation of macro-level industry analysis: this kind of analysis is always very descriptive by nature, and gives little practical advice on how to solve the problems identified. I believe, however, that it is important both for business people and researchers who are involved in the mobile business to pay attention also to the structure and dynamics of the mobile services market where they develop and market their services. Experiences from the Japanese and the UK market indicate that a key player with enough market power (e.g., NTT DoCoMo in Japan or Vodafone in the UK) can succeed in reshaping the entire industry towards a more favorable industry configuration. There is, however, one precondition for this: the regulatory framework must not prohibit this kind of natural evolution of the industry, otherwise there is a risk that the development of mobile market is halted. The challenge for national regulatory authorities is to have the wisdom to see beyond the wishes and demands of the dominant players that may be stuck with the business paradigms of the past. In Japan the key question is how to make sure that the regulation of the mobile services industry evolves as the mobile phone market is becoming mature and new 3G technologies are being implemented. For the Finnish authorities a major challenge is to analyze whether regulatory framework that worked well in the highly standardized GSM voice market is still viable as the business moves towards more complex mobile multimedia services.

References

Baker, G., & Megler, V. (2001). *The semi-walled garden: Japan's i-mode phenomenon.* IBM Red Paper.

Barabas, A.-L. (2000). *Linked.* New York: Penguin Books.

Berger, S., Sturgeon, T., Kurz, C., Voskamp, U., & Wittke, V. (1999, October 8). *Globalization, value networks, and national models.* Memorandum prepared for the IPC Globalization Meeting.

Fine, C.H. (1996, June 24–25). Industry clockspeed and competency chain design: An introductory essay. *Proceedings of the 1996 Manufacturing and Service Operations Management Conference*, Dartmouth College, Hanover, NH.

Fine, C.H. (1998). *Clock speed: Winning industry control in the age of temporary advantage*. Perseus Books.

Funk, J. (2004). *Mobile disruption: The technologies and applications driving the mobile Internet*. Hoboken, NJ: Wiley-Interscience.

Hoshi, T., Kashyap, A., & Scharfstein, D. (1991). Corporate structure, liquidity, and investment: Evidence from Japanese industrial groups. *The Quarterly Journal of Economics, 106*(1), 33–60.

Kodama, M. (2002). Transforming an old economy company into a new economy success: The case of NTT DoCoMo. *Leadership & Organization Development Journal, 23*(1), 26–29.

Matsunaga, M. (2001). *Birth of i-mode: An analogue account of the mobile Internet*. Singapore: Chung Yi Publishing.

Mendelson, H., & Pillai, R.R. (1999). Industry clockspeed: Measurement and operational implications. *Manufacturing & Service Operations Management, 1*(1), 1–20.

Mitchell, W., & Singh, K. (1996). Survival of business using collaborative relationships to commercialize complex goods. *Strategic Management Journal, 17*(3), 169–195.

Mizukoshi, Y., Okino, K., & Tardy, O. (2001, January 15). *Lessons from Japan*. Retrieved November 27, 2004, from http://telephonyonline.com

Natsuno, T. (2003). *The i-mode wireless ecosystem*. John Wiley & Sons.

Vargo, S.L., & Lusch, R.F. (2004). The four service marketing myths: Remnants of a goods-based, manufacturing model. *Journal of Service Research, 6*(4), 324–335.

Vesa, J. (2003, August 23–24). The impact of industry structure, product architecture, and ecosystems on the success of mobile data services: A comparison between European and Japanese markets. *Proceedings of the ITS 14th European Regional Conference*, Helsinki, Finland.

Vesa, J. (2004a, March). *The impact of industry structure and product architecture on the success of mobile data services*. Austin Mobility Roundtable, University of Texas, Austin.

Vesa, J. (2004b, Septmber 5–7). Regulatory framework and industry clockspeed: Lessons from the Finnish mobile services industry. *Proceedings of the ITS 15th Biennal Conference*, Berlin, Germany.

Vesa, J. (2005). *Mobile services in the networked economy*. Hershey, PA: Idea Group.

Chapter XVIII

Digital Multimedia Broadcasting (DMB) in Korea: Convergence and Its Regulatory Implications

Seung Baek, Hanyang University, Korea

Bong Jun Kim, Korea Telecommunications (KT) Marketing Laboratory, Korea

Abstract

The launch of portable Internet, alongside mobile Internet technology and cellular technology, is a new milestone, converging wireless with wired technology. Along with these new technologies, a new telecommunication service has been introduced and has received much attention from the Korean public. This is the Digital Multimedia Broadcasting (DMB) service. DMB is a digital multimedia service combining telecommunications with broadcasting technologies. DMB enables users to watch various multimedia contents on their phone screens while they are on the move. Since DMB services in Korea are the first in the world, the Korean Government has much interest in DMB services. However, the repeated failures in establishing a regulatory framework for DMB and ill-defined roles of players in the DMB industry interfere the diffusion of DMB in the Korean market. As the convergence of broadcasting and telecommunications makes progress, proper modifications of existing regulatory frameworks should be made in order to guarantee success of DMB service in Korea. This chapter reviews DMB technology, its business model, its market structure, and its policy. In particular, it explores business opportunities around DMB services and identifies major issues that must be solved to launch DMB services successfully.

Introduction

In Korea, the number of Internet users has been growing rapidly, nearly doubling each year since 1997. What is even more interesting is that most Internet users subscribe to high-speed Internet service. In 2001, the number of subscribers per 100 people was 21.8 people in Korea (about 40% of all Internet users), 4.5 people in the United States (about 9% of Internet users), and 2.2 people in Japan (5% of Internet users). This dramatic expansion of the high-speed Internet service has even received worldwide attention. The International Telecommunication Union (ITU) and the Organisation for Economic Co-operation and Development (OECD) announced that Korea ranked first in the diffusion of high-speed Internet service. Ninety-seven percent of all households in Korea have some way of connecting to the Internet and 60% of all households in Korea access the high-speed Internet.

In Korea, the phenomenal growth of ownership of cellular phones was not a government initiative, rather a private industry-driven one. Due to the highly efficient electronics industry which was able to manufacture low-cost/high-capacity cellular phones, the Korean public quickly adopted the use of cellular phones in their everyday lives. According to statistics, almost 90% of all adults now own a cellular phone, which makes Korea the country with highest ownership of cellular phones in the world. Recently, the use of mobile Internet through various handsets, such as cellular phones and personal digital assistants (PDAs), has become popular.

Now, many Korean users have utilized the Internet for personal communications (e.g., e-mail) and information searching. As the user population of the high-speed Internet service is growing quickly in Korea, many users are more inclined to use the Internet for multimedia entertainment, such as games, movies, and music. The high-speed Internet service shifts its main usage to entertainment. In terms of mobile Internet, its main usage is also concentrated on entertainment, such as ring/avatar downloads.

The launch of portable Internet, alongside mobile Internet technology and cellular technology, is a new milestone, converging wireless with wired technology. Along with these new technologies, a new telecommunication service has been introduced and has received much attention from the Korean public. This is the Digital Multimedia Broadcasting (DMB) service. DMB is a digital multimedia service combining telecommunications with broadcasting technologies. DMB enables users to watch various multimedia contents on their phone screens while they are on the move. By combining telecommunication and broadcasting technologies, DMB adds tremendous value to broadcasting. Traditionally, broadcasting technology is used to transmit information to many unspecified persons (one-to-many, one-way communication), and it is very difficult to watch TV while on the move. Whereas telecommunication technology allows individual communications (one-to-one, two-way communication) and it is easy to provide personalized services.

DMB, selected as one of the 10 new-growth engine sectors—in other words, one of the 10 most promising industries to propel Korea toward its goal of passing the $20,000 mark in GDP per capita—opens up a vast new horizon for broadcasting, making the most of strengths and specificities of different media, including terrestrial, cable, and satellite broadcasts. Economic and social ripple effects to be expected from DMB also are certainly substantial.

The question remains, however, as to whether DMB can indeed stake out its own market in Korea, as CDMA or the high-speed Internet has done in the past. The answers to this question are so far mutually contradictory even among experts. Figure 1 illustrates the respective positions of different types of telecommunication services in Korea, from which the place of DMB can be roughly deduced. More optimistic onlookers hold the view that DMB, by overcoming the one-way services thus far provided by the mobile Internet or the HSDPA-based mobile Internet, and by wooing over customers with more competitive prices, will be able to create its own niche. The opposing view holds that the market strategy of putting faith in DMB as a killer application is blunted by the fact that many of its supposedly killer service features are already offered by mobile Internet or HSDPA-based mobile Internet services. This redundancy has the implication that consumers constituting the demand source for DMB may also overlap with those making up the existing portable multimedia service market. In other words, DMB may turn out to be merely a complementary service, remedying some of the weaknesses of competitor services.

As predictions on the market viability of DMB remain divided among onlookers, what place, then, will DMB actually occupy within Korea's telecommunications market and how will it evolve within it? And what would be the response strategies by communications carriers? The objective of this chapter is to offer answers to some of these essential questions regarding the prospect of DMB in the domestic market.

Figure 1. Telecommunication services positioning in Korea

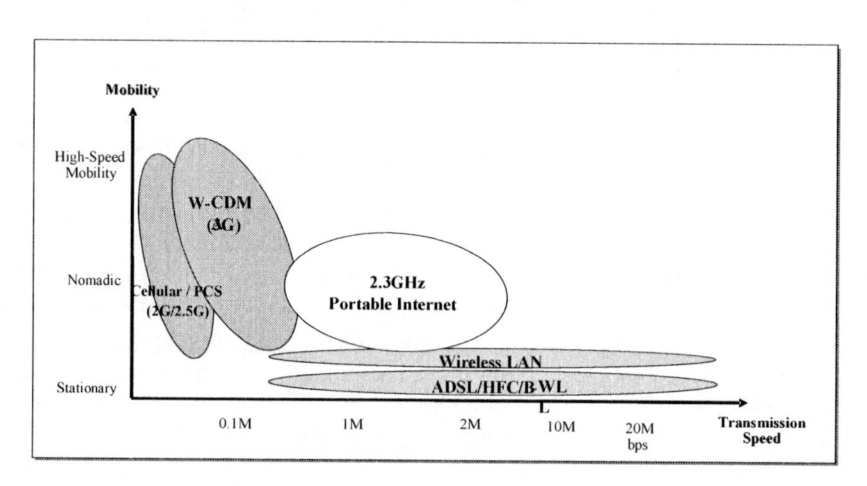

Digital Multimedia Broadcasting (DMB)

DMB Technology

Broadcasting is quickly moving into the era of digitalization by replacing traditional analog broadcasting technology with digital broadcasting technology. In 1997, the Ministry of Information and Communications (MIC) created a committee for terrestrial digital broadcasting. This committee mainly focused on the shift from analog radio broadcasting to digital radio broadcasting. At that time, the committee investigated various ways to launch the terrestrial Digital Audio Broadcasting (DAB) service into the Korean market. As telecommunication companies expressed their interests in DAB, many companies became interested on satellite DAB as well as terrestrial DAB. DAB services enable customers to receive CD-like quality radio programs, even in the car, without any annoying interference and signal distortion. Aside from distortion-free reception and CD-quality sound, DAB offers further advantages as it has been designed for the multimedia age. DAB can carry not only audio, but also text, pictures, data, and even videos. In 2002, by adding multimedia components to DAB, MIC announced a new service, called DMB (digital multimedia broadcasting), as a way to accelerate the convergence of telecommunication and broadcasting services. DMB services can be categorized into terrestrial DMB and satellite DMB.

Terrestrial DMB

Transmission of TV and radio signals using ground-based transmitters in a terrestrial network is traditionally the most used and known distribution form. End users can receive

Figure 2. Terrestrial DMB

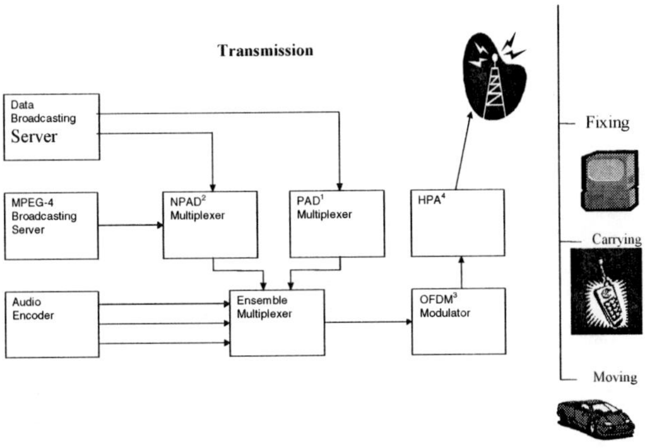

[1] PAD: Program Associated Date.
[2] NPAD: Non-Program Associated Data.
[3] OFDM: Orthogonal Frequency Division Multiplex.
[4] HPA: High-Power Amplifier

the signals through reception roofs or in-house antennas. This distribution form can also be used for digital TV. As shown in Figure 2, once terrestrial DMB integrates various multimedia contents by using multiplexers and orthogonal frequency division multiplex (OFDM) modulators, it transmits these integrated multimedia contents through ground-based transmitters. In 2001, MIC chose Euraka-147 as a standard for terrestrial DMB.

Satellite DMB

Unlike terrestrial DMB, satellite DMB transmits signals through satellites. Terrestrial DMB normally covers limited areas, whereas satellite DMB can transmit signals far away from a country's border. In addition, satellite DMB enables the signals to be transmitted with high technical quality. The high capacity of satellite DMB provides the opportunities to broadcast more channels and various digital services. However, compared with terrestrial DMB, satellite DMB is not cost efficient. As shown in Figure 3, satellite DMB transmits signals through satellites. For fringe areas, it retransmits the signals by using gap fillers. This is a major difference between regular satellite broadcasting and satellite DMB. The world's first DMB satellite, Hanbyol (MBSat), was successfully launched in March 2004. The satellite was launched by a joint venture between SK Telecom affiliated TU Media and Japan's Mobile Broadcasting Corporation (MBCo). SK Telecom and MBCo are responsible for 34.66% and 65.34% of the total cost for launching the satellite, respectively. Besides SK Telecom, KT, a giant Korean telecommunication company, also expressed its interests in satellite DMB and jumped into the DMB business. From ITU (International Telecommunication Union), KT already occupied frequency band that it is going to use for its satellite DMB service in future. Since MIC announced DMB services in 2002, two giant Korean telecommunication companies, SK Telecom and KT, competed with each other to gain a competitive advantage in the evolving satellite DMB industry. Since two companies plan to use different standards for their own DMB services (SK Telecom uses System E and KT uses System A), their competitions were fierce, rather than collaborative. In 2003, MIC selected System E as a standard for satellite

Figure 3. Satellite DMB

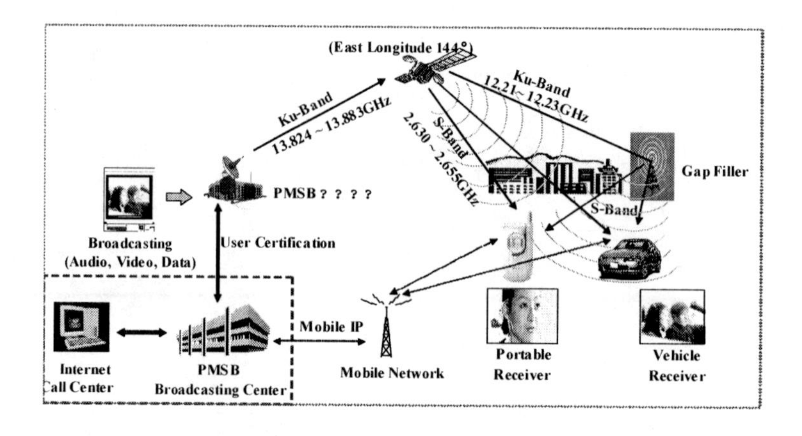

Table 1. IT 8-3-9 strategy

Telecommunication Services	Infrastructures	Applications
WiBro Service DMB Service Home Network Service Telemetries Service RFID Service W_CDMA Service Terrestrial DTV VoIP	BCN U-Sensor Network Ipv6	Next Generation Mobile Telecommunications Digital TV Home Network IT SoC Next Generation PC Embedded Digital Contents Telemetries Intelligent Robot

Source: MIC, 2004.

DMB. At that moment, by focusing more on wireless Internet business, KT gave up the satellite DMB business, and SK Telecom-affiliated TU Media has led the satellite DMB business ever since.

As the voice communication market has been saturated, telecommunication companies and MIC have looked for ways to make profit by using existing infrastructures. Recently, MIC announced the IT 8-3-9 Strategy. The IT 8-3-9 Strategy proposes eight new telecommunication services, three telecommunication infrastructures, and nine application areas as concentrated items that the Korean Government has decided to foster (see Table 1). It is an ambitious plan by the Korean Government to promote its competitiveness in the global market. Since DMB services in Korea are the first in the world, the Korean Government has much interest in DMB services. On the other hand, it imposes new political and regulatory challenges and makes rethinking and redesigns of the existing regulatory framework.

DMB Services and Business Models

DMB can be defined as a digital broadcasting medium, capable of providing stable transmission of CD-quality audio, data, and video content to fixed or portable reception devices. It is divided into two categories depending on the signals used: terrestrial DMB and satellite DMB. Figure 4 shows the differences between existing broadcasting services and DMB services. Both terrestrial DMB and satellite DMB allow customers to watch TV on their various handsets, such as PDAs or cellular phones, while on the move. Furthermore, by integrating existing telecommunication infrastructures and the Internet, DMB makes interactive, and personalized broadcasting possible.

In principle, terrestrial DBM is a free service for customers. Major income sources of terrestrial DMB operators heavily depend on advertisement sales, whereas satellite DMB operators generate income based on subscription charges. Therefore, it is very critical for satellite DMB to differentiate its contents from terrestrial DMB and to provide its

Figure 4. DMB service areas

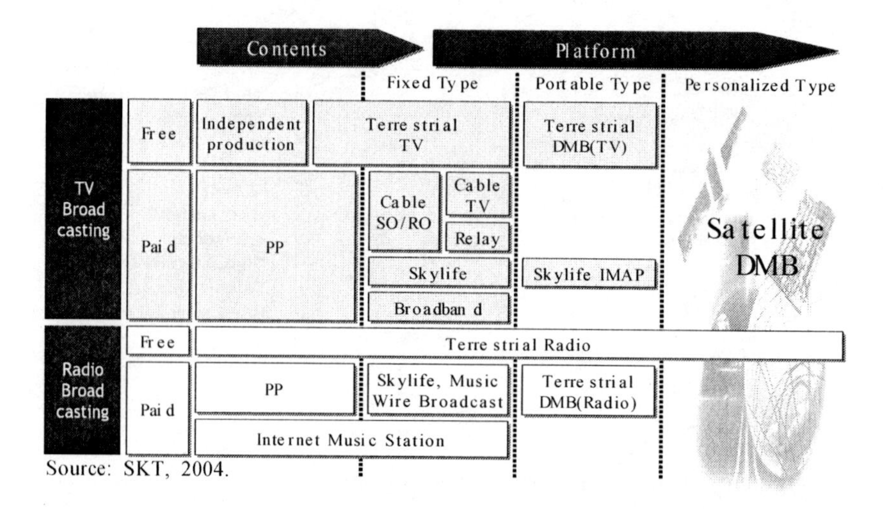

Source: SKT, 2004.

Table 2. DMB service economic spread effect

	2005	2006	2007	2008	2009	2010
Production Effect (Billion Won)	5,924	11,995	20,496	27,886	36,917	43,660
Employment Effect (Thousand People)	6.8	13.7	23.1	31.1	40.9	48.0

services for a reasonable price. On the other hand, it is crucial for terrestrial DMB to acquire various contents in order to compete with satellite DMB.

The DMB service of Korea is one of the national new-growth powers that can provide domestic telecommunication and broadcasting markets with vital energies. Furthermore, DMB, as a national exporting industry, is expected to have strong ripple effects on the Korean economy. Table 2 summarizes the economic effects of DMB service.

The industry of terrestrial DMB is supposed to be led by three major Korean terrestrial broadcasting companies (KBS, MBC, and SBS). Compared with the satellite DMB industry led by SK Telecom, terrestrial DMB industry does not have many business leaders who are able to or intend to invest large amount of money into terrestrial DMB. From all aspects, including service variety, service coverage, and service quality, terrestrial DMB is inferior to satellite DMB. However, to diffuse DMB service into Korean market at the beginning, the role of terrestrial DMB is crucial due to free service. Since many companies, especially broadcasting companies, cannot guarantee the returns to their investments, they hesitate to participate in terrestrial DMB business actively. At this time, KBS, the nation's representative public broadcaster, is preparing for a terrestrial DMB service, from developing technology of terrestrial DMB to experimenting media. Since KBS has a monopoly in the terrestrial DMB market, other competing broadcasting companies, such as MBC and SBS, recently decided to invest quota in

Figure 5. Equity composition of TU Media

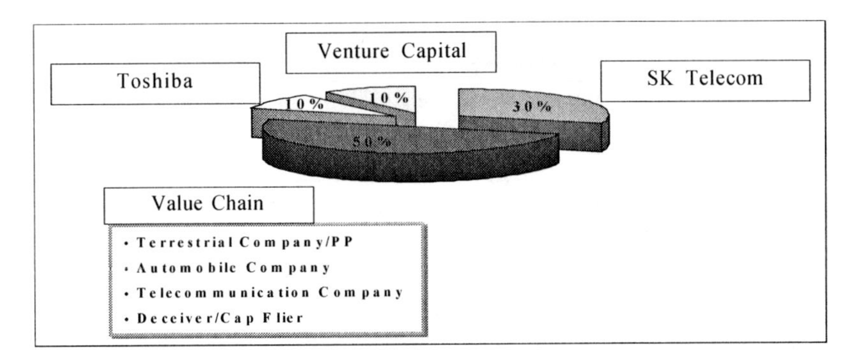

satellite DMB which can be a competitor of terrestrial DMB. It causes conflicts among the three major terrestrial broadcasting companies. From a public broadcaster's point of view, KBS cannot participate in launching charged services like other private broadcasters.

In terms of satellite DMB, SK Telecom, a market leader of the mobile communication, and KT, a market leader of wire communication, were preparing for satellite DMB enterprise together. However, the recent withdrawal of KT from satellite DMB enterprise enables SKT-affiliated TU Media to prepare a satellite DMB enterprise by itself (see Figure 5).

Market Issues

The Relationship Between Terrestrial DMB and Satellite DMB Players

With the exception that terrestrial DMB and satellite DMB do not share the same frequency band and use different transmission media/paths, they are virtually identical services. Regarding to revenue sources, the main income source of terrestrial DMB is advertising, whereas the income source of satellite DMB is a subscription charge. In terms of coverage, unlike satellite DMB providing nationwide services from the beginning, terrestrial DMB service is restricted to the Seoul and several metropolitan areas early in its commercialization. As DMB service becomes popular, it will offer national coverage in conjunction with DTV migration. Considering the traditional reticence exhibited by the Korean public toward fee-based services, without retransmission by terrestrial TVs of satellite DMB, satellite DMB service may find itself at a significant competitive disadvantage.

With new developments taking place in the telecommunication and broadcasting markets, it might be difficult to clearly define roles of terrestrial DMB and satellite DMB and their mutual relationship. If policy assigns and regulates the respective territories of these two services, it may help to avoid the overlapping of these two services, but it may also prove to be an overintervention, hurting DMB market takeoff. On the other hand, by entirely depending on the market and by lifting or loosening restrictions, the two services are likely to develop a competing relationship with each other. It is quite a difficult task for the Korean Government to balance between restrictions and free competition.

Another potential problem is passive investment of existing terrestrial broadcasting companies into terrestrial DMB. Unlike satellite DMB being expected continuous large-scale investments by major telecommunication companies and others, terrestrial DMB might experience difficulty in obtaining such large investments, due to the lack of well-defined business models. If such scenarios turn out to be true, this will certainly be a hindrance to continuous content development. It can ultimately result in the stagnation of the terrestrial DMB market. Although the fact that terrestrial DMB offers free services might be an appealing formula for the market in the short run, terrestrial DMB might be beaten out of the competition with satellite DMB, owing to the lack of resources for developmental activities. Moreover, to use network effects, satellite DMB players may reduce subscription prices in time. In the beginning, the Korean government might need to provide strong incentives that enable terrestrial broadcasting companies to invest huge amounts of money into terrestrial DMB.

From the viewpoints of service coverage, diversity of contents, and investment size and intention, things are not going to get any brighter for terrestrial DMB. Because the Korean government has a tendency to implement a market-centered policy in the DMB market, the relationship between the two services may inevitably turn out to be competitive. Under this situation, there is high possibility that the terrestrial DMB market will collapse. Accordingly, a policy helping the two services to form a complementary relationship with each other will be welcome for long-term development of this market.

The Relationship With Other Competitive Services

In Korea's communications service market, besides standard mobile wireless data services, there are various services. Among them, WiBro service recently received much attention from the public. It was launched in February 2005. WiBro service is an official name of Korea's portable Internet service. It is a new-concept, wireless high-speed Internet service using the 2.3GHz band, enabling indoor and outdoor reception by devices moving even at nomadic speed. Many Korean telecommunication companies, including SK Telecom, KT, and Dacom, are applying for its license. Many experts forecast that its ripple effects are almost comparable to those caused by voice mobile communications services half a decade ago. However, a simultaneous launch of two similar services, DMB service and WiBro service, can reduce the possibility of successful market penetration in the short run. In spite of the risk, these new services will play a crucial role for the industry. By identifying the relationship between DMB service and WiBro service, we can investigate ways that make the two services more competitive. This session examines their relationships from three different areas of interest.

Service Coverage and Content Variety

In the case of WiBro, it is practically impossible to offer a nationwide coverage during the initial stage after the service launch. Although it has the national coverage at later stages, it still has shadow zones in urban downtown sections. In regards to DMB, satellite DMB can easily offer national coverage, whereas terrestrial DMB is limited to Seoul and some metropolitan areas. By using special transporters, both terrestrial DMB and WiBro can cover shadow zones (including downtown districts and underground) even during early stages of commercialization.

In terms of the variety of content provided, while WiBro service offers a full spectrum of contents, ranging from information to entertainment, DMB service focuses exclusively on multimedia entertainment programs, with only a limited segment dedicated to information such as weather forecast service in collaboration with the Korea Meteorological Administration. In terms of live broadcasting, WiBro, unable to provide real-time TV broadcasting, appears less competitive than DMB, transmitting live broadcasting and contents. So far, terrestrial TV content is only supplied through terrestrial DMB. Also, satellite DMB service is going to retransmit terrestrial TV contents. If both terrestrial DMB and satellite DMB retransmit regular terrestrial TV contents, WiBro service will be inferior to DMB service. According to a survey on DMB demand conducted by ETRI in 2004, 41.1% of sampled respondents said they would be interested in subscribing to a DMB service, if DMB retransmitted TV contents in addition to DMB's own programs. Only 17.1% of the respondents were interested in DMB, in the case where DMB provided only its own programming.

Customers

As can be seen in Figure 6, in the case of WiBro, the respondents of the ETRI survey (2004) expressed a higher degree of interest for information searching, e-mail, and news. By age group, the younger the respondents, the stronger the appeal for contents such as chatting, games, and messaging. In comparison, older age groups showed more interest in value-added services which could yield practical benefits.

Contrasting responses were obtained regarding DMB service (see Figure 7). Drama series, music, and news were the preferred types of content that attracted respondents toward DMB. E-finance and shopping-related contents appeared to weigh in relatively little in their interest in DMB. By age group, teenagers exhibited a strong preference for music and entertainment, while respondents in their 20s and early 30s mostly favored films and animations, and those in their mid-30s or older, were interested in sports broadcast.

Receiving Devices

Figure 8 indicates that PDA and notebook computers were the preferred receiving devices for WiBro service across all age groups.

Figure 6. Preferences of WiBro service contents

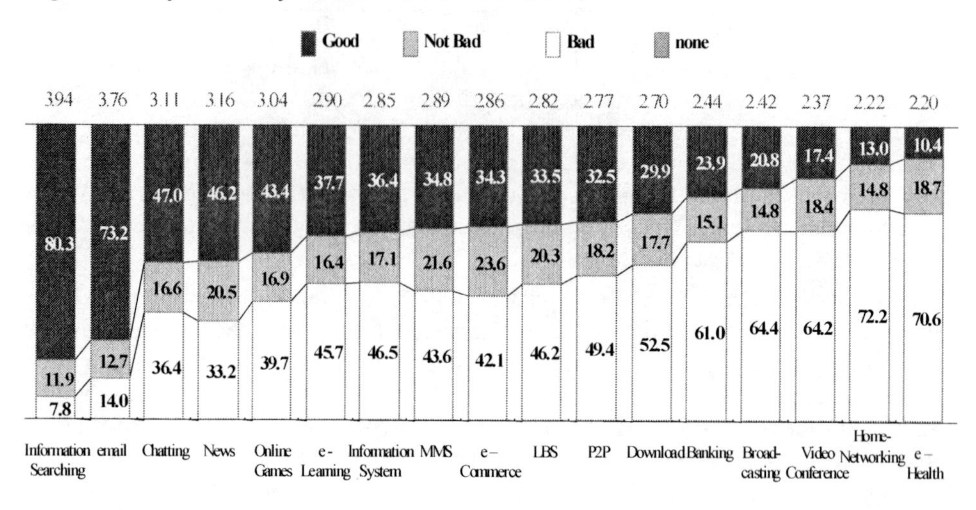

Regarding DMB, handheld devices (61.2%) and car-mounted devices (23.8%) were the preferred devices. Respondents showed only a low level of interest in stand-alone devices (4.4%). Meanwhile, an overwhelming majority of respondents favored a dual-reception device enabling the reception of both terrestrial and satellite DMB (80%). Moreover, as many as 74% answered they would wait with their DMB subscription until the dual-reception devices became available.

This survey finds several common features shared by the two services. First of all, they are both multimedia services, and both services target teens as their principal customer group. Moreover, they are portable services focusing on the customer group primarily interested in using the services under an outdoor and mobile environment. On the other

Figure 7. Preferences of DMB service contents

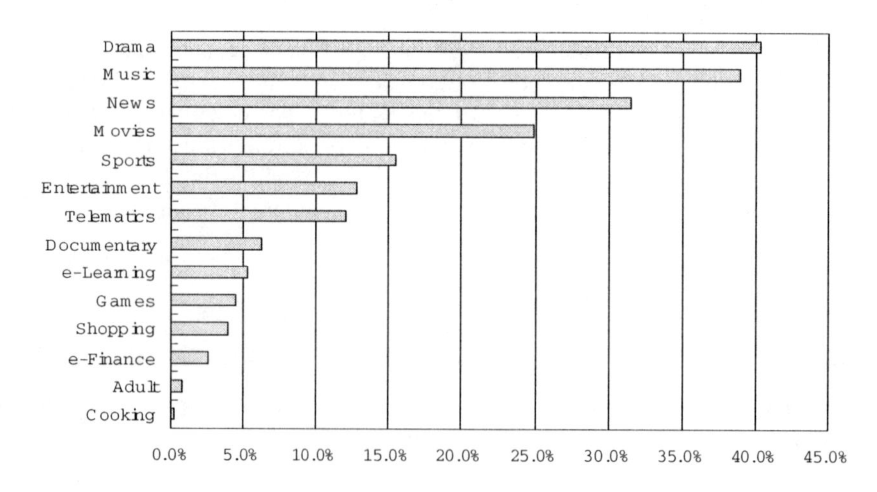

Figure 8. WiBro service receiving device preference

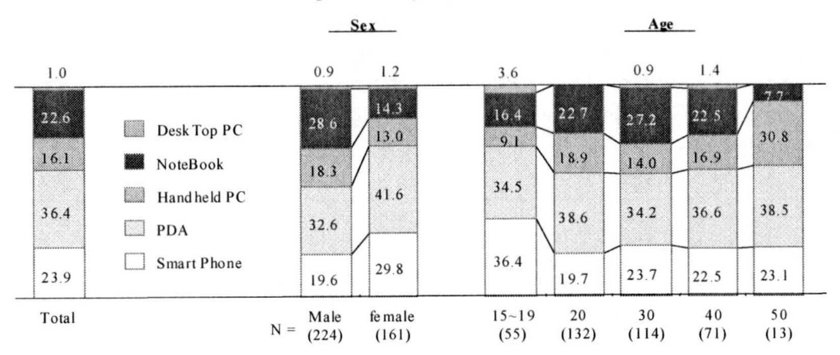

Table 3. WiBro vs. DMB

		Category	
		WiBro	DMB
Service Coverage/ Content Variety	Coverage	Initial stage: 7 major cities Thereafter: Nationwide Preferred reception locations: Car, bus, train	Terrestrial DMB: Seoul and metropolitan areas Satellite DMB: Nationwide Preferred reception locations: Car, bus, train
	Content Variety	Information, e-commerce, chatting and communication, multimedia services	Multimedia-centered services
Customer		Prefer information and shopping services	Prefer multimedia services
Receiving Device		PDA, notebook computer	Mobile phone, car-mounted devices
Price		30,000 won	7,640 won

hand, the survey results also indicate a number of differences. First, WiBro service is perceived as an information and e-commerce-oriented service, in contrast to DMB, which is extensively associated with multimedia services. The preferred receiving devices for WiBro and DMB services were also clearly distinct. The second major difference has to do with service coverage. While WiBro service coverage is limited during the early stages of its implementation, satellite DMB can offer nationwide coverage from the outset. Table 3 compares WiBro with DMB.

If terrestrial and satellite DMB launch as competing services, it is likely to reduce their appeal to customers and to ultimately weaken their competitiveness vis-à-vis other services. Furthermore, although two services partially overlap with each other in service content and target customer groups, it is not a major problem in building a collaborative relationship between the two services. This phenomenon has occurred in many other products and services. In fact, there is a high possibility that entirely new services or products turn out to be ones that fail to reflect market needs. In other words, when the DMB service is introduced to the market as a complementary service, rather than a competing service, to other similar services, its ripple effect over the communications market will be stronger and more substantial. Accordingly, for the successful launch of DMB and WiBro, instead of forming a rival relationship with WiBro service, DMB must be seen as complementing the perceived shortcoming of WiBro, namely its weakness in multimedia content offering. For long-term interest of the two services, it is wise for them to position themselves as each other's companion services.

Policy Issues

Monopoly Market

Due to the withdrawal of bid by KT, the satellite DMB market has been monopolized by SK Telecom-affiliated TU Media. However, in the case of terrestrial DMB, many experts are still debating whether monopoly market or free competition market is appropriate. To prevent large broadcasting companies from monopolizing the terrestrial DMB market, the Broadcasting Act proposed by the Korea Broadcasting Commission limits the number of multiplex to one per operator. By using one multiplex, broadcasting companies can operate one video, three audio, and one data channels. If they want to operate only one multiplex, they can operate up to two video channels. However, KBS (Korean Broadcasting System, www.kbs.co.kr), the nation's representative public broadcaster, plans to operate two multiplexes for retransmission of KBS 1 TV, KBS 2 TV, and EBS (Education Broadcasting System). Other private broadcasting companies, including MBC (www.mbc.co.kr) and SBS (www.sbs.co.kr), are strongly protesting KBS's plan, by saying that if KBS operates two multiplexes and others operate one multiplex, it will constitute an unfair preferential treatment. Industry insiders are expressing apprehension as to the possibility of this discord leading to an outbreak of large controversy on an alleged inequity. The Korean Government is still investigating appropriate market structures of terrestrial DMB.

Retransmission of Terrestrial TV

In June 2004, at a public hearing, the Korea Broadcasting Commission disclosed a regulatory decision prohibiting the retransmission of terrestrial TV programs by satellite DMB, with the exception of KBS 1 TV and EBS. TU Media, a satellite DMB operator, was fiercely protesting against the prohibition of terrestrial TV retransmission. Numerous consumer surveys indicate that potential viewers are substantially less interested in subscribing to satellite DMB, if this service does not transmit terrestrial TV. Accordingly, with this decision by the Government, the market prospect of satellite DMB business as a whole will be put into question. Recently, Skylife, a Korean satellite broadcaster, finally received the authorization to retransmit terrestrial TV (MBC, SBS) starting next year. This restriction has been a true obstacle for satellite DMB providers' efforts to expand subscriber pools, and continues to limit and mitigate the outlook for this business area.

Because the Korean Government is trying to prevent satellite DMB market from becoming a monopoly, its decision will not change anytime soon.

On the other hand, if the retransmission of terrestrial TV by satellite DMB is permitted, it will cancel out the key competitive elements of terrestrial DMB. Accordingly, it is advisable that the Government, rather than making an unconditional lifting of the retransmission restriction, instead wait and see how the market plays out and adjust its policy accordingly, so that these two DMB services can develop a complementary relationship.

Channel Compositions

The 25MHz frequency band used by satellite DMB has a transmission capacity of 7Mbps, and it is sufficient for operating up to 13 video channels. As public interests in satellite DMB are concentrated on video contents rather than audio contents, video content transmission is a priority for satellite DMB providers. However, because the Korean government strongly believes that terrestrial DMB service is inferior to satellite DMB, it supports terrestrial DMB and imposes many restrictions on satellite DMB. Given that situation, if the relationship between satellite DMB and terrestrial DMB turns out to be competitive rather than complementary, the restrictions on satellite DMB in the number of video channels allowed would be unfair and inappropriate. The overall DMB service market will be better served by loosening regulatory, directly geared toward stimulating terrestrial DMB business, rather than burdening the competing services with further restrictions such as a video channel restriction for satellite DMB.

Conclusions

This chapter reviews DMB technology, its business models, its market structures, and its policies. In particular, it explores business opportunities around DMB services and identifies major issues that must be solved to launch DMB services successfully. While the convergence of broadcasting and telecommunications is still in its early stages, it is nevertheless useful to draw a few preliminary conclusions from this survey in order to forecast trends in the worldwide telecommunication market, as well as the Korean telecommunication market.

In 1995–1997, many European countries, such as the UK, Sweden, France, and Germany, introduced DAB being similar to DMB into their markets. However, their markets have not expanded quickly. Experts point out that this market failure is caused by inability of existing terrestrial broadcasting companies in marketing and new service developments, and their conservative investments. Because telecommunication companies that have ample funds and are experienced in aggressive investment and marketing, play important roles in the DMB business in Korea, they are expected to overcome the problems encountered in European countries. The future of DMB in Korea is unpredictable, as it introduces a range of interrelated political, economic, and technical challenges. Particularly, the repeated failures in establishing a regulatory framework for DMB and ill-defined roles of players in the DMB industry might interfere the diffusion of DMB in the Korean market. As the convergence of broadcasting and telecommunications makes progress, proper modifications of existing regulatory frameworks should be made in order to guarantee success of DMB service in Korea.

References

ETRI. (2004). *Survey for DMB service* (Working paper).

Simpson, S. (2004). Universal service issues in converging communications environments: The case of the UK. *Telecommunication Policy, 28*, 233–248.

SK Telecom. (2003). Satellite DMB business plan (Working paper).

Tadayoni, R., & Skouby, K.E. (1999). Terrestrial digital broadcasting: Convergence and its regulatory implication. *Telecommunication Policy, 23*, 175–199.

Wu, I. (2004). Canada, South Korea, Netherlands and Sweden: Regulatory implications of the convergence of telecommunications, broadcasting and Internet services. *Telecommunications Policy, 28*, 79–96.

Yu, B. (2003). Policy directions for introduction of terrestrial digital multimedia broadcasting. KT Telecommunication Market.

About the Authors

Stuart J. Barnes is chair and professor of management at the University of East Anglia, UK. Barnes has been teaching and researching in the information systems field for more than a decade. His academic background includes a first class degree in economics from University College London and a PhD in business administration from Manchester Business School. He has published three books and more than 70 articles including those in journals such as *Communications of the ACM*, *International Journal of Electronic Commerce*, *e-Service Journal*, *Electronic Markets*, and *Journal of Electronic Commerce Research*. Two more books are in progress for 2005.

Eusebio Scornavacca is a lecturer of electronic commerce and information systems at the School of Information Management, Victoria University of Wellington, New Zealand. Before moving to Wellington, he spent two years as a researcher at Yokohama National University, Japan. He has published and presented more than 40 papers in conferences and academic journals. He is on the editorial board of the *International Journal of Mobile Communications* and the *International Journal of Electronic Finance*. His research interests include mobile business, electronic business, electronic surveys, and IS teaching methods. He recently received a prestigious award at the MacDiarmid Young Scientist of the Year Awards. He is the founder of M-lit – the mobile business literature database Web site (www.m-lit.org).

* * *

Seung Baek is an assistant professor with the College of Business Administration, Hanyang University, Seoul, Korea. He earned an MBA and a PhD from the College of Business Administration, The George Washington University. Before joining Hanyang

University, he was an assistant professor at Georgia State University and Saint Joseph's University. His main research interests include business intelligence and e-service in the telecommunication industry.

Stefan Berger holds a degree in business administration from the Universität Passau, Germany, and a PhD from the University of Regensburg, Germany. His research interests are in the areas of knowledge management and mobile computing.

Liz Burley is a lecturer in the Faculty of Information and Communications Technology at Swinburne University of Technology, Australia. She holds a Master of Business Systems from Monash University, Melbourne. Prior to accepting the lecturer position at Swinburne, Burley spent six years as an IT consultant mainly doing project management for three professional services companies in Melbourne, Canberra, and California. The projects Burley has managed have covered the broad spectrum of IT projects, from managing a project team for a dot.com in California to managing the development of a contract management system for the Department of Defence in Canberra to managing an offshore development team in Hyderabad, India for a Web-application development project based in California. Burley is currently pursuing PhD studies with SIMS, Monash University.

Amy Carroll is Equities Research Associate in the Global Markets division at Citigroup, New Zealand. Her academic background includes a bachelor degree in commerce and administration in information systems and marketing, and an honours degree in information systems from the School of Information Management at Victoria University of Wellington, New Zealand. Her interests include marketing, finance, and mobile technologies.

Guillermo de Haro is a professor with the Information Systems Department, Instituto de Empresa (Madrid, Spain). He has earned a European Doctor in Business Administration in the UPV/EHU (Bilbao, Spain), a Master in Business Administration *cum laude* at the Instituto de Empresa (Madrid, Spain), and a Telecommunication Engineer by the Superior School of Engineers in Bilbao (Spain). His research interests are in customer relationship management, key success factors for the implementation of new technologies, and new multimedia channels. Professor de Haro has been working in the consulting and education fields since 1998, having launched entrepreneurial projects like AvanGroup.

Albrecht Enders is an assistant professor of strategic management at Nuremberg University, Germany. He previously worked as a consultant with the Boston Consulting Group and as a research fellow at INSEAD where he conducted research in mobile and electronic commerce. He has authored articles on e-business and strategy which have been published in the *European Management Journal*, *International Journal for Electronic Business* and *Zeitschrift für Betriebswirtschaftliche Forschung*. In 2005 he co-authored a textbook on e-business titled *Strategies for e-Business: Creating Value through Electronic and Mobile Commerce* (FT/Prentice Hall). Enders holds a PhD in strategic management from the Leipzig Graduate School of Management and a BA in economics from Dartmouth College.

José María García is general manager at Telecinco.es and a professor with the Information Systems Department, Instituto de Empresa (Madrid, Spain). He earned a Master in Business Administration at INSEAD, and graduated in law, and economical and entrepreneurial sciences from the Universidad Pontifica de Comillas - ICADE (Madrid, Spain). He has formerly held executive positions in the Telecinco Group, International Herald Tribune, the new technologies holding Net-Juice, and the consulting firm McKinsey & Co. He cofounded Net-Juice. His research interests are marketing, strategy and entrepreneurship in innovative environments, and impact of technology in consumers. He has published the Buongiorno! MyAlert and the Coca-Cola cases in the Instituto de Empresa.

Mark Gaynor holds a PhD in computer science from Harvard University, USA, and is an assistant professor in the Graduate School of Management, Boston University. His research interests include ATM, TCP/IP over high-speed ATM networks, packet classification for Quality of Service, standardization in the IT area, designing network based-services, and wireless Internet services. He is on the research board of Telecom City (a regional technology development project) and is a co-PI on an NSF grant studying virtual markets on a wireless grid. He is technical director and network architect at 10Blade. His first book, *Network Services Investment Guide: Maximizing ROI in Uncertain Markets*, is in press with Wiley (2003).

Steve Goldberg is the founder of INET International Inc., Canada, a highly specialized IT management consulting firm. The firm: (1) researches ways to accelerate healthcare delivery enhancements using wireless technology, (2) conducts workshops for IT client account/project managers on how to engage and deliver mobile e-health projects, and (3) produces and facilitates INET Mini-conferences. In 2002, INET completed a study at Hamilton Health Science (HHS) to define the requirements for a standardized wireless IT environment (infrastructure.) HHS is one of Canada's largest teaching hospitals. This study has been included in collaborative papers for presentation at international conferences.

Barbara Haddon is an aged care consultant and registered nurse with 20 years experience in residential aged care facilities as a charge nurse, director of nursing, and consultant. This was followed by 11 years with the Commonwealth Department of Health and Aged Care Standards Monitor and Review Officer (auditor) and for the last five years as a program manager of outcome standards, resident classification scale and complaints/ user rights. Haddon has done post basic studies in gerontology, health education, management, training in negotiation skills, team dynamics, train the trainer, presentation skills, plus workshops and seminars. She regularly presents papers on aged care programs and associated issues to the Victorian University of Technology, Deakin University, Mayfield Education Centre, peak industry organizations, conferences and seminars and facilitated departmental workshops. She co-authored a 'how to' RCS manual for staff in residential aged care facilities and facilitated workshops to promote and demonstrate the use of the manual particularly for 'hands on' staff in facilities. She then spent two and a half years working as a consultant in residential aged care facilities, assisting staff and management to improve documentation systems, work processes and practices. In the past three years she has been involved in the development of an electronic care management system for the residential aged care industry and was directly responsible for the development of the content of the system.

Shengnan Han is a researcher in the Institute for Advanced Management Systems Research, Åbo Akademi University, Finland. She received her MSc in economics from Renmin University, China. Her research interests are focus on mobile commerce, user adoption of mobile products and services and industry foresight for mobile commerce. She has published more than 10 scientific articles and international conference papers on topics related to these interests. She is completing a doctoral degree at Turku Centre for Computer Science (TUCS).

Nada Hashmi is the leading director for Web services development at 10Blade, Inc. Prior to joining 10Blade, she received her Master in Science from the University of Maryland, College Park (UMd). While at UMd, she interned for a year with Fujitsu Laboratories of America, Inc. Her research interests include network services and techniques.

Tawfik Jelassi is dean of the School of International Management, Ecole Nationale des Ponts et Chaussées (Paris) and a professor of e-business & information technology. He was previously an associate professor at INSEAD and coordinator of its technology management area. Dr. Jelassi holds a PhD from New York University (Stern School of Business) and graduate degrees from the Université de Paris-Dauphine. His research focuses on e-business and the strategic use of IT. It has appeared in his books: *Competing through Information Technology: Strategy and Implementation* (Prentice Hall, 1994), *Strategic Information Systems: A European Perspective* (Wiley, 1994), and *Strategies for e-Business: Creating Value through Electronic and Mobile Commerce* (Financial Times/Prentice-Hall, 2005). Dr. Jelassi has also published more than 80 research articles in leading academic journals and conference proceedings, and was awarded several teaching, case writing, and research excellence awards. He is also a member of the editorial board of several international journals. Dr. Jelassi has taught extensively on MBA and executive education programs in more than a dozen countries internationally and has served as an advisor to several international corporations and government organizations.

Vaida Kadytë is currently a PhD candidate at Åbo Akademi University and the Turku Centre for Computer Science (TUCS), Finland. As a researcher at the IAMSR Mobile Commerce Research Centre, she has been working on producing foresight scenarios for mobile technology applications in global markets and leading one of the projects with the National Technology Agency of Finland (TEKES), which is aimed at improving B2B communications with number of industrial customers and B2E internal logistics in the fine paper industry. The focus of her recent scientific publications is on how customer relationship management theory can be adapted to and implemented on mobile technology platforms for industrial customers, covering broad range of issues from product inception phase to usability testing and & pilot release. An MSc graduate of Kaunas University of Technology, she has a background in management and industrial engineering.

Markku Kallio is a senior consultant in Duodecim Medical Publications Ltd., Finland. He received his MD and PhD from the University of Helsinki. He also received an international MBA from the Helsinki School of Economics and California University. He has started medical practice since 1982 as a full time medical doctor in several hospitals, mostly at the children's hospital at the University of Helsinki. Kallio has published more than 40 papers in international journals, mostly of cholesterol metabolism and infection diseases. Since 1998 he has been intensively involved in developing professional databases (for example, Evidence-Based Medicine Guidelines) for medical doctors, which run both in the Internet and on mobile platforms.

Bong Jun Kim is a research associate at Division of Telecommunication and Broadcasting, Korea Information Strategy Development Institute (KISDI), Seoul, Korea. He received a BA and an MA from Hanyang University, Seoul, Korea. He is interested in telecommunication policy.

Kyoung-Joo Lee is a PhD candidate in the Graduate School of Commerce and Management, Hitotsubashi University, Japan. He earned a master's degree in Japan studies from the Graduate School of International Studies, Korea University (1999). His current research interests include comparative institutional studies, institutionalism in economics and politics, relation between organization and technological innovation, B2B electronic marketplace, and innovation in mobile communication and computing.

Steven Moulton is an associate professor of surgery and pediatrics director, Pediatric Surgery & Trauma, and the surgical director, Pediatric Intensive Care Unit, Boston University, School of Medicine, USA. Dr. Moulton's principal areas of research include: pediatric trauma, clinical pathways, pediatric neurotrauma, and the role of angiogenesis in intra-abdominal adhesion formation.

Pekka Mustonen is the managing director of Duodecim Medical Publications Ltd., Finland. He received his MD and PhD from the University of Helsinki. He has served as Sigrid Juselius research fellow at the Institute Jacques Monod, Paris. He was an associate editor of the *Duodecim Medical Journal* from 1996-2000. He has been author or co-author of 18 research papers in international journals and a large number of articles in domestic medical journals.

Dan Myung is the director of application development at 10Blade, Inc. He graduated with an AB in computer science from Harvard University, where his main interests were in network communications. While at Harvard, Myung worked at several software companies, developing network optimization and diagnosis tools for Opticom, Inc., as well as alerting services for America Online. Prior to joining 10Blade, Myung was a software engineer at IBM providing solutions for their messaging and groupware products. He resides in Cambridge, Massachusetts and is an avid photographer.

Marissa Pepe is currently a student at the Boston University School of Management, USA, where she is pursuing an MBA and an MS in information systems. Prior to graduate school, she worked for Harvard University in the University Information Systems Department. She has also worked for a dot.com and a startup venture capital firm. She received an AB in psychology from Cornell University.

Ulrich Remus studied management information systems (MIS) at the University of Bamberg, Germany. From 1996 to 1998 he worked for a large IT consulting company. He received his PhD in MIS from the University of Regensburg, Germany in the field of process-oriented knowledge management (2002). He is currently a senior lecturer for MIS at the University of Erlangen-Nuremberg, Germany. His research interests focus on knowledge management and portal engineering & management.

Helana Scheepers is a senior lecturer in the School of Information Management and Systems, Faculty of Information Technology, Monash University, Australia. She holds a Masters in Commerce from Pretoria University, South Africa and a PhD in computer science from Aalborg University, Denmark. For the last 15 years, Scheepers has worked in the area of information systems in universities in South Africa, Denmark, and Australia. Prior to joining Monash, she was a visiting researcher at Aalborg University, Denmark. Scheepers has also worked in industry as a systems developer and her research has mainly been conducted in the area of systems development. She is currently conducting research in the area on mobile computing and mobile systems development and how these technologies can be applied in business.

Matti Seppänen is the product manager of mobile solutions by Duodecim Medical Publications Ltd., Finland. He received his MD, Licentiate in Medicine from the University of Tampere. He has a 10-year career in Duodecim, including several years as a member of the editorial team of the *Evidence-Based Medicine Guidelines* (Finnish edition) and also as a deputy editor. He has been a member of the Finnish Association of Science Editors and Journalists since 1997. He is a part-time practicing physician. His current research interest has a focus on mobile medical solutions.

Clarry Shchiglik is a researcher at the School of Information Management, Victoria University of Wellington, New Zealand. He holds BCA, Honors and Graduate Diploma in professional accounting qualifications from Victoria, and is currently a candidate for the MCA degree. Shchiglik's research interests include evaluating the quality of e-commerce Web sites and mobile services, particularly games. He has published and presented in the *Journal of Computer Information Systems* and *Australian Conference on Information Systems*. In both 2002 and 2004 he was awarded ITU fellowships to represent New Zealand at conferences in Hong Kong and Hawaii respectively.

Akira Takeishi is a professor at the Institute of Innovation Research, Hitotsubashi University, Japan. He received his PhD in management from the Sloan School of Management, Massachusetts Institute of Technology (1998). His current research interests include management of the inter-firm division of labor for innovation, architecture of business systems, history of the music business, and innovations in mobile communication and computing. His work has appeared in the *Strategic Management Journal* and *Organization Science*. Recently, he has written a book, *The Division of Labor and Competition: Strategic Management of Outsourcing*, published by Yuhikaku (in Japanese).

William Tollefsen is a medical student at New York Medical College pursuing an MD. He is also a recent graduate of the Boston University School of Medicine's division of Graduate Medical Sciences and Connecticut College. Beginning his career in medicine as an emergency medical technician, Tollefsen has had the opportunity to work for many EMS systems in both urban and rural settings giving him a unique perspective to the challenges facing EMS providers today. His research interests include the use of technology to streamline data capture and to improve patient care and management in the pre-hospital setting.

Pablo Valiente holds an MSc in industrial management from the Royal Institute of Technology, Stockholm. He is currently a research assistant at the Stockholm School of Economics and a PhD candidate working with a thesis on organizational process innovations. Previous research projects include the analysis of the Swedish 3G-license allocation process for the International Telecommunications Union (ITU) and the study of mobile information systems at a number of companies in Sweden and The Netherlands in cooperation with Vrije Universiteit, Amsterdam. Valiente holds previous work experience from Ericsson Marketing and Product Management where he worked with customer business case models, a middle-size Internet company in Stockholm as project manager for the implementation of enterprise resource system (ERP) and Saab Automobile.

Jarkko Vesa is a research fellow at Helsinki School of Economics in the Information Systems Science Department. In the past five years, he has worked as a freelance journalist for various business and IT magazines and newspapers in Finland. He works frequently as a management consultant for leading Finnish companies. His research concentrates on multichannel services and technologies. A special topic of interest in Vesa's research is the transformation of mobile services industry where he has analyzed the differences in the structure and success of mobile service in Japan and in Europe. Vesa has presented his research papers in various international conferences. He has written books and contributed to edited books on topics such as strategic planning of e-business, global customer relationship management and the evolution of mobile services industry. His latest book that focuses on mobile services in a networked economy will be available in January 2005 (Idea Group Publishing). He received his MSc (Econ.) at the Helsinki School of Economics.

Brett Walker is a research professional at the Centre for Interuniversity Research and Analysis on Organizations (CIRANO), Montreal, Canada. He holds a BCA with Honours from Victoria University of Wellington, New Zealand. In 2003, he was awarded the Ericsson Scholarship in Information Technology and Telecommunications while studying for his Honours degree. His research interests centre on information systems in the enterprise, including the role of wireless technologies. The project he is currently engaged with at CIRANO is developing a measure of business process integration in organisations.

Matt Welsh is an assistant professor of computer science at Harvard University (USA). Prior to joining Harvard, he received his PhD from UC Berkeley, and spent one year as a visiting researcher at Intel Research Berkeley. His research interests span many aspects of complex systems, including Internet services, distributed systems, and sensor networks.

Alf Westelius is an assistant professor in information systems and management at Linköping University and an associate researcher at the Stockholm School of Economics, Sweden. His research teaching and consulting explores people's images and understandings of their work, their organisations and of information systems in organizational change projects where IT is an important factor. He works with commercial enterprises as well as non-profit organizations and government. Westelius's publications include books, book chapters, conference papers and articles appearing in journal including *Scandinavian Journal of Management* and *Reflections* (MIT Press).

Nilmini Wickramasinghe currently researches and teaches in several areas within information systems including knowledge management, e-commerce and m-commerce, and organizational impacts of technology. In addition, Dr. Wickramasinghe focuses on the impacts of technologies on the healthcare industry. She is well published in all these areas and regularly presents her work throughout the U.S. as well as Europe and Australia. Dr. Wickramasinghe is the U.S. representative of the Health Care Technology Management Association (HCTM), an international organization that focuses on critical healthcare issues and the role of technology within the domain of healthcare. Dr. Wickramasinghe is the associate director of the Center Management Medical Technologies (CMMT), a unique research think tank that focuses on groundbreaking issues in the healthcare domain and holds an associate professor position at the Stuart Graduate School of Business.

Index

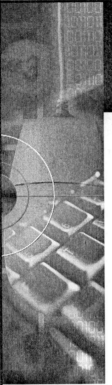